Formal Descriptions of Developing Systems

NATO Science Series

A Series presenting the results of scientific meetings supported under the NATO Science Programme.

The Series is published by IOS Press, Amsterdam, and Kluwer Academic Publishers in conjunction with the NATO Scientific Affairs Division

Sub-Series

I. Life and Behavioural Sciences	IOS Press
II. Mathematics, Physics and Chemistry	Kluwer Academic Publishers
III. Computer and Systems Science	IOS Press
IV. Earth and Environmental Sciences	Kluwer Academic Publishers
V. Science and Technology Policy	IOS Press

The NATO Science Series continues the series of books published formerly as the NATO ASI Series.

The NATO Science Programme offers support for collaboration in civil science between scientists of countries of the Euro-Atlantic Partnership Council. The types of scientific meeting generally supported are "Advanced Study Institutes" and "Advanced Research Workshops", although other types of meeting are supported from time to time. The NATO Science Series collects together the results of these meetings. The meetings are co-organized bij scientists from NATO countries and scientists from NATO's Partner countries – countries of the CIS and Central and Eastern Europe.

Advanced Study Institutes are high-level tutorial courses offering in-depth study of latest advances in a field.
Advanced Research Workshops are expert meetings aimed at critical assessment of a field, and identification of directions for future action.

As a consequence of the restructuring of the NATO Science Programme in 1999, the NATO Science Series has been re-organised and there are currently Five Sub-series as noted above. Please consult the following web sites for information on previous volumes published in the Series, as well as details of earlier Sub-series.

http://www.nato.int/science
http://www.wkap.nl
http://www.iospress.nl
http://www.wtv-books.de/nato-pco.htm

Series II: Mathematics, Physics and Chemistry – Vol. 121

Formal Descriptions of Developing Systems

edited by

James Nation
Department of Mathematics,
University of Hawaii, Honolulu, U.S.A.

Irina Trofimova
Collective Intelligence Lab,
McMaster University, Hamilton, Canada

John D. Rand
Department of Physiology,
University of Hawaii, Honolulu, U.S.A.

and

William Sulis
Department of Psychiatry and Behavioural Neurosciences,
McMaster University, Hamilton, Canada

Kluwer Academic Publishers

Dordrecht / Boston / London

Published in cooperation with NATO Scientific Affairs Division

Proceedings of the NATO Advanced Research Workshop on
Formal Descriptions of Developing Systems
Manoa, Hawaii, U.S.A.
2–6 October 2002

A C.I.P. Catalogue record for this book is available from the Library of Congress.

ISBN 1-4020-1567-4 (HB)
ISBN 1-4020-1568-2 (PB)

Published by Kluwer Academic Publishers,
P.O. Box 17, 3300 AA Dordrecht, The Netherlands.

Sold and distributed in North, Central and South America
by Kluwer Academic Publishers,
101 Philip Drive, Norwell, MA 02061, U.S.A.

In all other countries, sold and distributed
by Kluwer Academic Publishers,
P.O. Box 322, 3300 AH Dordrecht, The Netherlands.

Printed on acid-free paper

TABLE OF CONTENTS

PREFACE

It is early May in Moscow. The last vestiges of a long winter have passed into memory. Seeds, deposited into the soil the previous fall, have suffered a seemingly endless cycle of frost and thaw, their hard outer coats torn asunder by their life giver, water. Warming slowly under the mild spring sun, their metabolic machinery comes to life, ingesting and expelling, transforming dead nutrients into living cells. In the course of a few days, a population explosion will burst forth from the now dead husk, and rapidly flow towards the surface, diversifying along the way into roots, stems, leaves, and flowers. Beauty will come to the city for a brief summer before the cycle repeats itself anew.

The cycles of birth and death, of procreation and reproduction, of growth and maturation occur all around us. They inform our lives. In many ways, they are our lives. They provide some of the deepest mysteries of life, and they call out to us, capturing our attention, firing our imagination, and stimulating our thoughts. How can all of this occur? What drives it? What shapes it? How are we to describe and, hopefully, understand this?

Artists and philosophers have thought about these questions for millennia. In recent decades, scientists have also turned their attention to matters of reproduction, development, and emergence. Like the proverbial drunkard looking for his lost watch on the empty pavement beneath the streetlamp solely because the light is there, scientists have used what tools have been at hand, regardless of how ill fitting they might be. Some progress has been made, but it has been hampered by a lack of appropriate methods, theories, and models.

Development poses deep problems to all who seek to understand it. Development never occurs in isolation. There is always an environment, a context, and quite often that context plays a critical role in determining whether or not the developmental processes play out as necessary, or whether they go astray, leading to malformation or death. Developmental systems do not stand still. They grow in size and number. Their basic components are constantly in flux. Nutrients flow through them, rather than residing within them. In one sense they are never the same from moment to moment. In another sense they unfold with great regularity. Any description of a developing system must change as the system develops yet respect the deep regularities. It is not possible to fix one description and then observe how it goes under different conditions. The description itself must change, and how it changes will depend upon the conditions that the systems encounters along its way.

Worse, development does not follow a prescribed path. Happenstance also plays a role. We might study an ideal case of development in which everything unfolds like an origami flower, but in reality little goes as planned, and these quirks of fate are necessary for the appearance of adaptation and creativity. Without them, our world would lack diversity.

On that day in May, these thoughts were uppermost on our minds. Later that same month, at a complex systems conference in Boston, the idea for this workshop was born. We asked ourselves: What features of developing systems appear most critical for our understanding? If the tools currently available are inadequate for the study of developing systems, then what properties are needed to make them more effective? What tools presently at the cutting edge might prove worthy of application to the study of development?

We decided to bring together some of the most innovative researchers currently creating formal approaches to the study of developing systems to tackle these questions, to share their ideas and experience, and to try to offer a survey of some of the more promising approaches for future researchers to explore.

This book represents the outcome of our NATO Workshop, "Formal Descriptions of Developing Systems", which was held at the University of Hawaii at Manoa, October 2-6, 2002. The book reflects the ideas of both theorists and experimentalists, for it is our strong belief that the study of developing systems will proceed only through a close collaboration of both kinds of specialists. The study of developing systems is inherently multidisciplinary, and so a multidisciplinary and multinational approach was staken in selecting reserachers and topics. The book is divided into four principal sections, which can be read independently. They deal with Global Systems, Biological Systems, Emergence, and Modelling.

In his overwiew, James Nation revisits our principal questions from the viewpoint of the mathematician.

In the first section dealing with global developmental systems, renowned social theorist and scientific educator, Sergei Kapitza examines the use of a stochastic model for the growth of populations. Ed Laws tackles some of the inter-relationships that exist within food webs, a basic construct in ecology. Victor Sergeyev then shows how a thermodynamic approach can address some importnat problems in the study of economic systems.

The second section deals with various aspects of development in biological systems, the preeminent source for models of development. Vladimir Skulachev examines the nature of programmed death phenomena at various levels of development of the living systems. Jes Stollberg studies the development of motor control in vertebrates. Beata Zagorska-Marek explores dislocations in the repetitive unit patterns of biological systems. Jerzy Nakielski and Zygmunt Hejnowicz present a description of the growth of plant organs based on the growth tensor. Finally John Rand examines developmental aspects of sleep.

Emergence is a fundamental feature of all development and some of its aspects are examined in the third section. Jack Cohen, one of the leaders in the field of complexity studies, and a strong proponent of contextualism, tackles the problem head on, asking How Does Complexity Develop? Jean-Pierre Aubin presents an approach to the modelling of the dynamics of phase spaces in his paper on the adaptive evolution of complex systems under uncertain environmental constraints using the viability approach that he pioneered. William Sulis presents an original approach to the description and modelling of emergence in general based upon his concept of archetypal dynamics.

The fourth section deals with modelling specifically. Irina Trofimova examines the concepts of sociability and diversity as control parameters in formal models of developing systems. Wendy Brandt uses simple formal models to illuminate complex phenomena in Tetrahymena and ant colonies. Olga Melekhova and Yuri Liberov argue for the use of embryogenesis as a natural model of a developing system. H.S. Bracha and his colleagues examine another natural model system for the development of psychopathology in post traumatic stress disorders in their work on objective histomarkers of stress sensitization and high allostatic load during early brain development.

In the last section we include two papers submitted by workshop participants to illustrate the contributions of the next generation of developmental researchers.

Felix Arion discusses limits to developing a national system of agricultural extension, while Göksel Esmer proposes some ideas about the development of black holes.

William Sulis   October 2002
Irina Trofimova
James Nation
John Rand

ACKNOWLEDGEMENTS

The Organizers of this workshop would like to thank the NATO Science Committee for their generosity and support in providing the necessary funding, and the University of Hawaii at Manoa for serving as host institution. The lushness and diversity which is Hawaii served as a wonderful and stimulating backdrop for this meeting. We would also like to thank Protea Travel for ensuring that everyone arrived safely and in a timely manner. We thank Annelies Kersbergen at Kluwer Academic Publishers, for her patience and assistance in typesetting the manuscript. Finally we thank the Departments of Psychiatry and Psychology at McMaster University, and especially Gary Weatherill, for all of their support with our NATO activities.

On the photo:
> lower row, from the right to the left: Goksel Esmer, Przemyslaw Prusinkiewicz, , Irina Severina, Vladimir Skulachev, Irina Trofimova, Felix Arion,;
> second row: Enriko Coen, Beata Zagorska-Marek, John Rand, Wendy Brandts, Jerzy Nakielski, Paulo Neves;
> upper raw: Victor Sergeev, Leonid Levkovich, Stefan Bracha, George Wilkens, Jesse Stollberg, James Nation, Jack Cohen, John Head.

The list of participants of NATO ARW "Formal Description of Developing Systems", October 2-6, 2003, Honolulu, Hawaii:

Arion, Felix	University of Agricultural Sciences and Veterinary Medicine, Cluj-Napoca, Romania
Aubin, Jean-Pierre	Centre de Recherche Viabilite, Jeux, Universiti Controle, Paris
Bracha, Stefan	Department of Veterans affairs, National Center for PTSD, Honolulu, USA
Brandts, Wendy	Physics Department, University of Ottawa, , Canada
Coen, Enrico	Department of Cell & Developmental Biology, John Innes Centre, Norwich, UK
Esmer, Goksel	Physics Department, University of Istanbul, Turkey
Cohen, Jack	Ecology and Epidemiology Unit, Biology Department, University of Warwick, UK
Head, John	Department of Chemistry, University of Hawaii, USA.
Kapitza, Sergey	Institute for Physical Problems, Russian Academy of Sciences, Moscow, Russia
Levkovich-Masluk, Leonid	Keldysh Institute of Applied Mathematics, Moscow, Russia
Melechova, Olga	Department of Biology, Moscow State University, Russia
Nakielski, Jerzy	Department of Biophysics and Cell Biology, University of Silesia, Poland
Nation, James	Department of Mathematics, University of Hawaii, USA
Neves, Paulo	School of Management, Algarve University, Portugal
Prusinkiewicz, Przemyslaw	Computer Sciences Department, University of Calgary, Canada
Rand, John	Department of Physiology, University of Hawaii, USA

Severina, Irina I.	Belozyorski Institute of Physical-Chemical Biology, Moscow State University, Russia
Skulachev, Vladimir	Belozyorski Institute of Physical-Chemical Biology, Moscow State University, Russia
Sergeyev, Victor	Center for International Studies, Moscow State Institute for International Relations, Moscow, Russia
Sulis, William	Department of Psychiatry and Behavioural Neurosciences, McMaster University, Canada
Stollberg, Jesse	Pacific Biomedical Research Center, Hawaii, USA
Trofimova, Irina	Collective Intelligence Lab, McMaster University, Hamilton, ON, Canada
Wilkens, George	Department of Mathematics, University of Hawaii, USA
Zagorska-Marek, Beata	Institute of Plant Biology, University of Wroclaw, Poland
Shevyakov, Sergey	Lomonosov State Academy of Fine Chemical Technology, Russia

FORMAL DESCRIPTIONS OF DEVELOPING SYSTEMS:

AN OVERVIEW

J. B. NATION
Department of Mathematics
University of Hawaii
Honolulu, HI 96822, USA

Abstract. In this talk we will survey some significant results, methods and trends in mathematical modeling, in an attempt to define some of the questions to be explored in this workshop. We are particularly concerned with the formal description of biological systems. The examples, however, will be primarily drawn from the *mathematical* literature. A large and counterproductive gap remains between mathematics and the other sciences. Certainly one goal of this workshop is to narrow that gap.

1. The Scope of the Workshop

The official purpose of this workshop is

> to delineate the fundamental questions relevant to the study of the formal description, analysis, and modeling of developing systems, to critically examine approaches to these questions, to identify those approaches which seem to be most significant and fruitful, and to propose plausible methods and approaches to those questions that remain outstanding.

My assigned task is to parse that: to expand upon this charge by discussing some of the topics and problems that we will be addressing over the next few days.

The unofficial purpose of the workshop is to gather scientists from different fields - and different countries - for an exchange of ideas and expertise. Our success in this area will be measured not only in the improved perspective that we will each bring individually to our research, but also in the scientific cooperation and collaborations that will hopefully result.

The topic which brings us together is the *formal description of developing systems*. We should begin by determining what we mean by a *developing system*. A proper definition of the term might prove difficult, because its boundaries are diffuse. Our situation may be likened to an ancient kingdom: we know some areas that it definitely contains, and others that it definitely does not, but there is no carefully demarcated border. The following subject areas fall within the realm of developing systems. The references given are either to talks at this workshop (most of which are included in this volume), or to survey papers, or to recent work of particular interest.

— Population growth (Kapitza).

1

J. Nation et al. (eds.), Formal Descriptions of Developing Systems, 1–7.
© 2003 *Kluwer Academic Publishers. Printed in the Netherlands.*

- Epidemiology (Hethcote [6]).
- Embryonic development (Melekhova, Stollberg).
- Biological growth (Coen, Nakielski, Zagórska-Marek).
- Evolution (Cohen, Skulachev, McNamara *et. al.* [10]).
- Ecology (Laws, Neuhauser [11]).
- Cosmology and geology (Esmer).
- Neurology and psychology (Bracha, Rand).
- Theory of learning and intelligence (Cucker and Smale [1], Grossberg [4]).
- Neural networks and artificial intelligence.
- Social institutions (Arion, Neves, Sergeev, Trofimova).

Surely this list is not complete, and there is considerable overlap between fields. An argument could be made for including proteomics and genomics (Karp [7]), which is certainly an important tool for the study of developing biological systems.

By a *formal description* we customarily mean a mathematical model or, more generally, a mathematical analysis. The mathematical tools which may be used in the description themselves constitute a developing system. Our toolbox includes the following items.

- Discrete dynamical systems and iterative models (Nation).
- Ordinary and partial differential equations (Wilkens).
- More general dynamical systems (Aubin, Brandts, Kopell [8], Sulis).
- Thermodynamics (Sergeev, Lieb and Yngvason [9]).
- Chaos and catastrophe theory (Rand).
- Statistical analysis.
- Fuzzy systems (Yager and Zadeh [13]).
- Ordered systems and closure operators (Ganter and Wille [3]).
- Simulation (Levkovich-Maslyuk, Prusinkiewicz).
- Genetic algorithms (Head).

This toolbox, like *The Magic Pudding*, is continually replenishing itself, adapting to its required applications.

At this point, an interesting question raises its head. How much of the above material can one person know? How much do you know? How about your colleague? Clearly, teamwork in science is increasingly necessary and beneficial. Promoting scientific cooperation, both international and interdisciplinary, is a major goal of this workshop.

2. The Use and Abuse of Models

The object of the modeling process is to find a mathematical structure, whose logic we understand, that mirrors the workings of a complex biological, chemical, physical or social process. A good model should be both

- *descriptive* or *explanatory*, that is, its behavior should reflect qualitatively the process in question, increasing our understanding of it, and
- *predictive*, in the sense that it can be supported or denied by experiment.

A model may, or may not, include a *simulation* of the process.

Each of us has seen useful, and not-so-useful, models. A colleague once facetiously described the modeling process as follows.

1. Measure n parameters.
2. Find a family of curves with n parameters that look roughly like your data.
3. Fit the data.
4. Declare that you now understand what is going on.

In fact, each of these steps occurs in modeling, but without an overall rationale nothing much will be accomplished.

Let us give some more useful guidelines for modeling. This list was distilled from conversations with Brian Fellmeth and Jes Stollberg of the Pacific Biomedical Research Center. (See also Fellmeth [2]).

- Let the data directly suggest model construction.
- Adopt provisional schemes to suggest experiments to test the scheme.
- Reject or revise models that fail the test.
- Don't overcommit to models that seem to work. In particular, be wary of carrying the model beyond the limits of its applicability.

One would hope that these caveats were unnecessary with this august crowd, but it never hurts to reiterate the basics.

3. A Simple Example

Let us digress to consider a familiar class of models. Furthermore, let us look at the example as a mathematician would, which is probably a different standpoint than that of say a biologist. Again, some of the ideas in this section originated in discussions with Brian Fellmeth.

Suppose we have a fixed quantity of a substance or substances which can exist in n different states or phases. The system has a given initial distribution of states, and is then allowed to approach equilibrium. There is a standard way to describe the evolution of this system under suitably general conditions (e.g., constant temperature). Let the vector $s(t) = \langle s_1(t), \ldots, s_n(t) \rangle$ denote the amount of material in each state at time t, and for $i \neq j$ let k_{ij} denote the relative rate at which elements in state i transform to state j. The evolution of this system is described by the initial value problem

$$\frac{ds}{dt} = Ms$$

$$s(0) = s_0$$

where

$$M_{ij} = \begin{cases} k_{ji} & \text{if } i \neq j \\ -\sum_{l \neq i} k_{il} & \text{if } i = j. \end{cases}$$

From physical considerations, we want each $k_{ij} \geq 0$ for $i \neq j$, and we have built in the conservation of mass in the diagonal elements, by making the column sums zero. Moreover, because the Gibbs free energy satisfies $G_i - G_j \propto \log k_{ij} - \log k_{ji}$, our system satisfies the following condition.

(∗) For any cyclic permutation $(i_1 i_2 \ldots i_m)$ on $\{1, \ldots, n\}$, we have
$$k_{i_1 i_2} k_{i_2 i_3} \ldots k_{i_{m-1} i_m} k_{i_m i_1} = k_{i_1 i_m} k_{i_m i_{m-1}} \ldots k_{i_3 i_2} k_{i_2 i_1}.$$
This condition says that the product of the rate constants around loops is the same in both directions, e.g., $k_{12} k_{23} k_{31} = k_{13} k_{32} k_{21}$.

The solution to our system is of course given by $s(t) = e^{Mt} s_0$. However, this elegant form of the solution is not very instructive. The following version, most of which is standard, gives a better description of the solution.

THEOREM 1. *If M is as above, then the following hold.*

1. *M is singular, so that zero is an eigenvalue.*
2. *Every eigenvalue of M is real and nonpositive, i.e., $\lambda_i \leqq 0$.*
3. *If M has no repeated eigenvalues, then the solution of $\frac{ds}{dt} = Ms$ is of the form*

$$s(t) = v_1 + \sum_{j=2}^{n} e^{-\mu_j t} v_j$$

where $\lambda_j = -\mu_j$ for $2 \leqq j \leqq n$ are the eigenvalues of M, and the v_j's are the corresponding eigenvectors, with v_1 corresponding to the eigenvalue 0. (A slight modification is required for repeated eigenvalues.)

Hidden in this solution is the fact that the scaling of the eigenvectors v_j depends on the initial vector s_0. On the other hand, it is transparent that the solution tends to v_1, which will represent equilibrium, and that it does so without oscillation (by part 2).

Proof. The matrix M is singular because its column sums are zero.

To see that the eigenvalues are real, we define the symmetric matrix B by

$$B_{ij} = \begin{cases} \sqrt{m_{ij} m_{ji}} & \text{if } i \neq j \\ m_{ii} & \text{if } i = j. \end{cases}$$

Using a theorem of H. Sachs [12] and the condition (∗), one can show that M and B have the same characteristic polynomial. Hence M has real eigenvalues. The Gerschgoren Circle Theorem states that each eigenvalue λ of the matrix M satisfies $|\lambda - m_{ii}| \leqq \sum_{j \neq i} |m_{ji}|$ for some index i. For our matrix this translates to

$$-2 \left(\sum_{j \neq i} k_{ij} \right) \leqq \lambda \leqq 0$$

The remaining claims of the theorem are standard.

Typically, all we can measure experimentally is one component of the solution, say the first:

$$s_1(t) = v_{11} + \sum_{j=2}^{n} e^{-\mu_j t} v_{j1}.$$

From this we want to determine

1. What is n?
2. What are the time constants μ_j?

The current methods for doing so, using Laplace or Fourier transforms, are not very robust or satisfactory (see, e.g., [2]). They do allow you to obtain at least a lower bound for n, the number of states, which is an important component of any model.

On the other hand, for complex systems we might want to allow an infinite number of discrete states, or a continuous state-space. Moreover, it is not clear that the condition (∗) applies to developing systems. For that condition corresponds to a form of conservation of energy, which need not apply locally in developing systems. The same comment could be made regarding the conservation of mass.

The form of the most direct continuous analogue of this situation is straightforward. Given a non-negative function $m(x, y)$ on some domain, the corresponding equation would be

$$\frac{\partial s(x, t)}{\partial t} = \int m(x, z)s(z, t)\, dz - s(x, t) \int m(u, x)\, du$$

where $s(x, 0)$ is given. It is again easy to write down a formal solution to this equation, but what is required is an analysis of its behavior. This is an example of an old problem where basic, and potentially useful, work remains to be done.

What has been described is a class of models, based on rather minimal assumptions, which can be used in a variety of situations. Some modifications may be appropriate for developing systems. Nonetheless, I would claim that there is some virtue in this "top-down" approach, as opposed to inventing separate models for each particular application. Or, put more plainly, general mathematical systems may have wide applicability.

4. A Plug for Ordered Structures

Another example of this type is the application of thermodynamical principles to different situations. Professor Sergeev is speaking in this tradition, and the survey article by Lieb and Yngvasson [9] gives a nice exposition of the formal development of entropy and the second law of thermodynamics. However, the theory outlined in [9] includes a strong, and to me rather strange, assumption. The theory is based on a quasi-order (a reflexive, transitive relation) on a space of states. The relation that state X precedes state Y means that it is possible for the system to go from state X to state Y. The assumption called the *Comparison Hypothesis* is made: All pairs of states are comparable, i.e., either X precedes Y or Y precedes X (or both). In other words, the quasi-order induces a total order on the equivalence classes of states under precedence.

The total order assumed in this Comparison Hypothesis seems somewhat artificial and archaic. Nature may contain partial orders, which need not be linear orders. Note the successful use of lattice theory in data structures (as in Ganter and Wille's Formal Concept Analysis [3]) and social choice theory (see Johnson and Dean [5] for a recent update and references).

There seem to be two orders of business.

1. Assumptions of linear order should be questioned, and where appropriate, replaced.
2. A dynamic theory of ordered sets or lattices may be necessary to describe evolving and developing systems.

Just as we have learned the limitations of using only linear dynamics, so we must learn the limitations of linear order.

5. Mathematics and Science: C. P. Snow Revisited

Finally, let us consider the interaction of two largely distinct and non-overlapping cultures: mathematics and science.

The mathematics establishment is conservative in nature.

- Graduate education in mathematics focuses heavily on 19th and early 20th century mathematics. Much of this material is irrelevant to a career in modern mathematics, whether pure or applied.
- Research support is comparatively minimal, especially in pure mathematics. Research results remain a major factor in academic promotion and tenure, and it is usually a factor in personal and professional satisfaction. However, most mathematicians make their living teaching undergraduate mathematics.
- Most mathematicians work in specialized areas, with very little incentive to diversify or consider applications. There is merit in an individual pursuing a long-term, very specific, research program. But when an entire mathematics department becomes isolated, then it is time to re-establish communication with the rest of the world.
- Publication standards are very different. Mathematics papers must be precise and correct. You must say exactly what you mean, be precise and define every term. Papers with major errors, undefined terms or no proofs are not allowed in principle. (This can lead to frustration when mathematicians try to read papers from other fields.)
- This is reflected in publication policy. Mathematicians often wait six months to a year for referee reports; publication takes at least another additional year. On the other hand, a good mathematics paper may remain relevant for decades.

Moreover, it is safe to assume that most mathematicians have only minimal training in the sciences.

Of course, there is an active research community in biomathematics, biostatistics, biophysics and bioinformatics. This community is well-represented here. Similarly, there are active groups in mathematical social sciences and economics. Unfortunately, these fields remain isolated from the mainstream of mathematics. This will not be changed by retraining older mathematicians: we need to train a new generation. This process must include attracting some of our best students to applied mathematics at a relatively early age. The new Mathematical Biosciences Institute funded by the National Science Foundation at Ohio State University is a big step in the right direction.

We have left a large factor out of this discussion: computers and computer science. The effect of increased computing power and availability for both mathematics and science is evident. However, historically, computer science has developed largely independently of mathematics and science. In many universities, computer science represents a distinct and isolated department, to the detriment of all concerned. This should not be allowed to happen with biomathematics and modeling, or to mathematics for the social sciences.

The fault is not all with mathematics. Scientists from various disciplines need to learn the mathematical language. The mathematical training of scientists, outside of chemistry and physics, is often minimal. Most scientists learn mathematics up to about the year 1700 plus a little statistics. That won't cut it. Future generations of scientists must have a better, and more thorough, mathematical education.

The interaction between mathematics and science is a developing relationship. It would be interesting to try to model the developing relation between mathematics and the other sciences. Our task here is simpler, merely to promote that relationship. I wish you all a stimulating and successful workshop.

References

1. Cucker, F. and Smale, S. (2002) On the mathematical foundations of learning, *Bull. Amer. Math. Soc.* **39**, 1–50.
2. Fellmeth, B. (1981) *Kinetic study of ionic currents in the crayfish giant axon using the voltage clamp technique*, Ph.D. Dissertation, University of Hawaii.
3. Ganter, B. and Wille, R. (1999) *Formal Concept Analysis*, Springer-Verlag, Berlin.
4. Grossberg, S. (2000) Linking Mind to Brain: The Mathematics of Biological Intelligence, *Notices Amer. Math. Soc.* **47**, 1361–1372.
5. Johnson, M. and Dean, R. (2001) Locally complete path independent choice functions and their lattices, *Math. Soc. Sci.* **42**, 53–87.
6. Hethcote, H.W. (2000) The mathematics of infectious diseases, *SIAM Rev.* **42**, 599–653.
7. Karp, R.M. (2002) Mathematical challenges in genomics and molecular biology, *Notices Amer. Math. Soc.* **49**, 544–553.
8. Kopell, N. (2000) We Got Rhythm: Dynamical Systems of the Nervous System, *Notices Amer. Math. Soc.* **47**, 6–16.
9. Lieb, E.H. and Yngvason, J. (1998) A guide to entropy and the second law of thermodynamics, *Notices Amer. Math. Soc.* **45**, 571–581.
10. McNamara, J., Houston, A. and Collins, E. (2001) Optimality models in behavioural biology, *SIAM Rev.* **43**, 413–466.
11. Neuhauser, C. (2001) Mathematical challenges in spatial ecology, *Notices Amer. Math. Soc.* **48**, 1304–1314.
12. Sachs, H. (1964) Abzählung von Wäldern eines gegebenen Typs in regulären und biregulären Graphen, I, *Publ. Math. Decebren* **11**, 119–134.
13. Yager, R. and Zadeh, L., eds. (1992) *An Introduction to Fuzzy Logic Applications in Intelligent Systems*, Kluwer Acad. Pub., Boston.

Part 1.

Global Systems

THE STATISTICAL THEORY OF GLOBAL POPULATION GROWTH

SERGEY P. KAPITZA
P.L. Kapitza Institute for physical problems
RAS, Moscow, Russia

> The sciences do not try to explain, they hardly even try to interpret, they mainly make models. By a model is meant a mathematical construct which ... describe observed phenomena. The justification of such a mathematical construct is solely and precisely that it is expected to work.
>
> *John von Neumann*

Abstract. Of all global problems world population growth is the most significant. The growth of the number of people expresses the sum outcome of all economic, social and cultural activities that comprise human history. Demographic data in a concise and quantitative way describe this process in the past and present. By applying the concepts of nonlinear dynamics and synergetics, it is possible to work out a mathematical model for a phenomenological description of the global demographic process and project its trends into the future. Assuming self-similarity as the dynamic principle of development, growth can be described over practically the whole of human history, with a growth rate proportional to the square of the number of people. This is due to a collective interaction, responsible for the growth of human numbers, as the result of the informational nature of development. The large parameter of the theory and the effective size of a coherent population group is of the order of 105. As the microscopic parameter of the phenomenology, the human life span is introduced into the theory. Estimates of the beginning of human development 4 – 5 million years ago and of the total number of people who have ever lived since then ≈ 100 billion are made. In the framework of the model, large scale cycles defined by history and anthropology are shown to follow an exponential pattern of growth, culminating in the demographic transition, a veritable revolution when population growth is to stabilize at 10 to 12 billion.

J. Nation et al. (eds.), Formal Descriptions of Developing Systems, 11–35.
© 2003 *Kluwer Academic Publishers. Printed in the Netherlands.*

1. The demographic problem

The main object is to present the demographic problem in mathematical terms. The treatment will be as simple as possible, without sacrificing the meaning and content of the problem, and at the same time in no way attempting generalizations that tend to obscure, rather than help in analyzing growth. The point is to focus on the physics of demographic growth, and then find the appropriate mathematical concepts to express these processes in terms of a phenomenological theory of many particle systems. In this methods the tradition of theoretical physics will be followed, rather than the approach practiced in demography, where the growth rate is described in terms of the birth and death rates, not taking into account that the global population is really a non-equilibrium evolving dynamic system of nonlinear interacting entities [10, 11].

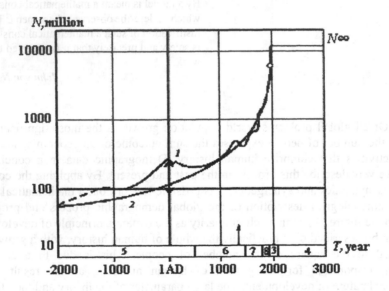

Figure 1. World population from 2000 BC to 3000 AD. Limit $N_\infty = 12$ billions 1 – data of Biraben [2], 2 – blow up growth, 3 – demographic transition, 4 – stabilized population, 5 – Ancient world, 6 – Middle Ages, 7 – Modernity, 8 – Recent history, ↑ – the Plague, ○ – 2000. As the demographic transition is approached, the time of history and development is compressed.

In developing this approach one has to keep in mind both the complex nature of the demographic system, and, what complicates these studies, the subjective desire, often subconscious, to offer immediate explanations by obvious and seemingly meaningful explanations. For the subject of life and death, of procreation and numbers of people deeply engage any student of human affairs and makes it difficult to be personally sufficiently detached from these issues. This leads to an often involuntary urge for reductionism, suggesting partial explanations having no real value, which can hardly help in reaching an in depth understanding. Much in the final treatment may seem paradoxical, confronting conventional wisdom, when the general and somewhat remote laws are sought for. But this is as an attempt to describe social phenomena by the methods of sciences that arrogantly call themselves natural and exact.

The population of the world at any given moment of time T will be characterized by a function $N(T)$. N is an additive variable and is the only one taken into account, to describe the global population. All other variables – the distributions of population in countries, towns or villages, by age and sex, by income etc., in the first approximation are not taken into account. This is the outcome of the slaving principle, formulated in synergetics by Herman Haken [8]. At this point it should be mentioned that in a demographic context the term 'slaving' is probably in a way out of place. What is meant that N is the senior variable, subordinating all others, and is the parameter of order. For the demographic problem this corresponds to and is an expression of the principle of the demographic imperative, and is introduced in discussing these issues. The idea of singling out the principle variable is based on asymptotic methods, used for solving problems of great complexity. In this case all secondary variables are averaged over so as to obtain an abridged equation of growth, retaining only the senior variable N. This is a significant simplification, as all rapidly changing variables are excluded and in the 'short' equation only two variables T and N are left:

$$\frac{\partial N}{\partial T} = F(N,T,X,Y,K,\tau,\nabla^2 N,...) \rightarrow \frac{dN}{dT} = f(N,K,\tau) \qquad (1)$$

Here K and τ are scaling factors for population and time. This exclusion of all other variables – spatial and internal systemic ones has become one of the main ideas in dealing with nonlinear problems and, as a result asymptotic methods are now widely used. But as with all approximate methods they have limits, limits set by the nature of the problem. For the demographic problem asymptotics means dealing with changes in time longer than that of a generation. In this case one may assume that asymptotically the birth and death rate, the duration of a lifetime will not explicitly enter into the formulas, and, likewise, changes in distributions in age and sex, and all other rapidly evolving variables of a social and economic nature shall not matter. Only in the next approximation the processes of birth and death, those of concern for demography can be accounted for by bringing in our reproductive period and a finite human lifetime, which become significant during the population transition.

In dealing with systems, having many degrees of freedom, where many factors affect growth it can be assumed that this multitude of processes could be treated not only statistically, but that the nature of these statistics does not change. On these conditions it may be assumed, that systemic growth will be self-similar, and, as a consequence of dynamic self-similarity, scale. This main hypothesis can be expressed mathematically by stating that the relative rate of change in the demographic system is constant:

$$\lim_{\Delta N,\Delta T \to 0} \frac{\Delta N}{(N-N_a)} \frac{(T-T_b)}{\Delta T} = \frac{d \ln\left| N-N_a \right|}{d \ln\left| T-T_b \right|} = \alpha \qquad (2)$$

N_a and T_b are the initial values for population and time. In most cases $N_a = 0$ and N is always positive. The expression means that self-similar growth is necessarily described by a power law:

$$N = C(T_1 - T)^\alpha \tag{3}$$

C and α are constants, and time is reckoned from T_1. In self-similar growth the ratios in the change of time and population is a constant. But this self-similar pattern of growth is not unbounded and can happen only within certain limits. These limits are set by the discrete nature of population – it cannot be less than one person and by the shortest unit time – by the length of a generation. The simplest case of self-similar development is linear growth, when $\alpha = 1$. In the framework of the model this corresponds to the initial stages of development. But for the main epoch **B** of blow-up growth a different law of growth rate is valid, being proportional to N^2, with $\alpha = -1$.

2. The case of quadratic growth

The 'short' equations are asymptotic and, as a statistical average, describe self-similar growth. In other words, we are dealing not only with average numbers, but also with statistically averaged functions, in treating growth and development. A good example is the physics of gases. The kinetic theory of gases is a well developed chapter in theoretical physics, and as a classical many-particle system, gases provide useful analogies for population studies. But if in thermodynamics the phenomenological, macroscopic, parameters of a gas are its temperature, pressure and density and the system is in equilibrium or changing slowly, in the case of population we are dealing with a rapidly evolving nonadiabatic dynamic system, far from equilibrium.

In an ideal gas all states are self-similar, for in this case there is no specific volume, pressure and temperature to refer to. The gas laws are, indeed, a simple proportion, intuitively easy to understand: for example, by doubling the pressure the density of a gas is doubled, hence the gas laws scale. But if the finite volume and the interactions of molecules are taken into account, as it is done in the theory of a non-ideal gas, the limits of the asymptotics of the ideal gas equations of state may be determined. This is done much in the same way as the finite human life time sets the limits of the asymptotics of hyperbolic growth. In the case of a non-ideal gas, as in the Van der Waals model, at certain temperatures (and pressure), a phase transition occurs, just as in the case of population growth the demographic transition happens at a certain time.

In thermodynamics and kinetics of population growth temperature and time are the parameters, determining the changes of these systems, one closed and conservative, the other open and evolving. The non-ideal gas model helps in yet another way – the Van der Waals interaction sums up all binary interactions between molecules and is proportional to the square of the density, the square on the number of particles in a given volume. In other words, it is similar to the collective interaction introduced to explain the growth of human numbers. This interaction is proportional to the square of total number of people on a finite Earth, regardless of their detailed distribution. It is suggested that this interaction is due to an exchange of generalized information throughout the whole development of mankind, and is peculiar for the mind of Homo.

For the demographic problem, as the first step, it is necessary to find the appropriate description of growth of the global population system, similar to the equation of state for an ideal gas. This was done by a number of authors, and probably the first, as it was indicated to me by Nathan Kheifitz, was MacCormic. In 1960 von Forster, an American engineer, suggested as an empirical formula (3) where the constants

$$C = (187 \pm 0.4) \cdot 10^9, \quad T_1 = 2027 \pm 5, \qquad \alpha = -0.99 \pm 0.09 \cdot$$

were obtained by a least square fit to world population data from B.C. to 1960 [6]. Although, in assessing the global population data, the accuracy in determining α is overestimated, this result indicates the robustness of hyperbolic growth. Later, in 1965 von Horner, discussing the possibility to deal with the run-away population explosion by escaping from the Earth to other worlds, suggested a similar expression:

$$N = \frac{C}{T_1 - T} = \frac{200}{2025 - T} \text{ (billions)} \qquad (4)$$

where $\alpha = -1$, that describes the growth of the global population over a very extensive time [9]. Independently this expression was obtained by the author, who right from the beginning recognized it as an asymptotic equation describing self-similar growth, limited in time in the past and present. This was a decisive step, as then the human life time was brought in. The necessities of taking into account the limits of asymptotics in scaling are discussed by Bahrenblatt [1].

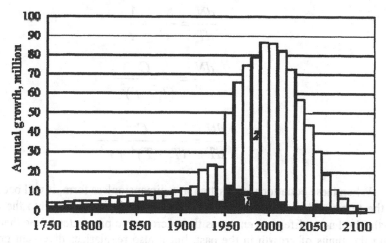

Figure 2. World demographic transition from 1750 – 2100 [16]. Annual growth averaged over a decade.
1 – developed countries, 2 – developing countries

3. Defining the model and excluding the singularities

The main point is how to establish the limits of scaling in equation (4). In the first place with the onset of the demographic explosion as the population is entering the population transition, the maximum rate of growth has to be limited. For the increase in the population in a effective generation cannot be expected to exceed the population of the world. At the other end, at the very beginning of development, when N is small, the growth rate cannot be less than one person, or rather a hominid, appearing in a generation. In chemical kinetics a similar limit is brought in at the initial stages of bimolecular reactions, where some growth has to be postulated to get the reaction started. For demography these conditions are

$$\left(\frac{dN}{dT}\right)_{\min\left|\begin{subarray}{l}N\to 1\\T\to T_0\end{subarray}\right.} \geq \frac{1}{\tau_0}, \frac{1}{\tau_1} \geq \left(\frac{1}{N}\frac{dN}{dT}\right)_{\max|T\to T_1} \tag{5}$$

and indicate, that the limits in (2) cannot be reached, assuming that the minimal growth rate of human numbers is limited by integers to one person appearing in τ_0, and at T_1 – by the maximum growth during the intrinsic time τ_1. Thus the inequalities (5) set the limits of scaling. The first limit is valid for the initial stages of development, and the other – for the present. The one for the past has a dimension [time×people], and the dimension of τ_1 is [time]. The equations for the growth rates are obtained by adding a term $1/\tau$ to the expression for quadratic growth. To introduce a cut-off during the population explosion a τ^2 term is added in the denominator to exclude the hyperbolic singularity during the population transition:

$$\frac{dN}{dT} = \frac{N^2}{C} + \frac{1}{\tau} \tag{6a}$$

$$\frac{dN}{dT} = \frac{C}{(T_1 - T)^2} \tag{6b}$$

$$\frac{dN}{dT} = \frac{C}{(T_1 - T)^2 + \tau^2} \tag{6c}$$

When these equations are written in a dimensionless form, it will become obvious that the terms added are of a similar nature. Having brought τ into the equations of growth, the human life time enters as the microscopic parameter of the theory, not only setting the limits of growth in the past, but it also regularizes divergent growth during the population explosion at T_1, growth that is then extended into the future, past the critical date.

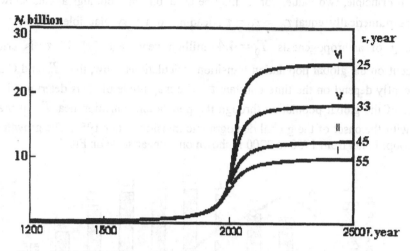

Figure 3. Global population growth models for different values of the time constant τ

N	N_∞, bil.	C, 10^9 year	τ, year	O_1, year	dN/dT,m. year	$1/N(dN/dT)$ max, %	T_{max}	N_{1990} s m.	$T_{0.9N}$ ∞ year	K, 10^4	T_0, m.year	P_0 bil.
I	10	160	55	1998	69	1.31	1964	5260	2157	5.72	4.9	99
II	12	173	45	2000	86	1.61	1981	5277	2136	6.2	4.4	95
III	14	186	42	2007	105	1.73	1989	5253	2138	6.66	4.4	103
IV	15	190	40	2010	119	1.01	1993	5259	2133	6.89	4.3	106
V	18	195	33	2017	180	2.18	2003	5230	2119	7.69	4.0	110
VI	25	200	25	2022	320	2.88	2011	5306	2009	8.94	3.5	114
VII	∞	200	(20)	2025	-	-	-	5713	-	(10)	(3.1)	115

Table 1. Models for global population growth

The constants appearing in these equations may be determined by integrating (6c)

$$N = \frac{C}{\tau}\cot^{-1}\left(\frac{T_1 - T}{\tau}\right)$$

(7)

and then fitting this equation to population data, describing the global population transition. A set of paths describing the transition, with parameters listed in Table 1, are presented in Table 2. The best fit is Model II with the following parameters:

$$C = (172 \pm 1)\cdot 10^9, \quad T_1 = 2000 \pm 1, \quad \tau = 45 \pm 1, \quad K = \sqrt{\frac{C}{\tau}} = 62000 \pm 1000.$$

(8)

In principle, two values for τ may be used, but anthropological data shows that both are numerically equal $\tau_0 = \tau_1 = \tau$, leading to a very plausible estimate for the beginning of anthropogenesis $T_0 = 4.4$ million years ago [7]. If τ is critically dependent on the global population transition, calculations show, that T_0 and C *do* not significantly depend on the time constant τ. The magnitude of τ is determined by the passage of the global population through the population transition near T_1, in practical terms with the onset of the global demographic transition after 1955. The growth of the global population from 1750 to 2200 is shown on a linear scale on Fig.7

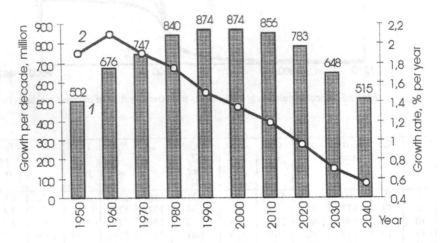

Figure 4. The global demographic transition. UN data for 1995. 1 – growth, averaged in a decade, (left scale). 2 – relative growth rate in % per year (right scale).

For this, apart from population data, it is useful to look into the absolute growth rate (6c) and to the relative rate:

$$\frac{1}{N}\frac{dN}{dT} = \frac{\tau}{\left[(T_1 - T)^2 + \tau^2\right]\cot^{-1}\dfrac{T_1 - T}{\tau}} \tag{9}$$

that reaches its maximum value

$$\left(\frac{1}{N}\frac{dN}{dT}\right)_{max} = \frac{72.5}{\tau}\% \tag{10}$$

per year at $T_m = T_1 - 0.43\tau$, and 1.6% in 1986, not taking into account a short time of postwar growth from 1950 to1970. The relative growth rate will pass in T_1 the high point of absolute growth

$$\left(\frac{1}{N}\frac{dN}{dT}\right)_1 = \frac{2}{\pi\tau} = \frac{63.6}{\tau}\% \tag{11}$$

per year, that is less than (10) and is 1.4% in 2000. In the rapidly changing circumstances of the population transition the maximum for the relative growth rate, expressed in per cent pr year precedes the high point for the absolute growth rate. For the global demographic transition the difference is $0.43\tau = 20$ years. During the demographic transition from $T_1 - \tau = 1955$ up to $T_1 + \tau = 2045$ the global population will grow on the average with

$$\frac{\Delta N}{\Delta T} = \frac{\pi K^2}{4\tau} = 67 \text{ million people a year} \tag{12}$$

added to the global population, not taking into account third order terms. The maximum growth rate in 2000 is $K^2 / \tau = 86$ million per year. The latest data UN is shown on Fig. 2. [16]. The hyperbolic run-away critical time is 2025, the shift to 2000 is due to the splitting of the singularity at 2025, and introducing a finite width of the population transition, an effect to be expected when renormalizing growth. For 1998 the world data the best fit is: $T_1 = 2000$, $N_1 = 6$ and $N_\infty = 12$ billion, with $\tau = 45$ years [16]. The UN data for global population and the calculated global population are given in Table 2 from 4-5 million year BP. The adjustments show how difficult it is to obtain valid data even for the current population of the world. In any case the revisions do not in any way change the principle result for an envisaged stabilization of the world at 10 to 12 billion.

Ãî ä	N, ì ëí	N_m, ì ëí	Ãî ä	N, ì ëí	N_m, ì ëí
-4,4·10⁸	(0)	0	1960	3019	3226
-1,6·10⁶	0,1	0,1	1965	3336	3467
-35000	1–5	2	1970	3698	3737
-15000	3–10	8	1975	4080	4037
-7000	10–15	17	1980	4450	4370
-2000	47	43	1985	4854	4735
0	100–230	89	1990	5292	5132
1000	275–345	181	1995	5765	5555
1500	440–540	362	2000	6251	6000
1650	465–550	515	2005	6729	6454
1750	735–805	715	2010	7561	6909
1800	835–907	885	2025	8504	8174
1850	1090–1170	1158	2050	10019	9683
1900	1608–1710	1661	2075	10841	10563
1920	1811	1998	2100	11185	11094
1930	2020	2218	2125	11390	11437
1940	2295	2486	2150	11543	11675
1950	2416–2515	2816	2200	11600	11980
1955	2752	3009	2500	UN data	12539

Table 2. Model for global population growth

4. Dimensionless variables of time and population

Most results are seen best if they are presented in a dimentionless form, with new variables for time and population:

$$t = \frac{T - T_1}{\tau} \text{ and } n = \frac{N}{K} \tag{13}$$

where time is measured in units of τ and N – in units of K. The constant K is the main dynamic parameter of the global population system that enters into all equations and results in the theory. As this parameter is large, it makes all asymptotics very effective, determining the scaling of temporal structures and population. In fact, even its natural logarithm $\ln K = 11.03$ is large. As with τ, it should be kept in mind that K *has* two dimensions: one is that of [people] in the case of dealing with scaling population, and in other cases K is a number. When these new variables are used the differential equations of growth:

$$\frac{dn}{dt} = \frac{n^2 + 1}{K}, \quad n = -\cot \frac{t}{K} \tag{14a}$$

$$\frac{dn}{dt} = n^2, \quad nt = -K \tag{14b}$$

$$\frac{dn}{dt} = \frac{K}{t^2 + 1}, \quad n = -K \cot^{-1} t \text{ or} \tag{14c}$$

$$\frac{dt}{dn} = \frac{t^2 + 1}{K}, \quad t = -\cot \frac{n}{K} \tag{14d}$$

become compact and the conjugate symmetry of time and population is seen, as well as how both of the singularities of hyperbolic growth are dealt with, if equations **a** and **d** are compared. In fact, near the singularity of the demographic transition at $t = 0$, population becomes, in a sense, the independent variable. Of course, physical time is invariant, but here time appears as a systemic variable, reciprocally connected with n as complimentary variables. In other words, the moment of the demographic transition is determined by population growth and is not explicitly causally dependent on time.
The growth of population and the appearance of the singularity may be illustrated by a simple geometric construction of the tangent function. The time of development from T_0 to T_1 is represented by the angle φ, going counterclockwise in units of τ, so that $K\tau = \pi/2$.

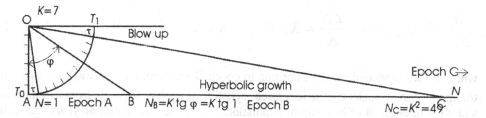

Figure 5. A tangent construction, showing the limits of asymptotics in growth $K=7$.

The radius $OA = K$ unit long sets the scale for population N, which is then plotted on the tangent AC. Then the connection between population and time becomes

$$\tan \varphi = \frac{AB}{OA} = \frac{N}{K} = n \qquad (15)$$

The beginning of epoch **A** of linear growth corresponds to a time change $\Delta \varphi = \tau$ and an increment of population $\Delta N = 1$. Linear growth will be pursued up to $\varphi_B = K\tau = 1$ and $N_B = K \tan 1$ at point B on the tangent line AC. All further quadratic growth will be determined by asymptotic hyperbolic growth of population as φ approaches $\pi / 2$. Population will rapidly grow as $N = K / \varphi$ up to the last interval of time τ. This is the point of blow-up, when the population has reached $N_c = K^2$. It is then, for times shorter than τ, the demographic transition has to set in, and from equation (14a) one has to pass to (14d). In this case the same construction may be used by substituting n into t, so as to describe growth from $N_1 \rightarrow N_\infty$ and $T \rightarrow \infty$. This is left to the reader. Fig. 5 shows that after the first linear epoch **A**, for the rest of development the time left is twice as short. The construction demonstrates the origin of asymptotics. In this case $K = 7$, the time from T_0 to T_1 is divided into 11 intervals, as $\pi/2 \cong 11/7$ and $N_C = K^2 = 49$, but even for $K = 7$ asymptotics work well and $\ln K = \ln 7 = 1.95$ provides an estimate of the number of demographic cycles equal to $3 = \ln 7 + 1$.

The growth rate in epoch **B** is described by the quadratic growth rate:

$$\frac{dN}{dt} = \frac{N^2}{K^2} \qquad (16)$$

that helps to understand the origin of growth and the meaning of K. It is really the first and simplest nonlinear term for a collective interaction. In this case nonlinearity is not a perturbation of linear growth, but the dominant term, and nonlinear growth only appears as a perturbation at the initial stage of anthropogenesis. From (16) its limits can be immediately established by assuming that, as $\Delta T = \tau$ or $\Delta t = 1$:

$$\frac{\Delta N}{\Delta t} = 1, \; N_{A,B} \approx K \; ; \; \frac{\Delta N}{\Delta t} = K, \; N_{1/2} = K\sqrt{K} \; ; \; \frac{\Delta N}{\Delta t} = N, \; N_{B,C} \approx K^2 \quad (17)$$

where N are estimated for the transition to epoch **B**, the Neolithic and the limit N_∞. In the model three epochs can be identified: epoch **A** of anthropogenesis from 4-5 million years BP to epoch **B** of blow-up of population, beginning 1.6-1.5 million years ago with the emergence of *Homo habilis*, and epoch **C**, which began with the global population transition in 1955 and is to lead to a future constant global population.

5. The limit of population and the number of people who ever lived

The solutions describing growth (7) immediately leads to the global population limit expected to be asymptotically reached in the foreseeable future:

$$N_\infty = \pi K^2 = 12 \text{ billion} \tag{18}$$

and the beginning of growth at $t_0 = -\pi/2 \cdot K$, or

$$T_0 = T_1 - \frac{\pi}{2} K\tau = T_1 - \frac{\pi}{2}\sqrt{C}\tau = -4.4 \text{ million years ago} \tag{19}$$

If growth is integrated from T_0 to T_1 the number of people, who ever lived is:

$$P_{0,1} = K \int_{t_0}^{t_{1/2}} \cot\frac{t}{K} dt + K \int_{t_{1/2}}^{0} K \cot^{-1} dt = \frac{1}{2} K^2 \ln K + \frac{1}{2} K^2 \ln(1+K) \cong K^2 \ln K$$

$$\tag{20}$$

For model II $P_{0,1} = 2.25 K^2 \ln K = 96 \approx 100$ billion people. This estimate is in good agreement with those made by Keyfitz [13] and Weiss [17], who obtained numbers from 80 to 150 billion. The multiplier 2.25 appears because in integrating the effective length of life is assumed to be $\tau = 45$ years, although the quoted numbers were obtained for an effective life span of 20 years. In the initial epoch **A** of anthropogenesis the number of early hominids approximately was:

$$P_A = 2.25 K \int_0^K \tan\frac{t'}{K} dt' = 2.25 K^2 \ln \cos 1 \cong 5 \text{ billion} \tag{21}$$

These estimates do not depend much on the model used, and are of general interest for anthropology and human population genetics. The larger number by Weiss is the result of assuming that the effective population during epoch **A** was 50 billion.

The general agreement of the model with most of the data available is consistent without any assumptions on the evolution of the constants appearing in the model. The crucial point is the appearance, some 1.6 million years ago, of a human with a brain that provided a critically new capacity for the human system to develop and multiply. This first period, which anthropologists associate with the Olduvai, lasted a million years leaving only half of that time – 500 000 years, for all of the future development of mankind up to the present crisis of the global demographic revolution.

During that eventful period there was not enough time (and generations) for evolutionary changes which could have markedly changed the human capacity to grow and develop. It is at this scale of time and growth that we should ascertain the significance of the changes through which mankind is now passing. During the demographic revolution the fundamental change in the paradigm of our development is of the same magnitude as the arrival of *Homo habilis* that gradually happened more than a million years ago. Today, for the greatest shock ever experienced by man since our origin, the change to a stabilized population is taking less than a hundred years.

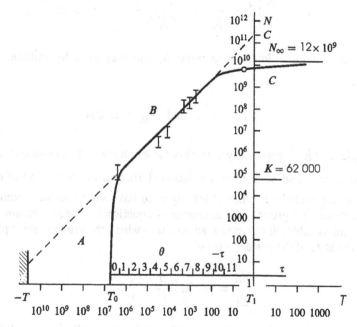

Figure 6. Global population growth from the origin of mankind and into the foreseeable future as described by the nonlinear model. $\theta = \ln t'$, $---$ (4), o – 1995

By means of mathematical modeling these calculations, based on an interpretation of the data of demography, anthropology and history lead to broad and challenging generalization. It should be stated that all data used are those generally acclaimed in the appropriate fields of study. They match the results of modeling right from the African origins of the genus *Homo* 4 to 5 million years ago within the accuracy of the numbers available [7].

Although for the initial stages only orders of magnitude of the population are known, the timing of events is known much better. In fact, for the demographic problem all asymptotic comparisons should be made in a $\log T - \log N$ space. In this space, matched to the nature of the dynamics, is where the development of mankind should be seen and interpreted. This will become evident when the long term global periods in human growth and development are described in Table 3.

6. Asymptotic solutions and autonomous equations

The asymptotic merging as growth, described as (14a) meets (14c), is best seen if the expansions for the appropriate functions for growth from T_0 and T_1 are compared:

$$n = -\frac{K}{t}\left(1 - \frac{1}{3t^2} + \frac{1}{5t^4} - \cdots\right), \quad t^2 \geq 1 \tag{22a}$$

$$n = -\frac{K}{t}\left(1 - \frac{t^2}{3K^2} - \frac{t^4}{45K^4} - \cdots\right), \quad t^2 \leq \pi K^2 \tag{22b}$$

These functions intersect at point A, half way on a logarithmic scale between T_0 and T_1

$$t_{1/2} = -\sqrt{K} \quad \text{and} \quad N_{1/2} = K\sqrt{K} \tag{23}$$

at an angle $2/3K^{-1}$ practically smoothly for any large K. It is obvious that time may be reckoned from T_0, so as to have a solution starting at the beginning of epoch **A** at t_0. Then one can exclude t from (14c), so as to have only one autonomous differential equation describing growth. An autonomous equation is an equation not containing the independent variable, in our case t, an equation where the growth rate explicitly depends only on the state of the system – on n:

$$\frac{dn}{dt} = K\sin^2\frac{n}{K} + \frac{1}{K} \tag{24}$$

This autonomous equation is valid throughout all times and has a solution describing the total human story:

$$n = K\tan^{-1}\left(\frac{1}{\sqrt{K^2+1}}\tan\frac{\sqrt{K^2+1}}{K^2}t'\right) \tag{25}$$

For a large $K \geq 1$ this solution may be simplified and then it becomes symmetric:

$$\tan \frac{n}{K} = \frac{1}{K} \tan \frac{t'}{K} \tag{26}$$

where time t' is reckoned from $t_0 = 0$, being the solution of the autonomous differential equation

$$\frac{dn}{dt} = \frac{K^2 - 1}{K} \sin^2\left(\frac{n}{K}\right) + \frac{1}{K} \tag{27}$$

that practically coincides with (25) with $n_0 = 0$, and the initial condition at $t_0 = 0$.

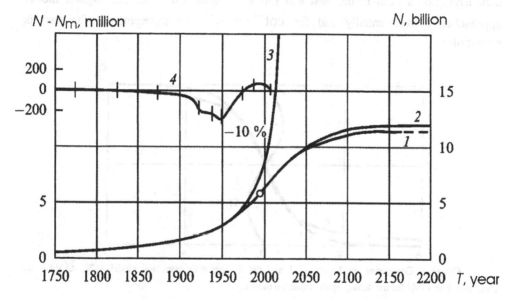

Figure 7. Population of the world from 1750 to 2200 near the singularity at 2000. 1 – recent projections of IIASA and UN, 2 – Model II, 3 – blow-up (4), 4 – difference between Model II and global population, enlarged five times, showing losses due to World Wars I and II, ○ – 1995.

Finally, the differential equation can be transformed into a finite difference equation

$$N_{m+1} = N_m + K^2 \sin^2 \frac{N_m}{K^2} + 1 \tag{28}$$

with discrete time $\Delta T = \tau = 1$ and population, which can be used for numerical modeling and stability studies. Other functions could be used to represent the transition and the appropriate polynomials have been examined in [11]. The transition is best described by the cut-off function introduced in (6) that corresponds to the observed data. In the case of a sharp transition function, where $N' \neq 0$ after the transition, an oscillatory pattern develops, which have been obtained in mathematical models. It is of interest to compare logistic growth with global population growth. The logistic is described by the differential equation:

$$\frac{dv}{dt} = v(1-v) \text{ with a solution } v = \frac{1}{1+e^{-t}} \tag{29}$$

where v is the normalized population that passes from 0 to 1, that can be compared with the demographic transition $\eta = (1/\pi)\cot^{-1} t$. These functions are anti-symmetric regarding the point (0, 1/2) and practically coincide during the transition, but their asymptotics both in the past and future are quite different. The logistic model approaches 0 exponentially and for $\cot^{-1} t \approx -t^{-1}$ the asymptote in the past is hyperbolic.

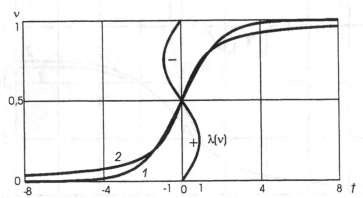

Figure 8. Comparing the logistic – 1 and the demographic transition model – 2. $\lambda(v)$ -- the Lyapunov index (30) for the demographic transition model.

7. Dynamic stability of growth

The dynamic stability of growth is determined by the growth of fluctuations and in a linear approximation any disturbance $\delta n = \delta n_0 \exp \lambda t$ grows or is damped, depending on the sign of λ. The Lyapunov index λ is determined by differentiating (24)

$$\lambda = \frac{\partial}{\partial n}\left(\frac{dn}{dt}\right) = \sin 2\frac{n}{K} \tag{30}$$

According to this criteria growth from T_0 to T_1 is unstable as $\lambda \geq 0$ and at $n = (\pi/4)K$, occurring at $t = -1$, or in 1955, when the Lyapunov index reaches its maximum value $\lambda_{max} = 1$. Only after $T_1 = 2000$, as λ changes sign, does growth stabilize and later stays asymptotically stable. See Fig. 8. Much in the same way the logistic is also unstable up to the transition point, as in this case $\lambda = 1 - 2\nu$.

The Lyapunov index deals with the linear stability of the 'short' equations for growth, where all internal processes are not taken into account. But it is exactly they that can stabilize, or destabilize for that matter, the system. Stabilization and that intuitively is plausible, can be expected from migration during the rising phase of the transition, when the instability is greatest. For example, in the 19th century in Europe there were massive migrations to the new world that certainly stabilized the global demographic system. On the other hand, the greatest disturbances during the 20th century – World Wars I and II – are close to the phase of maximum growth and the instability of growth to be expected in the developed countries. The instabilities were systemic and were due to the internal degrees of freedom in this part of the world, leading to a loss of global stability and a 8 – 10% decrease of population. See Fig.7. This is a nonlinear effect and happened because the systemic development in many degrees of freedom was to a great extent upset.

These considerations indicate that the hyperbolic growth is a path of maximum development – once this track is left, things can only get worse. In complex systems this can be seen as a general principle – as the system is evolving along an inherently stable path of maximum growth, most if not all changes, will nearly always make the system less effective. Analytically the stability of systems was studied by Haken [8] in developing the principles of synergetics and asymptotic methods, when the main motion is identified as $u(t)$ – unstable slow motion, in our case $N(T)$, and the fast stable $s(t)$ as motion in the internal degrees of freedom, that by non-linear coupling stabilizes the main motion. In mechanics this stabilizing effect of internal motion is seen, for example, when a spinning top becomes stable. A similar effect is observed, when rapid oscillations of the suspension point of a pendulum bring in a stabilizing force.

In the case of population growth a more fundamental understanding can be expected when the internal degrees of freedom will be taken into account in developing the theory of systemic growth. For the stability of the world population system the spatial distribution should then be taken into account. If diffusion is introduced into the kinetic equation one can expect a damping of systemic instabilities, as the eigenvalues of the Laplacian $\nabla^2 N$ are negative. This will require a more detailed analysis of fluctuations and instabilities of the solutions of partial differential equations, describing changes of population growth in space and time.

8. Structure of time and demographic cycles

An important property of the model is the change in the scaling of time in human history. The transformation of time can be expressed mathematically by the instantaneous exponential time of growth T_e. In exponential growth T_e is a constant, but in hyperbolic growth it changes in time:

$$T_e = \left(\frac{1}{N} \frac{dN}{dT} \right)^{-1} = \frac{1}{\tau} \left[(T_1 - T)^2 + \tau^2 \right] \cot^{-1} \frac{T_1 - T}{\tau} \qquad (31)$$

or for the past, before the demographic transition at T_1:

$$T_e(T) \cong T_1 - T \qquad (32)$$

describing the compression of the time of development as T_1 is approached. Exponential time of growth is obviously connected with the Lyapunov index $T_e = 2\tau / \lambda$. After the transition $\lambda \leq 0$, stability of growth rapidly rises, reaching at $T_1 + \tau \approx 2045$ its maximum value and then gradually decreases, but always retaining asymptotic stability, as exponential time of growth rapidly rises $T_e = (T - T_1)^2 / \tau$. The periodicity of the demographic cycles appear as a sequence of intervals

$$\Delta T(\theta) = K\tau \exp(-\theta) \qquad (33)$$

where θ is the number of the period and the integer part of $\ln t' = \ln |t - t_1|$, beginning with $\theta = 0$ and ending at $\theta = \ln K = 11$. In that case the 0-th period is $\Delta T_A = K\tau \approx 2.8$ million years and for the whole of the past we obtain:

$$T_1 - T_0 = K\tau \sum_0^{\ln K} \exp(-\theta) = K\tau \left[1 + \exp(-1) + \exp(-2) + \cdots + \exp(-\ln K) \right] \cong$$

$$\cong \frac{e}{e-1} K\tau = 1.582 K\tau \cong \frac{\pi}{2} K\tau = 1.571 K\tau \qquad (34)$$

In every one of the $\ln K = 11$ cycles of epoch B $\Delta P = 2.25 K^2 = 9$ billion people lived as the duration of the cycles changed from 1 million to 45 years. The slight difference in the total length of human development in (22) and (34) arises because in the first case development is described by growth following a $\tan t' / K$ path and in the second case growth is hyperbolic.

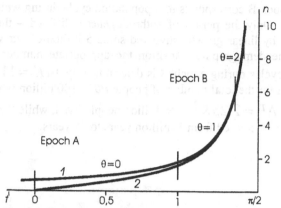

Figure 9. Comparing the initial stages of hyperbolic (1) and growth as $n = \tan t / K$.

The development at large is stabilized by internal degrees of freedom of cycles limited in their amplitude. They stabilize growth on a scale of T_e, but the origin of the demographic cycles, although they match anthropological data, requires further analysis. It may be conjectured that these cycles are due to bifurcations, taking into account the birth and death cycle, or their origin can be traced to the periodicity of the logarithmic function in the complex domain, corresponding to integer values of θ. A clarification of the origin of these cycles, resembling in many ways Kondratieff cycles, as well as the inferences of quadratic growth for economic theory are of interest [12].

For much of classic economic theory is based on the concept of local equilibrium and of reversible exchanges of goods or services. These processes correspond to the principle of detailed equilibrium of thermodynamics, implying that in a closed system changes are slow and adiabatic. In the case of population growth the rate of changes are determined by the propagation of information both vertically between generations, and horizontally, synchronizing development in space. As information is propagated it is irreversibly multiplied and is not conserved in an open and evolving system. This is the fundamental property of quadratic growth due to a collective interaction. In other words, the informational and cultural superstructure of society drives the economy. What the model suggests is that information is not a minor component of macroeconomics, but in the framework of metaeconomics it is the controlling factor. As Francis Fukuyama has very properly noted, "The failure to understand that the roots of economic behavior lie in the realm of consciousness and culture leads to the common mistake of attributing material causes to phenomena that are essentially ideal in nature."

The beginning of the Neolithic is right in the middle of epoch **B** of quadratic growth, seen on a logarithmic chart. This supports the point of view that the Neolithic really belongs to history, rather than to prehistory. Although the identification of anthropologically and historically relevant periods is based on data other than population growth, these intervals should be called demographic cycles. For no direct demographic data can substantiate such inferences, as even for recent cycles there are no pronounced discontinuities in growth rates that are sufficiently significant to mark the onset of a new cycle. On the other hand, the appropriate periods are well identified and generally recognized by historians and anthropologists [11, 3, 5].

Each period of epoch **B** corresponds to a population cycle during which the same number of people lived. Only the period of anthropogenesis differed – the first three million years, described by linear growth, involved some 5 billion of our very ancient ancestors and during the demographic transition the appropriate number will be 12 billion. The number of cycles during epoch **B** is determined by $\ln K = 11$ and can be seen from the expression for the total number of people $P01 = 100$ billion who ever lived (20). During each cycle $\Delta P = 2.25 K^2 = 9$ billion people lived, while the duration of each period exponentially decreased from 1 million years to 45 years.

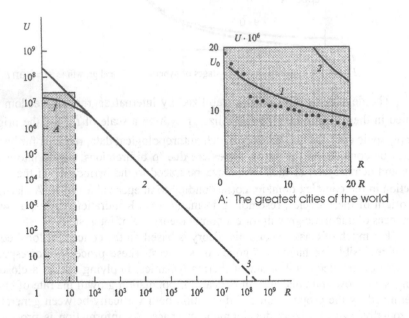

Figure 10. Ranking the global distribution of towns by size (1985). R – Rank.

$$1 - U(R) = \frac{U_0 \ln U_0}{R + \ln U_0}, \quad 2 - U(R) = \frac{U_0 \ln U_0}{R} = \frac{200 \cdot 10^6}{R}, \quad R \geq \ln U_0 = 17, \quad 3 - \text{vagabonds}$$

Insert A: $R=0$ – Tokyo, 1 – Mexico, 2 – San Paulo, 3 – New York, 4 – Shanghai, 5 – Calcutta, 6 – Buenos Aires, 7 – Rio de Janeiro, 8 – London, 9 – Seoul, 10 – Bombay, 11 – Los Angeles, 12 – Osaka, 13 – Beijing, 14 – Moscow, 15 – Paris, 16 – Jakarta, 17 – Tianin, 18 – Cairo, 19 – Teheran, 20 – Delhi

For describing the demographic transition Chesnais [4] introduced the demographic multiplier M, *defined* in the model as

$$M = \frac{N(T_1 + \tau)}{N(T_1 - \tau)} = \frac{\cot^{-1}(-1)}{\cot^{-1} 1} = 3.00 \qquad (35)$$

Epoch	Period θ	Date year	Number of people	Cultural period	ΔT years	Events in history, culture, and technology
C		2200	11×10^9	Stabilizing global Population		Global population limit 12×10^9 Changing age distribution
		2050	9×10^9		125	Globalization
	T_1	2000	6×10^9	World demographic Revolution	45	Urbanization Internet
B	11	1955	3×10^9		45	Biotechnology Computers
	10	1840	1×10^9	Recent	125	World Wars Electric power
	9	1500		Modern	340	Industrial revolution Printing
	8	500 AD	10^8	Middle Ages	1000	Geographic discoveries Fall of Rome
	7	2000 BC		Ancient World	2500	Christ, Muhammad Greek civilization India, China, Buddha, Confucius Mesopotamia, Egypt
	6	9000		Neolithic	7,000	Writing, Cities Bronze and iron metallurgy Domestication and agriculture
	5	29,000	10^7	Mesolithic	20,000	Microliths
	4	80,000		Moustier	51,000	America populated Shamanism *Homo sapiens*
	3	0.22 Ma	10^6	Acheulean	1.4×10^5	Language Speech, Fire
	2	0.6 Ma		Chelles	3.8×10^5	Europe and Asia populated Hand axes
	1	1.6 Ma	10^5	Olduvai	1×10^6	Choppers *Homo habilis*
A	T_0	4-5Ma	(1)	Anthropo-genesis	3×10^6	Hominida separate from Hominoids

Table 3. Growth and development of humankind, shown on a logarithmic scale

where $T_1 - \tau$ and $T_1 + \tau$ are the beginning and end of the transition. For the model the population of the world at points $T_1 - \tau$, T_1, $T_1 + \tau$ and T_∞ are in ratios $1 : 2 : 3 : 4$ in agreement both with past demographic data and the projections of demographers.

During the main part of the demographic transition from $T_1 - \tau$ to $T_1 + \tau$ population grows linearly, if third order corrections are not taken into account. Linear growth in the vicinity of the critical date T_1 is similar to the linear growth line at T_0 when gradual biological evolution and accumulation of a pool of genes led to the final appearance of a population of *Homo habilis,* the tool making man.

9. Scaling urban population and fluctuations

It has been noted that in a country the ranking of towns according to population can be described by a hyperbolic distribution and hence they scale [15]. It has also been remarked that large towns usually do not scale and this is attributed to the singular nature of these towns that by there sheer size cannot belong to a statistically valid local fractal set. But this may be expected for a global ranking of towns, as in this case such a set does exist, and it is the global population. For the population of a town $U(R)$ as a function of rank R the following expression is suggested:

$$U(R) = \frac{U_0 \ln U_0}{R + \ln U_0} \qquad (36)$$

that for $R \geq \ln U_0$ becomes a fractal hyperbolic distribution and as $R \to 0$ describes the population of the largest towns in the world U_0. By integrating this distribution from $R = 0$ to $R_{max} = U_0 \ln U_0$ and with $U_{min} = 1$, the population of the world becomes

$$N = \int_{R_{min}}^{R_{max}} \frac{U_0 \ln U_0}{R + \ln U_0} \, dR = U_0 \ln^2 U_0 \qquad (37)$$

By solving for U_0 it allows one to find the population of the largest town in the world and to obtain the ranking of towns. The distribution is based on the conjecture that $\ln U_0$ provides the natural scaling unit, so as to get rid of the singularity of the hyperbolic distribution at $R = 0$. It is obvious that this distribution is not applicable to any separate country. As (37) describes the global ranking of towns, it supports the idea of treating the global population as a single interacting system, where such a statistical approach, to be expected in an evolving system, is valid.

A crude estimate of fluctuations that can be expected in the population of the world has been made in [10]. The main point is that fluctuations are experienced not by N, but by n. In other words, one has to take into account the coherence and structures with a characteristic size $\approx K$, then relatively fluctuations grow as:

$$\delta n = \sqrt{n} \text{ and } \delta N = K\sqrt{N/K} = \sqrt{KN} \approx 20 \cdot 10^6. \qquad (38)$$

In this case on a relative scale the fluctuations will reach a maximum of the order of K at the beginning of epoch **B,** that is in agreement with data of anthropology on growth in early Paleolithic. These assumptions will have to be modified during the initial stages of anthropogenesis, as at that time the coherent population structures have yet to organize themselves and the whole approach is hardly valid for this stage of growth.

Apart from random fluctuations in the global population system population waves generated by a large coherent instability, propagate through the system. These waves are known as the demographic echo, the most prominent being the echo and the subsequent baby boom after the Second World War. The theory of the echo and of an effective width of the population transition could follow from a development of the model, to be dealt with by standard methods in theoretical physics for solving the Boltzman kinetic equation and its extensions. A detailed theory of fluctuations in the global population system has to be worked out, following the general principles of the theory of open systems far from equilibrium, where the demographic transition can be seen as a phase transition [14].

10. Conclusions and final remarks

It was when the newly acquired human capacity to develop mainly by intelligence began as the dominant feature of growth that the quadratic collective law of the growth of human numbers started in earnest. In other words, some 4-5 million years ago the evolution of man began as a predominantly evolutionary process that led to the appearance of intelligence and consciousness. By the end of that period, some half a million years ago when the human characteristics K and τ were attained, at values not that different to what we have now. It may be assumed that since the appearance of these fundamental, in the framework of the theory, constants they probably have not changed. It are these new qualities of *Homoce,* which determined the accelerating pattern of growth of human numbers ever since. This development, mainly social, technological and cultural happened so rapidly, without leaving much time for evolutionary, genetically determined and transmitted changes in human nature, simply because evolution is slow and quadratic growth is ever faster. That the dates and estimates of population fit the theory without bringing in any evolutionary changes of these constants is also a feature worth noting. The model provides for an opportunity to look into these processes from a quantitative point of view and suggests a basis for a better understanding of our origins, taking into account the dynamics of numerical growth, cultural development and biological evolution. The developed countries passed through the transition in some 50 years before the rest of the world. But although developing countries of the rest of the world are at very different stages of economic and social development the transition, a veritable worldwide demographic revolution is happening globally and simultaneously. It is a remarkable transformation, indicating the power of the global interaction, determining growth and development at this most critical stage for the whole history of mankind.

By passing to the 'short' equations of growth in equation (1), the spatial terms have been asymptotically excluded. This is possible because of the finite size both of the Earth and of the global population. In other words, we are dealing with a finite and discrete system and do not have to go to the limit of the continuum. This assumption leads to and is expressed in the synchronous pattern of large scale global systemic development, long noticed by historians, and justifies, in the first approximation, the exclusion of spatial terms in the short equation of growth, although these terms do contribute to the stability of growth.

In the model no external restraints appear and this leads to the conclusion that resources, at any rate up to now, do not limit growth, as it is assumed in Malthusian models, for example 'The limits to growth' model of the Club of Rome and other global models. Then of all the different processes taking place only those finally determining the gross features of systemic behavior will be left. This approach has been developed and formalized in synergetics, where methods for singling out the really significant variables have been developed. In the case of the global population system the variable that matters is the total population of the world. This point has been amplified by Mihaljo Mesarovich, who has indicated the importance of introducing the DRM principle – that of building dominant relations models and choosing the appropriate dynamic variables. What limits complicated models is not the lack of data, but the unknown coupling of all variables in complex systems. As Herbert Simon has noted "To address the 'curse of complexity', forty years of experience in modeling complex systems on computers, which every year have grown larger and faster, have taught us that brute force does not carry us along a royal road to understanding such systems... modeling, then calls for some basic principles to manage this complexity". Thus in our model we are restricted to only one variable N.

What limits growth is the generalized information transfer, the global interaction that determines the growth of human numbers, now outnumbering all other comparable creatures by a hundred thousand times. This interaction, expressed by N^2 growth, is peculiar to humankind that has acted as an information society right from its very beginning. In effect, the interaction leads to Lamarkian social evolution, when acquired information is transmitted between generations by education. Now growth is culminated by the demographic transition. In fact, it is the time devoted to bringing up and educating the next generation that at present limits the multiplication of human numbers. As a cooperative phenomenon the transition has all the features of a strong shock wave, limited in its duration by the microscopic time of the human life span. The discussion has brought out the main features of growth of the demographic system that in surprising detail can reveal in quantitative terms the human story, where previously only a descriptive, at best a chronological treatment was possible. The phenomenological theory of demographic growth is now open for further studies, as a well defined problem in complexity and evolution of nonlinear systems.

11. References

1.   Barenblatt G.I. (1996) Scaling, self-similarity and intermediate asymptotics, Cambridge University Press, New York.
2.   Biraben J.N. (1979) Essai sur l'evolution du nombre des hommes, Population **1**.
3.   Braudel F. (1980) On history. University of Chicago press, Chicago.
4.   Chesnais J.-C. (1992) The demographic transition. Stages, patterns, and economic implications. Translated from the French, Clarendon press, Oxford.
5.   Diakonoff I.M., (2000) The paths of history: From the prehistoric man to modern times. Cambridge University press, Cambridge.
6.   Forster, von. et al., (1961) Doomsday: Friday, 13 November, A.D. 2026, Science **132**, 1291, 1960. Discussion see vol.133, 942,
7.   Jones S. et al, (1994) The Cambridge Encyclopedia of Human Evolution Ed., **CUP**.
8.   Haken H. (1983) Advanced synergestics. Instability hierarchies of self-organizing systems and devices, Springer, Berlin.
9.   Horner von S. (1975) Journal of British interplanetary society **28**, 691.
10. Kapitza S. (1996) The phenomenological theory of world population growth, Physics-uspekhi **146 N1**, 63-80.
11. Kapitza S.P. (1999) The general theory of the growth of humankind. How many people have lived, live and are to live on Earth. Nauka, Moscow, (in Russian).
12. Kapitza S. (2000) A model of global population growth and economic development of mankind. Voprosy Ekonomiki, **N12**, (in Russian).
13. Keyfitz N. (1977) Applied mathematical demography. Wiley, New York.
14. Klimontovich Yu. L. (1995) Statistical theory of open systems. Kluwer academic publishers, Dordrect.
15. Mandelbrot B. (1983) The fractal geometry of nature, Freeman, New York.
16. Revision of the world population estimates and projections 1998, UN Population Division, New York, 1999.
17. Weiss K.M. (1984) On the number of the genus *Homo* who have ever lived and some evolutionary implications, Human biology **56**, 637 – 649.

MAXIMUM RESILIENCY AS A DETERMINANT OF FOOD WEB BEHAVIOR

E. A. LAWS
University of Hawaii
Honolulu, Hawaii 96822

Introduction

Ecologists have hypothesized for many years that the evolution of biological systems is driven by fundamental principles or forces. Lotka [1], for example, argued, "The first effect of natural selection thus operating upon competing species will be to give relative preponderence ... to those most efficient in guiding available energy ... The result ... is not a mere diversion of the energy flux through the system of organic nature along a new path, but an increase of the total flux through that system ... This may be expressed by saying that natural selection tends to make the energy flux through the system a maximum, so far as compatible with the constraints to which the system is subject."

Odum [2] expanded on Lotka's ideas as follows: "Lotka ... formulated the maximum power principle, suggesting that systems prevail that develop designs that maximize the flow of useful energy. The feedback designs are sometimes called autocatalytic. They maximize power, and theories and corollaries derived from the maximum power principle explain much about the structure and processes of systems. ... Examination of systems of many sizes with energy models suggests that there is a hierarchy of systems within systems, each with similar designs, differing mainly in scales of time and space. The possibility of generalizing and teaching the nature of the universe in this way with relatively fewer principles is most exciting...". This perhaps too inclusive characterization of the scope of applicability of the maximum power principle has drawn some sharp criticism. Fenchel [3], for example, has commented that Odum's approach, "had an appeal for some time, I suppose, only because it was sufficiently obscure and incomprehensible to appear profound." In reacting to such comments, Patten [4] observed, "The admissibility of virtually any system configuration, within given physical and resource constraints, does not mean the processes of ecosystem organization are lawless. The challenge is to find the laws, and that is the central thrust of Odum's work."

1. Comparison of Goal Functions

A number of authors have explored the implications of using various goal functions to constrain the behavior of model food webs. Cropp and Gabric [5], for example, used a genetic algorithm to simulate the adaptation of the biota in a simple linear food chain consisting of a limiting nutrient, autotrophs, and heterotrophs. Four

J. Nation et al. (eds.), Formal Descriptions of Developing Systems, 37–44.
© 2003 *Kluwer Academic Publishers. Printed in the Netherlands.*

basic selection pressures were formulated by considering thermodynamic (entropy, energy, and ascendancy) and ecological (sustainable biomass, primary productivity, and productivity per unit biomass) imperatives that influence ecosystems. They concluded, "Ecological systems exist within the constraints of thermodynamic laws that prescribe the transfer of energy. Ecologically defined "thermodynamic imperatives" such as entropy, energy and ascendancy provide whole-ecosystem selection pressures that constrain the evolution of individuals within an ecosystem. ...Our simulations suggest the hypothesis that, within the constraints of the external environment and the genetic potential of their constituent biota, ecosystems will evolve to the state most resilient to perturbation" [5].

In their simple linear food chain, the conditions associated with various selection pressures were easily identified, and the selection pressures suggested by Lotka [1] and Odum [2] led to essentially the same system behavior as did maximum resiliency. Fath et al. [6] likewise noted an equivalency of system behavior governed by seemingly disparate ecological goal functions. In more complex systems application of thermodynamic analogues as selection pressures to determine ecosystem behavior has met with mixed success. Månsson and McGlade [7], for example, pointed out that the flows of carbon in six marine ecosystems studied by Baird et al. [8], "were contrary to what Odum has suggested, in that the aggregate amount of cycling was an indication not of maturity but rather of the type of dynamics and levels of stress." On the other hand, Cropp and Gabric [5] observed that in some cases thermodynamic approaches have met with considerable success in estimating parameters to describe real ecosystems [9].

1.1. MAXIMUM RESILIENCY

May [10] provides an excellent introduction to the concept of ecosystem stability. Stability here is taken to mean local stability, i.e., the system will return to the steady state configuration if subjected to a small perturbation. Application of first-order perturbation theory shows that the steady state solution to a system of differential equations is stable if the real parts of the eigenvalues of the so-called community matrix are all negative. The community matrix is a matrix of partial derivatives of the differential equations describing the dynamics of the system evaluated at the equilibrium point. Assuming the steady state is stable; the time required for the system to return toward steady state is inversely related to the magnitude of the least negative eigenvalue. The most resilient steady state is the one that returns toward its equilibrium configuration most rapidly. Thus identification of the most resilient steady state can easily be made by examining the real parts of the eigenvalues of the community matrix associated with each steady state.

2. Application to Marine Pelagic Food Webs

The behavior of marine pelagic food webs, particularly with respect to their export of organic matter to the interior of the ocean, is a subject that has attracted increasing interest in recent years. The so-called biological pump is the complex set of processes by which inorganic carbon is fixed in the surface waters of the ocean and exported below the mixed layer through mechanisms such as sinking of fecal material and entrainment of dissolved organic matter. It is estimated that without this mechanism of biological sequestration, the concentration of CO_2 in the atmosphere would rise to more than 680 ppm, roughly 2.5 times higher than pre-industrial concentrations [11]. The response of the biological pump to climate change and to global warming in particular has been the subject of much speculation. The pump could provide either a positive or negative feedback to the accumulation of CO_2 in the atmosphere. Conventional wisdom regarding the mechanisms that control the biological pump is summarized in Fig. 1. In this figure, new production is the amount of primary production supported by the input of nutrients from outside the system, which in this case is taken to be the surface waters of the ocean. In the steady state, new production must be balanced by so-called export production, i.e., the rate of removal of organic matter from the surface waters of the ocean. Figure 1 is only slightly modified from the seminal paper by Eppley and Peterson [12], in which it was hypothesized that the ratio of new production to

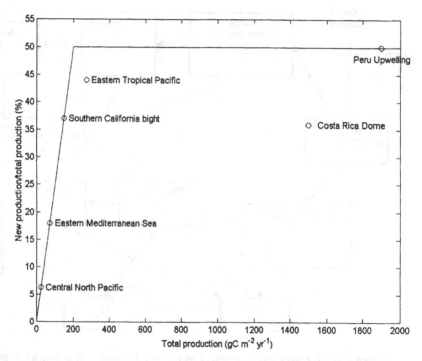

Figure 1. Total primary production versus the ratio of new production to primary production based on data summarized by Eppley and Peterson [12].

total primary production, the so-called f-ratio, was approximately a hyperbolic function of the primary production rate. This relationship was based on nothing more than empirical data, and since the publication of Eppley and Peterson [12] there has been increasing evidence that factors other than primary production control the f-ratio [13].

I have used the model food web depicted in Fig. 2 to explore the factors that regulate export production in pelagic marine communities and to determine whether application of the principle of maximum resiliency produces behavior that is consistent with field observations. In the model food web, photosynthetic production is partitioned between small and large phytoplankton cells. The food chain consisting of large phytoplankton → filter feeders → carnivores is the traditional grazing food chain [14]. The food web composed of small phytoplankton, flagellates, ciliates, bacteria, and dissolved organic matter (DOM) is the microbial loop [15]. Although the consumption of ciliates by filter feeders provides a mechanism by which organic matter produced by the small phytoplankton can be exported, the fact that organic matter consumed by predators is converted to biomass with only 30-35% efficiency [16] means that the microbial loop is an inefficient pathway for exporting organic matter. The characteristics of the system with respect to export production are therefore determined to first approximation by the relative amounts of organic matter synthesized by the large and small phytoplankton.

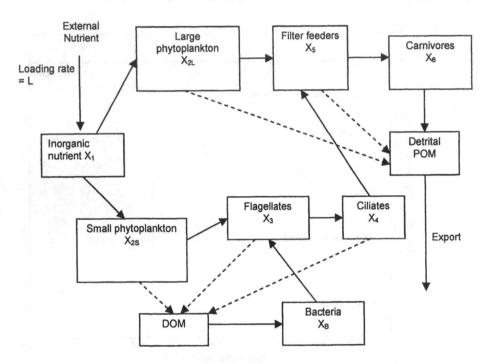

Figure 2. Feeding and excretion relationships in a model pelagic food web in which photosynthetic production is partitioned between small and large phytoplankton cells. Dashed lines indicate excretion. DOM = dissolved organic matter. POM = particulate organic matter.

Laws et al. [16] provide a detailed analysis of this model food web, and I will here provide only a brief summary. Most of the parameters in the equations describing the dynamics of the system can be estimated from information in the literature, including the temperature dependence of metabolic processes. What cannot be determined from the literature is the extent to which nutrient limitation reduces growth rates below nutrient-saturated values. The growth rates expressed as a fraction of their nutrient-saturated values are defined to be relative growth rates. Relative growth rates are dimensionless and must obviously lie in the range 0 to 1.

Imposition of the condition that the system be at steady state provides constraints but still leaves some relative growth rates unspecified, as well as the biomass of one trophic level. I used these degrees of freedom to identify the stable steady state with maximum resiliency.

2.1. MODEL RESULTS

Figure 3 summarizes the behavior of the model in Fig. 2 with respect to the export ratio (export production/primary production) when the condition of maximum resiliency is imposed. The export ratio (ef ratio) is uniformly low at low rates of primary production but rises abruptly as primary production rate increases. The value to which the ef ratio rises as primary production increases is temperature dependent. At temperatures close to 0°C the ef ratio plateaus at values of roughly 0.7, i.e., 70% of the organic matter synthesized by phytoplankton is exported to the interior of the ocean. However, at temperatures approaching 30°C the maximum ef ratio is much lower, roughly 15%.

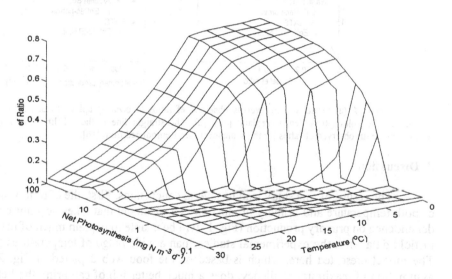

Figure 3. Calculated export ratios (ef ratios) as a function of temperature and net photosynthetic rate derived by applying the principle of maximum resiliency to the food web in Fig. 2.

A comparison between the predictions of the model with field data shows remarkable agreement (Fig. 4a). The observed ef ratios range from 0.07 to 0.68. The model accounts for 97% of the variance in the field data. When the same field data are plotted against the rate of primary production (Fig. 4b), there is no significant correlation.

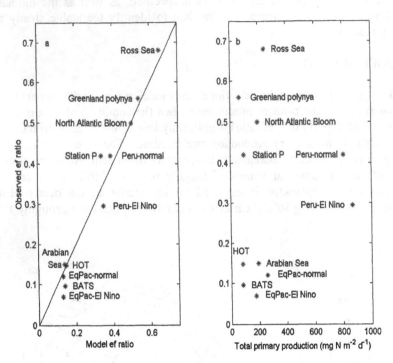

Figure 4. (a) model ef ratio versus observed ef ratios at oceanographic stations from the tropics (EqPac) to the poles (Ross Sea, Greenland polynya). The straight line is the 1:1 line. (b) Total primary production versus observed ef ratios for the same locations. Reproduced from [16].

3. Discussion

The results of this study lead to several conclusions. First, the ef ratio is a function of both temperature and primary production rate. A model that takes account only of the dependence on primary production (Fig. 1) will be unable to explain much of the variance in field data when the experimental studies span a wide range of temperatures (Fig. 4b). The model presented here, which is based on the food web depicted in Fig. 2 and the assumption of maximum resiliency, does a much better job of capturing the behavior of the ef ratio over a wide range of marine pelagic communities.

Qualitatively the dependence of the ef ratio on temperature and primary production is not hard to rationalize. Because of the positive correlation between heterotrophic metabolic rates and temperature, at high temperatures much of the organic matter produced by phytoplankton is respired before it has a chance to be exported. At low temperatures, it is relatively difficult for the heterotrophic community to decompose a large percentage of photosynthetically produced organic matter before the organic matter is exported via sinking, advection, or mixing processes. However, even at low temperatures, the ef ratio is low if the rate of primary production is low. Under these oligotrophic conditions, the most resilient steady state is the system that makes efficient use of photosynthetically fixed carbon and exports little to the interior of the ocean.

Application of the assumption of maximum resiliency in this model food web is a convenient way to constrain model parameters that would otherwise be free to vary. However, how sensitive are the results (e.g., Fig. 3 and 4) to the assumption of maximum resiliency? The answer depends on the values assigned to the independent variables, temperature and photosynthetic rate. Under some conditions, simply averaging the ef ratio over all stable steady states produces average ef ratios that are little different from the ef ratio associated with maximum resiliency. This is true, for example, at high temperatures and low photosynthetic rates. It is not the case, however, at low temperatures and high photosynthetic rates. Under these conditions the ef ratio associated with maximum resiliency can be more than twice the average ef ratio of all stable steady states. Thus merely requiring that pelagic marine food webs settle down into some random but stable configuration would not produce the kind of agreement between theory and observation apparent in Fig. 4b.

I am inclined to agree with Patten [4] that the processes of ecosystem organization are not lawless and that the challenge is to find the laws. I am also inclined to agree with Patten [4] that the central thrust of Odum's work was to find the laws. One may certainly argue over what the laws are. Odum suggested a number of thermodynamic analogues as possible goal functions of ecological systems. The results presented here suggest that maximum resiliency may be a useful construct for determining the behavior of marine pelagic food webs. However, some caveats should be noted. Planktonic communities are often regarded as being very dynamic communities. Algal blooms associated with upwelling events or seasonality are a common phenomenon in the ocean. The temporally variable behavior of planktonic communities might suggest that the assumption of maximum resiliency or indeed of steady state is inappropriate.

It should be obvious, however, that plankton have doubling times that are short compared to the time required for a bloom to develop, assuming of course that the bloom is the result of in situ growth and not some purely physical processes that concentrates cells. Laws et al. [16] have pointed out that because of the short generation time of microorganisms, "large changes in planktonic biomass and production may occur over a timescale of several weeks under conditions where there is only a small imbalance between production and consumption. As noted by Hutchinson [17], the seasonal cycle of planktonic organisms over the course of a year is equivalent to as many as 10,000 years in the successional history of a forest. In other words, what appear from the human perspective to be rapid changes in the biomass and production of planktonic communities are much more gradual when scaled to the division times of microorganisms." Maximum resiliency will undoubtedly fail to explain the behavior and structure of planktonic communities under all conditions, but it appears to be a useful construct for describing system behavior under a perhaps surprisingly wide range of conditions.

4. Acknowledgments

This work was supported by National Science Foundation grant OCE-97-25966.

5. References

1. Lotka, A. J. (1922) Contribution to the energetics of evolution. *Proc. Natl. Acad. Sci.* **8**, 147-150.
2. Odum, H.T. (1983) *Systems Ecology: An Introduction*, Wiley, New York.
3. Fenchel, T. (1987) *Ecology – Potentials and Limitations*, Ecology Institute, Oldendorf/Luhe
4. Patten, B.C. (1993) Toward a more holistic ecology, and science: the contribution of H. T. Odum. *Oecologia* **93**, 597-602.
5. Cropp, R. and Gabric, A. (2002) Ecosystem adaptation: Do ecosystems maximize resilience? *Ecology* **83**, 2019-2026.
6. Fath, B.D., Patten, B.C., and Choi, J.S. (2001) Complementarity of ecological goal functions. *J. theor. Biol.* **208**, 493-506.
7. Månsson, B.Å. and McGlade, J.M. (1993) Ecology, thermodynamics, and H T Odum's conjectures. *Oecologia* **93**, 582-596.
8. Baird, D., McGlade, J.M., and Ulanowicz, R. (1991) The comparative ecology of six marine ecosystems. *Phil. Trans. Soc. Lond. B* **333**, 15-29.
9. Jørgensen, S.E. and Straskraba, M. (2000) Ecosystems as cybernetic systems, in S.E. Jørgensen and F. Müller (eds.), *Handbook of ecosystem theories and management*, Lewis, Boca Raton, pp. 249-264.
10. May, R.M. (1974) *Stability and Complexity in Model Ecosystems*, 2nd ed., Princeton Univ. Press, Princeton, NJ.
11. EDOCC (2000) *Ecological Determinants of Ocean Carbon Cycle*, Final report from the EDOCC workshop held at Timberline Lodge, Oregon, March 13-16, National Science Foundation, Washington, D.C.
12. Eppley, R. W. and Peterson, B. J. (1979) Particulate organic matter flux and planktonic new production in the deep ocean. *Nature* **282**, 677-680.
13. Sarmiento, J.L. and Armstrong, R.A. (1997) *U.S. JGOFS Implementation Plan for Synthesis and Modeling The Role of Oceanic Processes in the Global Carbon Cycle*, U.S. Joint Global Ocean Flux Study Synthesis and Modeling Implementation Plan, U.S. JGOFS Planning Office, Woods Hole, Massachusetts. http://usjgofs.whoi.edu/mzweb/smp/smpimp.htm.
14. Steele, J.H. (1974) *The Structure of Marine Ecosystems*, Harvard University Press, Cambridge.
15. Azam, F. (1998) Microbial control of oceanic carbon flux: The plot thickens. *Science* **280**, 694-696.
16. Laws, E.A., Falkowski, P.G., Smith, W.O., Jr., Ducklow, H., and McCarthy, J.J. (2000) Temperature effects on export production in the open ocean. *Global Biogeochem. Cycles* **14**, 1231-1246.
17. Hutchinson, G.E. (1976) *A Treatise on Limnology*, Vol II, John Wiley, New York.

THERMODYNAMIC APPROACH TO THE PROBLEM OF ECONOMIC EQUILIBRIUM

V. M. SERGEEV
Center for International Studies
Moscow State Institute for International Relations
Moscow, Russia

1. Metaphors of equilibrium: deconstruction

Economic theory represents the unusual case when the ontological assumptions underlying mathematical models are somewhat mathematical models themselves, or more precisely, mathematical metaphors. A basic mathematical metaphor for the classical model of the market is mechanical equilibrium. The basic idea of such a model is that small deviations of the system from a point of equilibrium result in occurrence of "forces", which cause the system to return to the state of equilibrium. In some, very important sense, "the invisible hand" of the market in this model is equivalent to a mechanical force. The economy is considered as a dynamic system. Time stands as a key notion, and the mathematical structure of the economic models is represented by ordinary differential equations. Equilibrium is considered in the mechanical models as a state at which forces applied to the system counterbalance each other, and the potential energy achieves a minimum.[1] Consequently, for applications of the mechanical metaphor of equilibrium to economic theory, some analogues of the mechanical notions are needed. But this conceptualization is not completely harmless. It implies that, upon deviation from the state of equilibrium, the system, acting by itself, will return to this very state. It is well known that there are also some other approaches in physics to the conceptualization of the intuitive notion of equilibrium. In this context, thermodynamic equilibrium should be mentioned first of all. According to this concept, the system approaches a state of equilibrium not because it is being affected by "forces", but because this is the most probable state of the system, consisting of parts, each of which is characterized by its independent dynamics. This may apply to mechanical systems as well, subject to the laws of dynamics. But, in the case that such a system is very complex, its general behavior starts to be determined by absolutely different principles, very dissimilar to those of dynamics. This distinction in the mathematical description of the system's transformation of state is fundamental. In terms of the thermodynamic approach, the system, instead of evolving in time, simply

---

[1] It is worth remembering, by the way, that the subject of modern research in mathematical economy is, as a rule, the state of equilibrium in itself. The dynamics of the system is kept as a metaphor, pointing out the way in which this equilibrium state may be achieved. See Fisher [10], and such classical works as Wald [22], Debreu [7], and Arrow and Debreu [2]. In general, further evolution of the theory produced a state of affairs in which the system's dynamics in the vicinity of the equilibrium point was completely neglected by the researchers.

45

J. Nation et al. (eds.), Formal Descriptions of Developing Systems, 45–58.
© 2003 *Kluwer Academic Publishers. Printed in the Netherlands.*

changes its state in the space of microscopic parameters, remaining on certain surface, defined as the "equation of state." Time is not included into a set of parameters, important for the description of the system. The equation of state is set by linear dependencies between differentials of macroscopic variables, or the so-called Pfaffian equation. During the process of changing one or several macro-parameters of the system, it is simply moved over the surface of the equation of state. In essence, from the mathematical point of view, the investigation of equilibrium in such a model is a problem of differential topology of the surface, described by the equation of state [4].

Such a metaphor differs essentially from the mechanical one. Time occurs here not as an internal parameter of a system, determining its dynamics, but as an external one. Dependencies between differentials of changeable macro-parameters are determined by the internal system structure, for example, energy values of subsystems. In the beginning of the 20-th century, C. Caratheodory [6] proved that it is possible to develop thermodynamic theory logically, drawing exclusively on the assumption that "the equation of state", i.e. the surface in space of thermodynamic variables, corresponding to some Pfaffian equation, exists. In this case, the second law of thermodynamics is formulated as the existence, in any small vicinity of each state of the system, of such states, which cannot be reached without a change of entropy.

In other words, the simple assumption that the "surface of states" does exist appears to be logically sufficient to develop the system of thermodynamic equations. Accordingly, the idea of equilibrium will look completely different. Equilibrium, as against a mechanical metaphor, would mean in a thermodynamic metaphor not the presence of a singular point in the system of the differential equations or the extremum of a potential function, but moving on the integral surface of some Pfaffian equation, or the equation of state. There are most serious grounds to believe that Pfaffian equations (as the thermodynamic metaphor of equilibrium) are much more adequate for the description of economy than is the mechanical metaphor.

With some additional assumptions, for instance, on the thermodynamic system, described by the Pfaffian equations, the principle of Le Chatelieau [19] will be applicable. Namely, the system will demonstrate a behavior opposing influences exercised upon it as the result of changes of the external macroparameters. Economic systems can also demonstrate such behavior under certain conditions. For example, the increase of prices can be an incentive for production in order to support the consumption level; attempts to impose total control over levels of production or consumption trigger the process of corruption of executive bodies, which would diminish the effect of such control, etc.

The important discovery of the past was that the phenomena of such homeostasis in physics, sometimes appearing as almost reasonable behavior, can be explained on the basis of the thermodynamic metaphor, proceeding from the very simple assumption, namely, that for the most time the system will remain in its most probable state. In the case of economy, it may look as if the system is directed, citing A. Smith, by an "invisible hand." But, at the time when A. Smith was working on his book, the principles of thermodynamics were not known yet. Later the "invisible hand" was interpreted in terms of a mechanical metaphor, for which A. Smith's conceptual model, identifying human interests with "forces" of the market, really gave serious grounds.

Let's consider a bit closer the "mental experiment" by A. Smith [21]. Smith assumes that an increase of commodity prices brings about an increase in at least one of the price-making components - the rent, workers' salaries, or profit, forcing actors to change their behavior (to exploit more facilities, increase the number of jobs or to expand manufacturing), that finally leads to the reduction of price. However, there is one very weak spot in the mental experiment by A. Smith. The immediately accessible information for the participants of the market is only the price. Whenever it changes - up or down - the reasons for this shift are not immediately exposed to the observer. It is not quite clear to the latter what kind of changes in the price-making factors led to the fluctuations that have occurred.

In the studies of the Austrian school representatives we find serious arguments for a hypothesis that no complete information on the latent components of the price is available for the market participants. This, consequently, means that they simply cannot behave in the way described by A. Smith's "mental experiment." Wording it rigorously, this mental experiment was based on false assumptions. But that inconsistency necessitates that we then doubt the mathematical metaphor underlying the classical concepts of market dynamics, i.e., the mechanical metaphor. In other words, the ontology of the classical economy model is, indeed, not indisputable. F. A. Hayek [13] directly asserts that the market economy is based on the observance of moral principles, ensuring survival of the community in competition with the others, and that those principles not infrequently directly contradict mercantile interests. Similar views were also developed by M. Weber in his famous work about role of protestant ethics in the genesis of capitalist economy. He expressed conviction that the interests of the community members are satisfied in the course of communal observance of moral principles in a whole.

In neo-classical models of economic equilibrium, such as, for example, by Arrow-Debreu-McKenzie, it was supposed in the way of substantiating the existence of equilibrium, that the utility functions of economic actors are maximized. Thus, arguments about basic incompleteness of the information available to the actors of the market were ignored.

This postulate seems to be one of the basic reasons for the negation of applicability of mathematical methods in economics, claimed by representatives of the Austrian school. It looks as though they were not so much unsatisfied with mathematics itself, as with the mechanical metaphor used for construction of models of equilibrium - a metaphor in which, at the end, utility acted the same role as the potential in the classical dynamic system. However, dissatisfaction with such models of equilibrium was discernible not only on the part of the opponents of mathematical economy, but also among its supporters. The latter were anxious about the absence, within the framework of the mechanical metaphor, of a plausible stability theory of the economic equilibrium. Most disappointments were connected to the impossibility to consider, in all cases, the excess demand as a continuous function of the price. Controversial examples are well known (see, for example, the work by B. Arthur [3]).

It is, of course, possible to "improve" the theory, remaining in the confines of the mechanical metaphor. However, it seems that the arguments of the Austrian School are significant enough to be the reason to change the mathematical metaphor, not rejecting completely the applicability of mathematics in economics. And the arguments for this appear rather pressing. F. A. Hayek's concept of the market as a process of discovery emphasizes the key role of information in the market economy (as opposed to the priority of unobservable functions in neo-classical models, based on the mechanical metaphor of equilibrium). It looks plausible to make an attempt to develop the market equilibrium model proceeding from the mathematical theory of information. As has been proved by Brillouin [5], the mathematical theory of information is, in essence, identical to thermodynamics, if we identify information with negentropy.

The quantity of information, received at interaction of the subject with the system, is measured in information theory by means of the logarithm of the relative reduction of opportunities for choice, with respect to the choices enjoyed by the subject before the receipt of information. It is easily understandable that such an interpretation of information theory directly links it to the description of market actors' behavior, thus making entropy the most important parameter to describe the economic situation. This idea is not a novelty in itself.[2] Nevertheless, up to this time the idea of applying entropy was realized only for special purposes (mainly, in relation to the transportation tasks of economy), and was not used with the aim of building up the theory of economic equilibrium. Entropy alone would not be enough to create the thermodynamic theory of economic equilibrium. To this end, the entire spectrum of ordinary thermodynamic variables, such as temperature, pressure, chemical potential, free energy, etc., should be employed.

The basic idea of the thermodynamic approach to the analysis of economic equilibrium consists in the concept that, if the system is described on two ontological levels - "macroscopic" and "microscopic", and each macroscopic state is characterized by a certain number of microscopic states (called the statistical weight of macroscopic state), the system, generally speaking, will remain in the most probable state, i.e., in a state with the greatest statistical weight.

The conditions of applicability can be formulated for the thermodynamic approach in very general terms. It becomes clear thereby, that the applicability of thermodynamics goes beyond the framework of physical systems. The Large System can be imagined, which can be decomposed into a large number of Small Systems, having their independent dynamics. We admit, that the state of the Large System and its relatively large parts are described by some number of macro-parameters, which are additive, i.e. upon decomposing of the Large System, the total values of its macro-parameters appear to be the sum of values of the same macro-parameters of parts. Such a parameter in physics is energy, if one neglects the superficial interaction between the parts of the Large System. This forms the basis for application of thermodynamic methods in physics. It is possible also to admit that each of the considered systems is characterized, as well, by a set of micro-parameters, which can have various values at the same value of the macro-parameter. Their values are determined by the system's dynamics and, generally speaking, can be of interest in concern to the posed task only in one aspect. Having fixed them, an exact state of the system becomes known and it can be determined how many various micro-states correspond to one macro-state of the system. At this stage, the notion of statistical weight, as the number of micro-states corresponding to one macro-state, and the notion of

---

[2] At the end of the 1960's, A. Wilson suggested that entropy analysis should be used for the examination of transport flows [24]. Wilson mainly used the method of the maximization of entropy to tackle transportation problems. The use of the entropy approach is also well-known in economic studies in relation to the estimation of incomplete data on the basis of ideas by E.T. Jaynes; see Jaynes [14], [15], [16], Levine and Tribus [20], and Golan *et al.* [12]. Note, however, that these works do not include an analysis of economic equilibrium. For the precedent of the application of other methods from statistical physics to socio-economic studies, see for example Durlauf [8], [9].

entropy, as the extent of uncertainty of the macro-state of the system, which is a function of the number of micro-states, can be operated.

If the system is of such a kind that micro-states of the parts of the Large System are statistically independent, it is possible to look for the statistical weight of the Large System as a whole, based on the System components' statistical weights. For that purpose, the multiplying of statistical weights would be quite enough, so that the number of the Large System's micro-states can be found in a purely combinatorial way. If the extent of uncertainty (entropy) is implied to be additive too, we should regard it as the statistical weight logarithm. The logarithm of a product is equal to the sum of the logarithms of the terms, and the additive extent of uncertainty, or entropy, can be obtained for the Large System with statistically independent components.

Presuming that nothing is known about the dynamics of the system, except that it is very complex, the natural assumption for the probability value of any macro-parameter will be that it is proportional to the number of the appropriate micro-states, i.e., to the statistical weight. As the logarithm is a monotone function, the most probable state, i.e. the state with greatest statistical weight, will be at the same time the state with the greatest entropy, i.e., with the greatest extent of uncertainty. It is the core of the second law of thermodynamics - entropy tends to increase, that any part of the Large System goes in the most probable state at the interaction with the other parts. If we "isolate" some component out of the Large System in order to observe of the distribution of its probable states, it becomes clear that one needs to consider the rest of the Large System as a thermostat, that is a body, that ensures equilibrium (in thermodynamic sense) of the distribution of states within the subsystem, which has been isolated. To find the form of this distribution, it proves to be necessary to enter "temperature", equal to the reciprocal value of the derivative of entropy of the Large System on the macro-parameter, for which the conservation law is effective. The introduction of the parameter of equilibrium - i.e., temperature, is simply a result of the condition that there are no exchanges of flows between the parts of the system with respect to the keeping macro-parameter. If there are several macro-parameters of such kind, when there are as many parameters of equilibrium as the conservation laws.

It is worth noticing that nothing that was said above particularly relates either to physics, or physical laws and values. All given reasoning is valid in concern of the Large Systems of any nature, subject to the above-made assumptions. It also looks rather natural in relation to large economic systems, if we regard as macro-parameters the total income, manufacture or consumption of the goods, and as micro-parameters - ways of distribution of income, production or consumption of goods between the subjects of economic activity.

With this approach to the study of economic equilibrium, the necessity to explore subsystems' dynamics is avoided, so that the knowledge of institutional limitations on the goods production and distribution would be enough. Thus, *not only a new approach, appropriate to describe the economic equilibrium, is provided, but, also, the instruments to investigate the impact of institutional restrictions on the state of equilibrium.* Namely, we get the opportunity to build up the mathematical apparatus to analyze the transaction costs theory, to suggest quantitative methods to study the information asymmetry on the market agents' behavior, and to pursue solutions of many other problems as well.

Yet, it should be recognized that, for more than a century, the endeavors to prove thermodynamics within the framework of theoretical physics, by analyzing equations of motion, were not a great success. But no doubt the principles of thermodynamics work by themselves, regardless of possible reductionist interpretations. The two levels of description pose the problem of determination of certain parameters, that may be fully "inconspicuous", whose equality should constitute the condition for equilibrium, intuitively understood as the absence of significant flows between the parts of the system. If macro-parameters are functionally dependent to each other, and the surface of the developed equation of state is differentiable, the appearing linear functions of macro-parameter differentials produce Pfaffian thermodynamics equations.

Thus, the idea of thermodynamic equilibrium is quite appropriate in the framework of description of economic systems. They show observable flows of money, goods and people, and like physical systems, have two levels of description.

Consequently, the description of the economic system should be equivalent to the description of physical systems in thermodynamics, but the parameters of equilibrium that are used - "temperature", "pressure", "chemical potential", certainly will have quite different interpretations, connected with the particularities of economic systems.

At the same time, it is noticeable that thermodynamic terms were often used in relation to the economic systems in a "naive" manner, as metaphors of description: the stock exchange is "overheated", the national economic problems "boiled over" or "cooled down", stock exchange indices are associated with temperature degrees, etc. One of the aims of this work was, besides all, to provide the proof that, not rarely, there is more sense in the "naive" metaphors of such sort, than in the complex mathematical models based on mechanical metaphor of equilibrium, and that, proceeding from the thermodynamic metaphor, the thermodynamic theory in economics can be developed, not only catching hold of economic realities by no means less than the one built on the grounds of the mechanical metaphor, but also availing with the opportunity to account for the role of institutional restrictions for the establishment of economic equilibrium, - a task that has been infeasible to the theories, based on the mechanical metaphor.

2. Entropy and temperature in economic models

To start the creation of the thermodynamic model, the simplest example will first be taken. Let the economic system to be examined be made up of N economic agents, among whom the wholly invariable income is distributed. It will be assumed that there are a lot of modes of income distribution, and we are incapable of foreseeing all possible alternatives. (It is important to note that this assumption fully corresponds to the concept of market suggested by Hayek, where the market is described as a flow of discovery of new procedures and operations). For the sake of simplicity, it will be conceived that the income is quantum, i.e., presented in integers (which is natural, as the smallest currency unit is operational in economy). It is possible to find for each value of total income (E) the quantity of modes of income distribution between agents, as a characteristic of this value: $n(E, N)$, the statistical weight of the state with income E.

At this stage the concept of equilibrium can be introduced. The idea is that two systems (in above-mentioned terms) are in the state of equilibrium, if the distribution function of income is not changeable at their contact, so that there is no income flow between the systems. By "contact" here is understood an "open list" of possible modes of redistribution. It is remarkable that the calculation of statistical weight is feasible in concern to extremely various institutional limitations, imposed on the agents' income, so functions $n(E, N)$ may be different for different systems.

The given model is rather simple - we do not depend on the market at this phase of reasoning. Income restrictions may be sustained coercively, but, as such, it does not matter for this investigation. If two systems are in interaction, one of which has the total income E_1 and the number of economy agents is N_1, and another E_2 and N_2 respectively, then the integrated system will be characterized with the total income $E_1 + E_2$ and the number of agents $N_1 + N_2$. The point is in defining conditions, necessary for the state of equilibrium, or the state with no exchanges of income flows between the systems.

In order to build up the proper theory, we have to develop one more, extremely important, hypothesis on the nature of the systems under investigation. Namely, it should be presumed that all elementary states of income distribution have the same probability. The main ground for such assumption is symmetry of states. As in the theory of probability and in statistics, equal probability of elementary events is assumed just because there are no grounds to prefer one probability to another. Thus, it is of utmost importance for the theory, to include all possible states of distribution. The change of function $n(E, N)$, certainly, will change the results, obtained via the application of the model.

The problem to find the state of equilibrium consists in the determination of conditions under which the income is not to be redistributed between the systems in contact. For this purpose, let the case of redistribution of income be considered, while during the interaction certain part of income ΔE will pass from system 1 to system 2. So, the states and statistical weights of the systems will change and become equal to: $n_1(E_1 - \Delta E, N_1)$ and $n_2(E2 + \Delta E, N_2)$ respectively.

Proceeding from the principle of equal probability, maximum probability for the integrated system is the state in which its statistical weight is maximum, so we should seek the maximum of the function $n_{tot}(E_1, E_2, N_1, N_2)$, with the proviso that the total income $E_1 + E_2$ is constant. The statistical weight of the integrated system, if the first system has income E_1, and the second E_2, and the transfer of economic agents from one system to another is not possible, would be:

$$n_{tot}(E_1, E_2, N_1, N_2) = n_1(E_1, N_1)n_2(E_2, N_2), \qquad (2.1)$$

This means that the number of possible states of integrated system is obtained by means of multiplying of numbers of possible states of component systems. As $E_1 + E_2 = E$, then, at the process of the redistribution of income, $\Delta E_1 = -\Delta E_2$.

Instead of looking for the maximum of the function n_{tot}, we can look for the maximum of $\ln n_{tot}$, because logarithm is a monotone function. That $\ln n_{tot} = \ln n_1 + \ln n_2$ makes it more proper, and condition of maximum turns out to be very simple:

$$\frac{\partial}{\partial E_1} \ln n_1(E_1, N_1) = \frac{\partial}{\partial E_2} \ln n_2(E - E_1, N_2), \qquad (2.2)$$

or, as $dE_1 = -dE_2$,

$$\frac{\partial}{\partial E_1} \ln n_1(E_1, N_1) = \frac{\partial}{\partial E_2} \ln n_2(E_2, N_2). \qquad (2.3)$$

So, two systems are in the state of equilibrium, if they are characterized by the same value of the parameter $\frac{\partial}{\partial E} \ln n(E, N)$. In thermodynamics, the statistical weight logarithm is called the entropy of the system, and the entropy derivative with respect to energy is the reciprocal temperature $\frac{\partial}{\partial E} \ln n(E, N) = \frac{1}{\tau}$. In order to get on the equilibrium state, the contacting systems should have temperatures that are equal in value.

The statement given above can be found in any textbook on statistical thermodynamics [17]. We should analyze here the propriety of such an approach for the economic systems delineated above, at least under the constraints of the model. In terms of the above-made assumptions, it would be concluded that the economic system is inwardly in the state of equilibrium if it is rather homogeneous and there are no flows of income from one of its parts to another in case of separation into parts. It is supposed, certainly, that the homogeneity is kept only if there is no division into too meager parts, when significant fluctuations of flows are observed.

The same postulates are available in physical statistics. Small parts of the system produce significant fluctuations. In physics the question of an equilibrium of small parts of the system is solved on the basis of the assumption (and this assumption is closely connected to a postulate of equal probability of elementary states), that a long lapse of experiment over a small part of system leads to an adequate distribution probabilities of its state. This is one of the formulations of the so-called ergodic hypothesis.

A similar assumption can be made for the economic systems as well. We see that in the thermodynamic model, there are two extremely important characteristics - entropy and temperature. Ignorance of these parameters does not allow us to infer correctly the system's state of equilibrium, as the system (without partitions) is in the state of equilibrium only when its subsystems have identical temperature, and temperature cannot be calculated without knowledge of entropy.

The basic possible objection against the thermodynamic approach to economy is that the number of "particles" in the economic systems (in the given example, the number of the agents of economic activity) is much less than the numbers of particles usually considered in physical systems. If in the physical systems the number of particles, as a rule, is comparable to the Avogadro number, in the economic systems the number of economically active participants usually is approximately 10^3–10^8.

Dispersion has in statistical systems the order $\frac{1}{\sqrt{N}}$, where N is the number of particles. Thus, if the statistical errors connected to the number of particles in physical problems are completely insignificant, in economic problems it is necessary to expect considerably larger errors on the order of $\approx \frac{1}{\sqrt{3}}$, i.e. 3%, or less. It seems, however, that this may be reasonably regarded as a small discrepancy, taking into account the extreme roughness of economic models. One of the tendencies in physics has been to apply thermodynamic approach to systems with a rather small number of particles $10^3 - 10^8$ (nuclear physics, cluster physics, etc.) and the results appear quite valid not only in a qualitative, but also in a quantitative sense. See Frenkel [11].

Here it is necessary to make one important remark. The statistical models in physics show that the energy of a system has a peculiar quality that is responsible for the success of the thermodynamic theory: if some parameter of heterogeneity is a property of the system, then, subject to the condition of a large number of particles, the system has a very sharp entropy maximum, which is obtained when the parameter of heterogeneity tends to zero. Thus, not only is the maximally homogeneous state most probable, but even small deviations from it are *most improbable*.

To illustrate this thesis, it is possible to analyze the entropy of one very simple system. Let us suppose that the income distribution between the subjects of economic activity is arranged as follows: each subject has either zero income, or the fixed income A (presuming that the system of authority is organized in such a way that all deviations are annihilated through a special institutional redistribution mechanism). Such an example is not so much unreal, if one takes into consideration the economic experience of some countries aspiring to implement the "leveling principles" in distribution of income, stipulating that a part of population is generally excluded from economic activities, being allowed to have a very low level of income.) Then it is easy to find out that, if the total number of economic subjects is N, the number of the subjects with income A is L, so that the size of the total income is $L A$, then the number of possible states of this system would be equal to C_N^L Let us consider also this system to be in contact with a source of money. The conclusion can be reached, by comparing the statistical weights of the systems with the different levels of total income, that the biggest statistical weight is a characteristic of the state $E = LA$, where $L = N/2$ (to make it more simple, take N to be even). At the introducing of the parameter of heterogeneity $m = |L - N/2|$, the application of the Stirling formula for factorials will produce the following simple approximation to the dependence of statistical weight on parameter of heterogeneity:

$$n(N, m) = 2^N \left(\frac{2}{\pi N} \right)^{1/2} \exp\left(-\frac{2m^2}{N} \right). \tag{2.4}$$

This shows that statistical weight, (and, consequently, entropy) has an extremely sharp maximum depending on the parameter of heterogeneity, with the width $1/\sqrt{N}$. From the principle of equal probability of elementary states, it closely follows that the entropy of interacting systems will increase. The most probable state will be the state with maximum statistical weight, and, as a consequence, maximum entropy. As the number of possible states for the system decreases sharply with the increase of the parameter of heterogeneity, it would hardly be possible to find the system in such a state that the parameter of heterogeneity exceeds certain value, determined by the number of particles in the system (in the example above it was $1/\sqrt{N}$).

3. Thermostat and the function of income distribution

Now the question will be considered, in what fashion the income is distributed among the subjects of economic activity. Within the framework of the thermodynamic model, it appears that a universal function of distribution persists, whose configuration, if the number of the subjects is held constant, depends only on the temperature. This situation is well known in statistical thermodynamics, where such a distribution is called the Boltzmann distribution.

To examine this distribution, it can be imagined that the economic system X is very large, and out of it a small part Y is separated. The remaining large part of the system will be referred to as the reservoir. If we consider a certain state of the small subsystem with the income E_1, the probability for the small subsystem to have income E_1 would be proportional to the number of possible states of the reservoir with the range of income $E - E_1$, under the assumption that the total income of the system is constant. Then the ratio of the probability P_1 for the subsystem to have income E_1, to the probability of having income E_2, equals to the corresponding probability ratio for the reservoir:

$$\frac{P_1(E_1)}{P_2(E_2)} = \frac{n(E - E_1, N)}{n(E - E_2, N)}. \tag{3.1}$$

But, using the fact that $n = e^{S(E,N)}$, where $S(E, N)$ means the entropy of system, this can be put as:

$$\frac{P(E_1)}{P(E_2)} = e^{S(E-E_1, N)-S(E-E_2, N)} = e^{\Delta S}. \tag{3.2}$$

If the reservoir is far greater than the subsystem in question, we can make an expansion of ΔS in the first order on ΔE into a Taylor series, thus obtaining that $\Delta S = \frac{E_1 - E_2}{\tau}$, in which τ is the reservoir's temperature $\left(\frac{1}{\tau} = \frac{\partial S}{\partial E}\right)$.

It follows from this that:

$$\frac{P_1(E_1)}{P_2(E_2)} = e^{\frac{E_1 - E_2}{\tau}}. \tag{3.3}$$

Thus, the formula of probability distribution for a small system is derived.

If the subsystem Y at the incomes E_1 and E_2 has the number of possible states $n_Y(E_1)$ and $n_Y(E_2)$ respectively, the distribution formula will be:

$$\frac{W_Y(E_1)}{W_Y(E_2)} = \frac{n_Y(E_1)}{n_Y(E_2)} e^{\frac{E_1 - E_2}{\tau}}. \tag{3.4}$$

This formula supposes that we proceed from the state probability P to the probability of the level of income W. It should be noted that only two assumptions were made in these considerations: the system is homogeneous, so that it is liable to be separated into interacting parts without the effect of flows from one part to another, and the system's total income is constant. The development laid out above is actually a replica of the traditional distribution calculation by Boltzmann in statistical thermodynamics.

Hereby we have obtained something more than just a mental experiment for the verification of logical reasoning. When the appropriate data and procedure are at hand, it would be possible to verify this thesis in the context of a real economy, comparing, for example, the income distribution function of various sectors. But this type of verification lies beyond the purpose of this paper. Anyway, in economics the real situation cannot be regarded as a reference for the experiment, because it cannot be reproduced. Therefore, when deviations from the logical conclusions drawn on the basis of modeling, similar to that presented here, are observed in reality, they cannot serve as sufficient grounds for either acceptance, or rejection. It is necessary to study to what extent the postulates of the logical model are embedded in reality, and what would be the outcome of their realization in the especially framed situation.

The Boltzmann distribution function is realizable, with respect to income, in a system which is in contact with the reservoir at a certain temperature. In an isolated system, on the contrary, the temperature depends on income and entropy.

Boltzmann income distribution allows us to understand the relationship between the average income of an economy subject inside the system, and its "temperature". Let us suppose that there are no limitations on the economic agent's income and all levels of income are authorized. Such a situation is commonly associated with the market economy. Then, with the help of Boltzmann distribution function, we are equipped to easily calculate the average income of an economy agent:

$$E = \frac{\int_0^\infty E e^{-\frac{E}{\tau}} dE}{\int_0^\infty e^{-\frac{E}{\tau}} dE}. \tag{3.5}$$

Thus, in the case of no limitations on the economy agent's income, the calculation of equation (3.5) shows that an average income is equivalent to the temperature in the thermodynamic model. Integration here was carried out to ∞, despite the limits of total income present in the real system. But, due to the drastic exponential decrease, this does not matter, subject to the condition that the total income greatly surpasses the average income of a single agent, which is true for the major systems.

It will be seen, however, that under conditions of restrictions on income the relationship between "temperature" and average income may look quite different.

In conclusion of this part of the paper, it should be noticed that, as in statistical thermodynamics, in the thermodynamic model of economy the parameter called the "statistical sum" stands extremely useful:

$$Z_0 = \sum_E n(E, N) e^{-\frac{E}{\tau}}. \tag{3.6}$$

The sum here is taken over all possible values of E. The probability for the system, interacting with a thermostat with temperature τ, to have energy E is expressible through the statistical sum:

$$P(E) = \frac{e^{\frac{E}{\tau}}}{Z_0}. \tag{3.7}$$

With a fixed number of economic agents, the average income in the system at temperature τ would be:

$$\langle E_{cp} \rangle = \frac{\sum_E E n(E, N) e^{-\frac{E}{\tau}}}{Z_0} = \tau^2 \frac{\partial}{\partial \tau} \ln Z_0. \tag{3.8}$$

This means that, having the temperature-dependence of the statistical sum, it is possible to obtain via differentiation the value of the average income. Again, this is fully in agreement with the standard statistical thermodynamics techniques.

One paradoxical example will be tackled now, demonstrating that the use of a thermodynamic approach in application to economic systems can lead to extraordinary results.

4. The interaction of systems with restrictions on income and "free market" systems

In this section the "spin model", already addressed above, will be examined in detail, with two possible values of income: 0 and A. Despite its abstractness, this model is very useful, as it catches some important features of systems that are characterized with restrictions on income. We ask what the results would be, once such a spin system enters into interaction with a "pure" market system that does not suppose limitations on income. In the "pure" market system the entropy increases together with the increase of energy, and the temperature is always positive. It is not always like that for the restriction-on-income system. Actually, if the number of economy agents L with non-zero income surpasses the half of a total number of agents N, the number of probable states of the system n, which is equal to C_N^L, starts to diminish, and the reverse temperature, expressible as:

$$\frac{\partial}{\partial E} \ln n = \frac{1}{\tau}$$ (4.1)

is becoming negative. The question arises here, how the negative temperature could be explained. Phenomena of this kind have been investigated quite well in the course of the laser theory development [18]. Negative temperature implies an inverted population, i.e., a situation in which levels characterized by higher energy have a greater population than those with lower energy. In such a state, the system is imminently ready to give off energy upon contact with any system with positive temperature.

How can this be referred to the income distribution model of the economic system that has been described above? If two systems with positive temperature contact, it means that the redistribution of income goes in the direction from the higher to the lower-temperature system. As was seen in the case of the free market model, the temperature equals to the average income per economic agent. The redistribution of income from the "richer system" to the "poorer one" takes place in accordance with our intuitive understanding. In this case nothing that could be considered as something counter-intuitive is going on. However, at the interaction of the free market system and the income-restricted system, under certain conditions a kind of counter-intuitive process is observed, in which redistribution from the "poorer" system to the "richer" one takes place.

Let two systems, X and Y, be considered. Let X be a free market system, with its temperature equal to the average income per market agent. Let system Y be a spin model, where income is limited with the maximum income equal to A. It is conceived that the temperature of X is $\tau_x > NA$, and that the number of Y's subjects having non-zero income is more than half of their total number, so that a negative temperature occurs. What is going to happen at the interaction of the systems in question? The entropy of the joint system will increase, while the entropy of system Y will increase simultaneously, thereby precipitating a disordering of the system Y, which is effected by the transfer of part of its income to X. The latter is to take place, despite the fact that the average income of the system X is already higher than Y's average income.

As a result, system Y, with limited incomes, will get poorer, while the system X, with unlimited incomes, will get richer. This redistribution of income will not be an effect of coercion, but just a consequence of the fact that the combined system tends to acquire the most probable state. As for coercion, it may play a certain role, but not in the form of coercion of X towards Y, but rather as coercion inside Y, aimed at obstructing a rise of income above certain level.

Referring all this to real economic situations, our pattern model allows us to draw two important conclusions. 1)The policy of harsh restriction of income is dangerous, if the contact with free market surroundings is irreversible. 2) If such a policy was already embarked on, and the economy in which it was implemented is isolated from the free market, then the effect of the "market reforms" will depend on the sequence of two main steps - the release of incomes and the establishment of contact with the free market surroundings. For the "non-market" country, it would be necessary first to release the income from restrictions and then to "open" the economy to the market, only after the equilibrium had been achieved. Otherwise, the resources of the economy would be "sucked out" outwardly before the establishment of balance.

Of course, the arguments put forward here are based on a very simple idealized model. But it appears that our reasoning provides the grounds for explication of differences observed at the carrying out of market reforms in Eastern Europe, on the one hand, and in China and Vietnam, on the other. In Eastern Europe reforms were conducted by a "shock" mode: the national economies were "opened up" for outside activities without preliminary creation of the market institutions inside. The results mostly turned out catastrophic (flight of capital and collapse of production). In China and Vietnam there a reverse order was implemented - first the release of capitals inside and then the gradual opening up for outsiders, according to the process of creation of domestic market. Such a policy resulted in rapid economic growth.

Another case of interaction of the system with restricted incomes and the free market system is rather of interest as well. Let us consider the "spin system" with the fixed individual income A, having L wages for the community of N workers. It is quite understandable that, with a fixed value of A and an increase of the total value of wages (i.e., number of wages), the system's temperature would be negative at $L > \frac{N}{2}$. This means that in the state of equilibrium, the free market system stipulates that half of the workers should not get wages, or should be unemployed, as it is a far more probable state than the other ones with greater ratings of employment. So, this model includes unemployment as an inherent characteristic of the system in the state of equilibrium. Thus, the famous paradox, which stood as a stumbling block for the neo-classical school of economists, is dismissed. Namely, it reads: why is wages equilibrium not realizable in the case that the free labor force market is in operation? [3] It is clear that negative temperature may emerge in the system imposing restrictions on the salary growth. Really, to increase the wages' total value for the workers in the situation of a salary-restricted economy with the value of wage restricted to A, means that at the total value of wages NA the entropy will become zero. This implies that, at least for some interval $E_0 < E < NA$ temperature, expressible through $\tau = \frac{1}{\partial S/\partial E}$, will be negative, i.e. at any contact to free market system the total wages' value will fall, at least up to E_0. If, besides, there is a minimal wage value B, the number of workers will be determined by the following expression: $L < \frac{E_0}{B}$. That restriction will keep salary from turning down, providing for the appearance of a certain rate of unemployment, which will be more than $a = \frac{N-L}{N}$. In reality the wage level always is limited "from below" to the level of biological survival. Therefore, in the case when salary is restricted from above, the opportunity for unemployment is being created.

The analysis given above shows that unemployment emerges in the state of equilibrium at the positive temperature, if the wage value is limited simultaneously above and below. In our view, under certain circumstances it may serve as a proof of Keynesian arguments on the reasons of unemployment. Speaking about real situations, the businessmen aspire to limit the salary "from above", the trade unions "from below". This very

---

[3] The problem of unemployment is of utmost importance for the discourse on economic balance, as it represents a counter-example to the neo-classical theory of balance, stating that no excess demand is possible. This served as one of the greatest incentives for the emergence of the institutional approach to the economy. See Chapter 4 of Williamson [23] and Akerloff [1].

combination should, according to our model, result in the emergence of unemployment.

References

1. Akerloff, G.A. (1984) *An Economic Theorist's Book of Tales: Essays that Entertain the Consequences of New Assumptions in Economic Theory*, Cambridge Univ. Press, Cambridge.
2. Arrow, K.J. and Debreu, G. (1954) Existence of an equilibrium for a competitive economy, *Econometrica* **22**, 265–290.
3. Arthur, W.B. (1988) Self-reinforcing mechanisms in economics, in P. W. Anderson, K. J. Arrow, and D. Pines (eds.), *The Economy as an Evolving Complex System*, Addison-Wesley, Redwood City, CA, 9–31.
4. Born, M. (1920) Betrachtungen zur Traditionallen Darstelling der Termodinamik, *Physik Zschr.* **22**, 218–224, 249–254, 282–286.
5. Brillouin, L. (1956) *Science and Information Theory*, Academic Press, New York.
6. Caratheodory, C. (1909) Untersuchungen über die Grundlagen der Thermodynamik, *Mathematische Annalen* 67.
7. Debreu, G. (1959) *Theory of Value: an Axiomatic Analysis of Economic Equilibrium*, Wiley, New York.
8. Durlauf, S. (1993) Nonergodic economic growth, *Rev. Econ. Studies* **60**, 349–366.
9. Durlauf, S. (1996) A theory of persistent income inequality, *J. Econ. Growth* **1**, 75–93.
10. Fisher, F.V. (1963) *Disequilibrium Foundations of Equilibrium Economics*, Cambridge Univ. Press, Cambridge.
11. Frenkel, J. (1955) *Principles of the Theory of Atomic Nuclei*, Moscow (in Russian).
12. Golan, A., Judge, G. and Miller, D. (1996) *Maximum Entropy Econometrics: Robust Estimation with Limited Data*, Wiley, Chichester.
13. Hayek, F.A. (1988) The fatal conceit, in *Collected Works of F. A. Hayek*, University of Chicago Press.
14. Jaynes, E.T. (1957) Information theory and statistical mechanics, *Phys. Rev.* **106**, 620–630.
15. Jaynes, E.T. (1957) Information theory and statistical mechanics, *Phys. Rev.* **108**, 171–190.
16. Jaynes, E.T. (1983) Prior information and ambiguity in inverse problems, in *SIAM-AMS Proceedings* **14**, Amer. Math. Soc., Providence, pp. 151–166.
17. Kittel, Ch. (1970) *Thermal Physics*, Wiley, New York.
18. Klein, M.J. (1956) Negative absolute temperature, *Phys. Rev.* **104**, 589.
19. Landau, L. and Livschitz, E. (1963) *Statistical Physics*, Nauka, Moscow.
20. Levine, R.D. and Tribus, M. (1983) Foreword, in *The Maximum Entropy Formalism*, MIT Press, Cambridge.
21. Smith, A. (1776) *An Inquiry to the Nature and Causes of the Wealth of Nations*, London.
22. Wald, A. (1951) On some systems of equations of mathematical economics, *Econometrica* **19**, 368–403.
23. Williamson, O.E. (1975) *Markets and Hierarchies: Analysis and Antitrust Implications*, Free Press, New York and McMillan, London.
24. Wilson, A.G. (1970) *Entropy in Urban and Regional Modeling*, Pion, London.

Part 2.

Biological Systems

Part 2

Biological Systems

PROGRAMMED DEATH PHENOMENA AT VARIOUS LEVELS OF DEVELOPMENT OF THE LIVING SYSTEMS

VLADIMIR P. SKULACHEV
*Belozersky Institute of Physico-Chemical Biology and School of
Bioengineering and Bioinformatics,
Moscow State University*

Abstract: A concept is presented assuming that any complex biological system, from intracellular organelle (mitochondria) to multicellular organism is equipped with a program of self-elimination. Such a suicide program is actuated when the system in question appears to be unwanted for a system occupying a higher position in biological hierarchy. This principle called the "Samurai law of biology" ("It is better to die than to be wrong") will be illustrated considering defence of organelles, cells, organs and organisms against damaging effects of reactive oxygen species (ROS). In bacteria, DNA damage initiates (a) induction of synthesis of reparation enzymes, (b) arrest of the cell divisions and (c) autolysin activation resulting in the programmed death. In mitochondria, ROS can open the permeability transition pore, initiating in this way programmed death of mitochondria (mitoptosis), which can purify intracellular population of mitochondria from the ROS-overproducing organelles. In yeast, H_2O_2 induces some proteins causing programmed death. Inhibition of the protein synthesis prevents this effect. Also in yeast, high levels of a pheromone proved to cause ROS formation resulting in programmed death. In human HeLa cells, tumour necrosis factor α (TNF) initiates ROS formation and then programmed death (apoptosis). At supracellular level, the programmed death signal is transmitted from the TNF-treated to intact cells and such a transmission is arrested by catalase, indicating that H_2O_2 serves as an intercellular programmed death messenger. Conversion of a tadpole to a frog is shown to be mediated by thyroxine causing induction of an NO synthase in the tadpole tail cells. This results in strong increase in the H_2O_2 level due to inhibition by NO of catalase and gluthatione peroxidase. Moreover, NO causes antimycin A-like inhibition of mitochondrial respiratory chain, which strongly stimulates ROS production. Due to massive apoptosis, the tail disappears (organoptosis). It is suggested that ROS mediate aging which is considered as programmed death of organism (phenoptosis). The "Samurai law" is regarded as a mechanism preventing the great destructive potential of environment, as well as of the living systems per se, from being realised. It helps organisms to maintain intact their genomes that were developed during billion years of evolution but can be destroyed during one or several generations by a single mutation in one of thousands genome-composing genes.

1. Introduction: Weismann's Hypothesis Revisited

More than a century ago, August Weismann hypothesized that death of old individuals had been invented by biological evolution as a kind of adaptation. He wrote: «Worn-out individuals are not only valueless to the species, but they are even harmful, for they take the place of those, which are sound... I consider that death is not a primary necessity, but that it has been secondarily acquired as an adaptation. I believe that life is endowed with a fixed duration, not because it is contrary to its nature to be unlimited, but because the unlimited existence of individuals would be a luxury without any corresponding advantage» [1].

Weismann's hypothesis on aging as an adaptive mechanism was strongly

J. Nation et al. (eds.), Formal Descriptions of Developing Systems, 61–86.
© 2003 *Kluwer Academic Publishers. Printed in the Netherlands.*

criticized by Medawar [2], who assumed that aging could not have developed during the course of biological evolution. Medawar in fact assumed that, under natural conditions, the majority of organisms die before they become old. This assumption, however, cannot be applied to some periods of evolution of many species [3,4].

Moreover, Medawar did not take into account a possibility that individuals with changes in their genomes can dramatically affect the fate of a population even if they amount to only a very small part the population. Muir and Howard [5] recently published an excellent example illustrating this statement. A fish with an inserted human growth factor gene was studied. The transgenic animals have increased growth rate. In a mixed population of modified and unmodified animals, the larger modified males attracted four times as many mates as their smaller rivals.[1] However, only two-thirds of the modified animals survived to reproductive age.

Thus, the modification decreases the reproductive potential of the fish. Calculations showed that the whole population would become extinct within just 40 generations if 60 transgenic fish joined a population of 60,000 fish. Even a single transgenic individual could have such an effect, although extinction would take a longer time [5]. This study gave a quantitative description to practices applied for many years as a defense against some insects. To the insect population, some sterilized males were added, which resulted in extinction of the population since the balance between reproduction and death of insects proved to be shifted toward death.

Lewis wrote in his recent review: «It is quite possible that the main danger unicellular organisms face are not competitions, pathogens or lack of nutrients, but their own kin turning into «unhopeful monsters» causing death of the population» [7]. Muir and Howard's work clearly shows that this statement should be extended to multicellular animals for which an «anti-monster» defense is perhaps even more important than for a bacterium because of the much higher complexity of their organization. The very fact that a simple change in activity of a single gene in small minority of organisms composing a population can be sufficient to kill this population points to existence of a very well-organized system preventing such changes in the genome. It is this system that allows a species to exist for millions years in changing environmental conditions.

The system in question seems to include special measures (i) to prevent oxidative and any other damage to the genome, (ii) to repair a damage if it, nevertheless, appears, and (iii) to purify the population from genetically impaired individuals.

Among the type (iii) mechanisms, some suicide phenomena are involved. In this connection, I formulated the «Samurai law of biology»: It is better to die than to be wrong; or, in more detail: Any biological system from organelle to organism possesses a program of self-elimination. This suicide program is actuated when the system appears to be dangerous for any other system of higher position in the biological hierarchy [8]. In this review, I shall summarize possible role of such mechanisms in development of modern living creatures and in preservation of their genetic programs (see also [4,9]).

---

[1] Interestingly, tall men, like larger fish, were shown to have more reproductive success [6].

2. Mitoptosis, Apoptosis, and Organoptosis

2.1. MITOPTOSIS, PROGRAMMED ELIMINATION OF MITOCHONDRIA

In 1992, my coworker Dmitry Zorov and his colleagues suggested that mitochondria, the energy-transducing intracellular organelles, possess a mechanism of self-elimination [10]. This function was ascribed to the so-called permeability transition pore (PTP). The PTP is a rather large nonspecific channel located in the inner mitochondrial membrane. The PTP is permeable for compounds of molecular mass < 1.5 kDa. The PTP is usually closed. A current point of view is that PTP opening results from some modification and conformation change of the ATP/ADP antiporter, the protein normally responsible for exchange of extramitochondrial ADP for intramitochondrial ATP. Oxidation of Cys56 in the antiporter seems to convert it to the PTP in a way that is catalyzed by another mitochondrial protein, cyclophilin [11-13]. When opened, the PTP makes impossible the performance of the main mitochondrial function, i.e., coupling of respiration with ATP synthesis. This is due to the collapse of the membrane potential and pH gradient across the inner mitochondrial membrane that mediate respiratory phosphorylation. Membrane potential is also a driving force for import of cytoplasmic precursors of mitochondrial proteins. Moreover, it is strictly required for the proper arrangement of mitochondrially-synthesized proteins in the inner membrane of the mitochondrion. Thus, repair of the PTP-bearing mitochondrion ceases, and the organelle should perish [14-17].

It is noteworthy that the above scheme of elimination of a mitochondrion does not require any extramitochondrial proteins. It can be initiated by a signal originating from a particular mitochondrion, such as reactive oxygen species (ROS) produced by the mitochondrial respiratory chain. ROS seem to oxidize the crucial SH-group in the ATP/ADP-antiporter, thereby actuating the elimination process. This is why one can consider this effect as the programmed death of the mitochondrion (mitochondrial suicide). This event I coined *mitoptosis*, by analogy with apoptosis, the programmed death of the cell [14]. I also suggested that the biological function of mitoptosis is the purification of the intracellular population of mitochondria from those that became dangerous for the cell because their ROS production exceeded their ROS scavenging capacity. This situation may well be a particular case of the above-mentioned "Samurai law".

2.2. MASSIVE MITOPTOSIS RESULTS IN APOPTOSIS

Opening of the PTP leads to an osmotic disbalance between the mitochondrial matrix and cytosol, swelling of the matrix and, consequently, the loss of integrity of the outer mitochondrial membrane, thus releasing the intermembrane proteins into the cytosol. Among them, five proteins are of interest in this context: cytochrome *c*, apoptosis-inducing factor (AIF), the second *m*itochondrial apoptosis-*a*ctivating protein (Smac; also abbreviated DIABLO), procaspase 9 and endonuclease G [4, 18-20]. All these proteins are somehow involved in apoptosis.

AIF was the first mitochondrial component for which the ability to activate apoptosis was revealed [21]. AIF is a flavoprotein of sequence resembling that of dehydroascorbate reductase, an enzyme found in the intermembrane space of plant mitochondria [22]. The reductase regenerates ascorbate, the main water-soluble antioxidant, from dehydroascorbate, using NADH as the reductant. One might speculate that AIF was originally used by mitochondria as an antioxidant enzyme and later was employed by the cell as an apoptosis-activating factor [8]. When released from mitochondria, AIF goes to the nucleus and activates a nuclease that decomposes nuclear DNA [22].

The route of cytochrome c-mediated cell death was shown to be more complicated [23-26]. In cytosol, cytochrome c combines with very high affinity with a cytosolic protein called apoptosis-activating factor 1 (Apaf-1) and dATP. The complex, in turn, combines with an inactive protease precursor, procaspase 9, to form the «apoptosome». As a result, several procaspase 9 molecules are placed near each other, and they cleave each other to form active caspases 9. When formed, caspase 9 attacks procaspase 3 and cleaves it to form active caspase 3, a protease that hydrolyses certain enzymes occupying key positions on the metabolic map. This causes cell death. At least in certain tissues, procaspase 9 and some procaspases are also localized mainly in the intermembrane space [27].

Smac, the fourth proapoptotic protein of the intermembrane space, was recently shown to bind cytoplasmic inhibitors of apoptosis-activating proteins (IAPs) which suppress the activities of caspases 9, 3, and others. Smac•IAP complexes lack the antiapoptotic activity inherent in the free IAPs [28,29].

As to endonuclease G, it is also hidden in the intermembrane space. When releases from this space, endonuclease G goes to the nucleus and attacks DNA [18-20].

Considering these data, the following scenario of the final steps of the defense of a tissue from mitochondrion-produced ROS seems to be most likely. ROS induce PTP opening and, consequently, release of cytochrome c and other proapoptotic proteins from mitochondria to the cytosol. If this occurs in a small fraction of ROS-overproducing mitochondria, these mitochondria die. The cytosol concentrations of proapoptotic proteins released from the dying mitochondria appear to be too low to induce apoptosis. If, however, more and more mitochondria become ROS-overproducers, the concentrations in question reach a level sufficient for the induction of apoptosis. This results in purification of the tissue from the cells whose mitochondria produce too many ROS [9].

It should be stressed that dysfunction of mitochondria per se can be a reason for cell death in tissues where phosphorylating respiration is the major source of ATP. However, apoptosis caused by cytochrome c and other intermembrane «death proteins» occurs much earlier than the mitochondrial dysfunction results in exhaustion of ATP. On the contrary, dATP and/or ATP are required for apoptosis (e.g., to actuate the apoptosome formation, see above). It looks as if cytochrome c and other mitochondrial «death proteins», when they appear in large amounts in the cytosol, represent a signal for the cell that something is dramatically wrong with its mitochondrial population. Such a cell commits suicide, following the "Samurai law".

In 1994, I postulated a scheme in which mitoptosis is an event preceding apoptosis [15]. In the same year, Newmeyer and co-authors published the first indication of a requirement of mitochondria for apoptosis [30]. Quite recently,

Tolkovsky and her coworkers presented direct proof of the mitoptosis concept [31,32]. In the first set of experiments, axotomized sympathetic neurons deprived of neuron grow factor were studied. It was found that such neurons died within a few days, showing cytochrome c release and order typical features of apoptosis. However, the cells survive if a pan-caspase inhibitor Boc-Asp (O-methyl)-CH_2F (BAF) was added a day after the growth factor deprivation. The cell survival was due to that the mitochondrion-linked apoptotic cascade was interrupted downstream of the mitochondria. Electron microscopy showed that in majority of such cells *all the mitochondria disappear within 3 days* after the BAF addition (Fig. 1A) [31]. Later, the same group reported [32] that a similar effect could be shown using a classical experimental models of apoptosis, HeLa cells treated with staurosporine. Again, addition of BAF to the staurosporine-treated cells resulted in that (i) the cells lived longer and (ii) mitochondria disappeared in the time scale of days. This was shown to be accompanied by disappearance of mitochondrial DNA and as well as the cytochrome oxidase subunit IV encoded by nuclear DNA. On the other hand, nuclear DNA, Golgi apparatus, endoplasmic reticulum, centrioles, microtubules, and plasma membrane remained undamaged. Mitoptosis was prevented by overexpression of antiapoptotic protein Bcl-2, which is known to affect mitochondria upstream from the cytochrome c release.

Apparently, disappearance of mitochondria in the apoptotic cells without BAF could not be seen since the cells die too fast to reveal massive mitoptosis and subsequent autophagia of dead mitochondria. On the other hand, inhibition of apoptosis at a post-mitochondrial step prevented fast death of the cells so there was time for mitoptosis to be completed (Fig. 1B).

It should be mentioned that, besides the above described scenario, there are alternative and more delicate mechanisms of apoptosis, which do not abolish oxidative phosphorylation and do not kill mitochondria. They consists in modifications of porin localized in the inner mitochondrial membrane. Modified porin becomes permeable for proteins that release from the intermembrane space to cytosol. Modification in question can be carried out (i) by proapoptotic protein Bax migrating from cytosol to mitochondria in response to some apoptogenic stimuli [33,34] or (ii) by oxidation of porin by superoxide [35]. It seems probable that the mitoptotic (PTP-linked) and the non-mitoptotic (porin-linked) mechanisms are actuated by the mitochondrion-produced and extramitochondrial ROS, respectively.

Independently of the mechanism of release of mitochondrial intermembrane proteins, all the above mechanisms of apoptosis employ mitochondria as amplifiers of a suicide signal. A precedent is described when such a function, rather oxidative phosphorylation, appears to be the main one for these organelles. During differentiation of eosinophiles, mitochondria were shown to be required for apoptosis. As to the cellular respiration, it was cyanide resistant and could not produce membrane potential. All the energy-linked functions, including formation of membrane potential on the inner mitochondrial membrane, were supported hydrolysis of glycolytic ATP [36].

On the other hand, sometimes apoptosis was found to be mitochondrion-independent. This usually occurs when an apoptotic stimulus is very strong and can be realized without mitochondrion-mediated amplification [9].

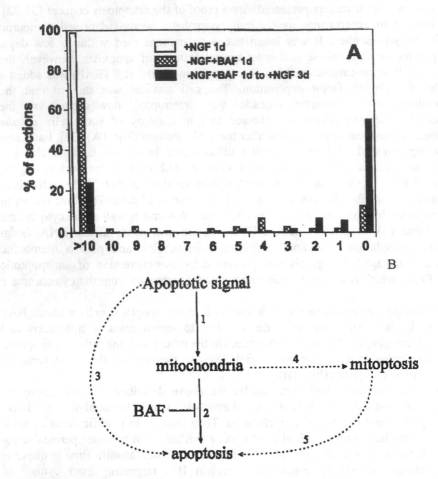

Figure 1. Mitoptosis in sympathetic neurons after nerve grow factor (NGF) deprivation and inhibition of apoptosis by pan-caspase inhibitor Boc-Asp(o-methyl)-CH_2F (BAF). (A) Up to 202 individual cell sections from each of the three treatments, namely (i) 1 day with NGF, (ii) 1 day without NGF and with BAF, and (iii) 1 day without NGF and with BAF to subsequent 3 days with NGF, were sorted into 12 groups (section containing 0-10 or > 10 mitochondria per cell section). Cells with NGF contained 59±10 mitochondria per section (From Fletcher et al. [31]). (B) A scheme explaining results of the above experiment. Apoptotic signal (NGF deprivation) reaches mitochondria (1), which multiply the signal via BAF-sensitive caspase activation and induce fast apoptosis (2). Moreover, slow apoptosis can be induced in BAF-resistant fashions. They can be mitochondrion-independent (3) or mitochondrion-dependent. In the latter case, mitoptosis takes place (4), an event entailed by a cell death (5). It should be stressed that the three types of apoptosis may differ not only in mechanisms but also final manifestations of the death program. Solid and dashed lines, fast and slow processes, respectively (From Skulachev [4]).

2.3. PROGRAMMED DEATH AT SUPRACELLULAR LEVEL: BYSTANDER EFFECT

Several cases were described when apoptotic cells formed clusters in tissues in vivo or in cell monolayers in vitro (so-called death of bystander cells) [37-49]. In 1998,

I suggested that it is H_2O_2 produced by an apoptotic cell that serves as an apoptogenic signal to the bystander cells surrounding the apoptotic cell [50]. In this way, I explained a role of H_2O_2 in antiviral defence of the tissue. At least in some cases, infected cells have been shown to produce H_2O_2 not only to commit suicide but also to induce apoptosis in nearly cells that are most probable targets for the virus during the expansion of the infection. As a result, a "dead area" is organized around the infected cell (for details, see [9,50]). In 2000, Bakalkin and coworkers [45] applied a combined experimental and mathematical analysis to this phenomenon. Monolayers of the human osteosarcoma Saos-2 cells were studied. Apoptosis was induced by serum deprivation. The randomness in spatial distribution of apoptotic cells was tested using the nearest neighbor distance and K-means clustering statistics as described for cell clustering in tissue section [40]. The nearest neighbor distance between apoptosis cells were analyzed using Clark-Evans statistics [51] The number of apoptoses around an apoptotic cell was found to be two to three times greater than around a viable cell. The bystander effect, in agreement with my suggestion [50], disappeared when catalase, the H_2O_2 scavenger, was added to the growth medium. In the same study, it was found that apoptosis of Saos-2 cells is accompanied by great increase in the H_2O_2 production. The H_2O_2 concentration proved to be strongly elevated in both the apoptotic cells and the serum-deprived medium where these cells were grown [45]. When H_2O_2 was added to intact Saos-2 cells, they entered apoptosis [45] (see also [52-57]).

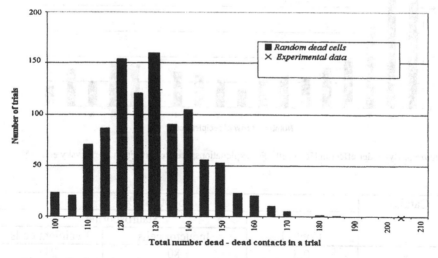

Figure 2. Total member of dead-dead contacts for a randomly selected dead cells (93 dead cells from 327 total in the graph) (From Alexeevsky et al. [58]).

Quite recently, we extended the Bakalkin et al. observation to human cervical carcinoma HeLa cells treated with another apoptogen, tumor necrosis factor α (TNF). The protein synthesis inhibitor emetine was added to prevent the NF-κB-mediated antiapoptotic effect of TNF. Experiments performed by O. Ivanova and L. Domnina clearly showed clustering of apoptotic cells. A. Alexeevsky and D. Alexeevsky created a computer program Clud ("ClusterDetector") to (i) calculate number of apoptotic cells contacting with a given apoptotic cell in the HeLa cell monolayer (an experimental value) and (ii) compare this value with a theoretical one obtained assuming occasional distribution of apoptotic cells in the monolayer. The result (Fig. 2) clearly showed clustering of apoptotic HeLa cell [58].

Another series of experiments with HeLa cells was performed in our group by O. Pletjushkina and E. Fetisova. The HeLa cells were grown on two cover glasses. One of them was treated with a protein kinase inhibitor staurosporine to induce apoptosis. Then this glass with the cells attached to it was removed from the staurosporine-containing medium and put to a fresh medium, being placed side by side to the second glass with the cells that were not treated by staurosporine. It was found that apoptotic cells appeared on the second glass, the number of such cells decreasing with increase in the distance from the first glass (Fig. 3). Such an effect was strongly suppressed by catalase (Table 1).

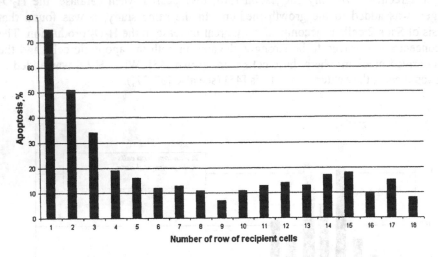

Figure 3. Bystander effect in HeLa cells. For explanations, see the text (From Alexeevsky et al. [58]).

Catalase	Apoptosis, %		
		Staurosporine	
	Control	Inductor cells	Recipient cells
-	0.5	80	21
+	0.5	82	6

Table 1. Catalase inhibits intercellular transmission of apoptotic signal of staurosporine (From Aleveevsky et al. [58]).

Thus, we can conclude that the hypothesis on H_2O_2-mediated bystander effect is experimentally proved. (Such a phenomena is apparently of great importance in plants where H_2O_2 and mitochondria are involved in so-called hypersensitive response [59,60]).

This does not mean that H_2O_2 is the only intercellular messenger of apoptosis, produced by apoptotic cell. $NO^.$, like H_2O_2, is often formed in large amounts during apoptosis. This small neutral molecule, also like H_2O_2, can (i) easily penetrate through the cell membrane and (ii) induce apoptosis [61,62]. Therefore $NO^.$ might be regarded as one more candidate to the role of intercellular transmitter of apoptotic signals. This was apparently not the case in Bakalkin's experiments since N^G-monomethyl-L-arginine, an inhibitor of $NO^.$ synthase, did not affect clustering of apoptotic Saos-2 cells whereas catalase completely prevented the clustering if a p53-containing cell line was used [45]. In our system, however, catalase did never completely suppress the bystander effect in HeLa cell. Interestingly, that (i) catalase failed to arrest completely clustering of a p53-negative Saos-2 cell line [45] and (ii) HeLa cells studied in our experiments are also p53-negative [51]. Possible involvement $NO^.$ is now under study in our group.

2.4. ORGANOPTOSIS, PROGRAMMED ELIMINATION OF AN ORGAN

It is obvious that massive apoptosis of cells composing an organ should eliminate the organ. This process can be defined as «organoptosis». For example, consider the disappearance of the tail of a tadpole when it converts to a frog. It was recently reported by Kashiwagi et al. [63] that addition of thyroxine (a hormone known to cause regression of the tail in tadpole) to severed tails surviving in a special medium caused shortening of the tails that occurred on the time scale of hours. The following chain of events was elucidated:

$$\text{thyroxine} \rightarrow NO^. \text{ synthase induction} \rightarrow [NO^.]\uparrow \rightarrow [H_2O_2]\uparrow \rightarrow \text{apoptosis} \quad (1)$$

Mechanisms of the $[H_2O_2]$ increase by $NO^.$ remains to be revealed. This may be inhibition by $NO^.$ of the main H_2O_2 scavengers, catalase and glutathione peroxidase (for refs., see [63]). One more possibility consists in affecting the mitochondrial respiratory chain. As was shown by Borevis and his colleagues, $NO^.$ inhibits the respiratory chain of mitochondria or submitochondrial particles in vitro at levels of cytochrome oxidase and cytochrome b. This was shown to be accompanied by H_2O_2 production which was as fast as in the presence of antimycin A [64,65]. Later the groups of Boveris and Cadenas showed the same phenomenon in isolated beating rat heart [66]. They also analyzed non-enzymatic mechanisms of $CoQH_2$ interaction with $NO^.$ as well as with $NO^.$ derivatives [67]. Among them three reactions of semiquinone formation were identified (see below, reactions 2, 5 and 6).

$$CoQH_2 + NO^. \rightarrow CoQ^{.-} + NO^- + 2H^+ \quad (2)$$
$$CoQ^{.-} + O_2 \rightarrow CoQ + O_2^{.-} \quad (3)$$
$$NO^. + O_2^{.-} \rightarrow ONOO^- \quad (4)$$
$$CoQH_2 + ONOO^- \rightarrow CoQ^{.-} + NO_2^. + H_2O \quad (5)$$
$$CoQH_2 + NO_2^. \rightarrow CoQ^{.-} + NO_2^- + 2H^+ \quad (6)$$

These events including $NO^.$ and such $NO^.$ derivatives as peroxinitrite ($ONOO^-$) and nitrogen dioxide ($NO_2^.$), may result in $CoQ^{.-}$ formation even under conditions when enzymatic oxidation of $CoQH_2$ by terminal span of the respiratory chain is inhibited by $NO^.$ (Fig. 4).

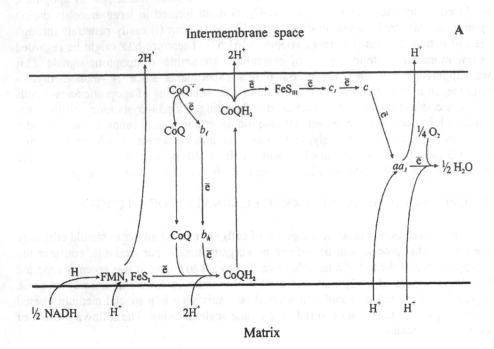

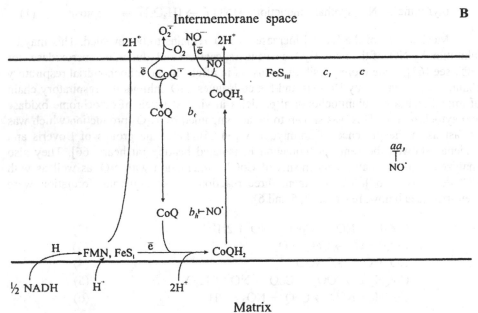

Figure 4. Respiratory chain of the inner mitochondrial membrane, producing H_2O in the absence of $NO^{\cdot}$ (A) or producing O_2^{-} in the presence of $NO^{\cdot}$ (B).

Thus the following mechanisms of $[H_2O_2]$ increase by $NO^{\cdot}$ may be postulated.

(1) $NO^{\cdot}$ increases O_2 flux to the tissue due to the well-known dilating effect of this compound on the blood vessels.

(2) $NO^{\cdot}$ decreases O_2 consumption in the tissues due to inhibition of the respiratory chain, which is the main mechanism utilizing O_2 inside the cell.

(3) $NO^{\cdot}$ converts (directly or via $ONOO^-$ and $NO_2^{\cdot}$) $CoQH_2$ to $CoQ^{\cdot}$, an excellent one electron O_2 reductant producing $O_2^{\cdot-}$ which is converted to H_2O_2 by superoxide dismutase.

(4) $NO^{\cdot}$ arrests decomposition of the formed H_2O_2 by means of inhibition of catalase and glutathione peroxidase.

3. Phenoptosis, Programmed Death of an Organism

Obviously, massive apoptosis in an organ of vital importance, resulting in organoptosis, must entail death of the entire organism. On the face of it, such an event should be regarded as a lethal pathology of no biological sense. However, this may not be the case if the organism in question is a member of a kin or community of other individuals. Here, altruistic death of individuals may appear to be useful for a superorganismal unit, being a mechanism for adaptation of the group of organisms to a changing environment. If it is the case, the chain of events: «mitoptosis → apoptosis → organoptosis» should be supplemented with one more step, i.e., the programmed death of an organism. I called such an event «*phenoptosis*». Phenoptosis can be defined as *a mechanism purifying kin or community from individuals that are not longer wanted* [9, 68-70].

3.1. PHENOPTOSIS AMONG BACTERIA

For unicellular organisms, programmed death of a cell is phenoptosis. It is already clear that the bacterial cell has suicide programs that are actuated when there is something dramatically wrong with the cell.

For instance, bacteria have a protein that monitors DNA damage. It is called RecA. DNA damage activates RecA, which hydrolyses the LexA repressor [71]. This results in derepression of genes encoding (i) proteins of SOS DNA repair and (ii) the short-lived protein SulA. The latter binds FtsZ, a protein that forms a division ring so cell division appears to be blocked. Surprisingly, the *sulA⁻* mutant is many-fold less sensitive to the same DNA damage (caused by quinolone antibiotics) than the wild-type bacterium [72]. Lewis [7] suggested that SulA, besides binding FtsZ, send a signal for cell suicide. He hypothesized that such a phenoptotic signal is realized by an autolysin, a peptidoglycan hydrolase causing decomposition of the cell wall and, in turn, phenoptosis of bacteria reaching the stationary growth phase or treated with some antibiotics (Figure 5). Another possibility is that a long-term arrest of the cell division per se is a suicide signal. Lewis concluded that *sulA⁻* mutants have an enormous survival advantage over the wild type, yet the immediate benefit of greater survival of *sulA⁻* cells is apparently outweighed by the longer-term disadvantage due to the loss of the ability to eliminate defective cells» [7].

The SulA story is very instructive in that a unicellular organism commits suicide due to DNA damage long before this damage per se becomes incompatible with the life of the organism. This means that a cell usually dies when DNA is damaged because of a suicide signal rather than because of a dysfunction of DNA. Such a strategy is consistent with the principle formulated above: «It is better to die than to be wrong». This insures the genetic program against dangerously modified DNA that can result from its accidental damage.

Lewis [7] listed in his excellent review some other examples of phenoptosis in bacteria. Among these are (i) active lysis of the mother cell of *Bac. subtilis* during sporulation, which is required to release spores; (ii) lysis of some cells of *S. pneumoniae* to release DNA which is picked up by other cells that did not lyse; (iii) lysis of *S. pneumoniae* caused by penicillin. In the latter case, a mutant was selected which was resistant not only to lysis by penicillin, but also by several other antibiotics acting on quite different targets. In spite of the absence of the *lysis* response, all the antibiotics tested were shown to inhibit *growth* of the mutant bacteria just as that of the wild type, a fact showing that the antibiotics were able to act normally against their targets in the mutant cells.

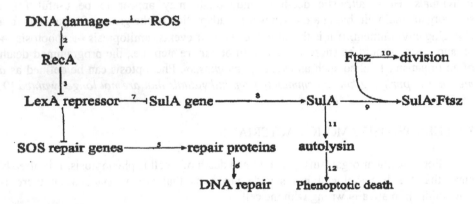

Figure 5. Regulation of DNA repair, cell division, and phenoptosis in bacteria. *1*, DNA damage caused by ROS (or by any other agents) activates RecA protein (*2*), which hydrolyses the LexA repressor (*3*), a protein that previously arrested the SOS repair genes (*4*), and the SulA gene (*7*); *5*, formation of the DNA repair proteins; *6*, DNA repair; *8*, formation of SulA protein, which binds the FtsZ protein (*9*) required for cell division *(10)*. Moreover, SulA is assumed to actuate the autolysin-mediated phenoptosis *(11, 12)*. (Modified from Lewis [7]).

Recently, the mechanism of this phenomenon was elucidated by Tuomanen and coworkers [73]. It proved to be related to the peptide involved in the so-called «quorum sensing» effect. It was found that bacteria (in this study, *S. pneumoniae*) always excrete a peptide composed of 27 amino acid residues (Pep-27). When the number of growing bacteria in a medium increases, this entails elevation of [Pep-27], which appears to be an agent inducing the autolysin-mediated lysis of the *S. pneumoniae* cells. In this way, phenoptosis regulates the level of bacteria in the environment, apparently optimizing their number under given ambient conditions. Pep-27 production was shown to strongly increase when something is wrong with the bacterium. This is why a wide range of antibiotics affecting various targets in the bacterial cell induce lysis—one more example of the operation of the "Samurai law"-linked phenoptosis.

One more bacterial suicide mechanism is shown to be composed of two proteins: (1) toxic protein MazF which is long-lived but present at low concentration and (2) antitoxic protein MazE which is sort-lived but present at high concentration. MazE binds MazF and neutralizes its toxicity. Under unfavorable conditions when rate of protein synthesis lowers, the short-lived MazE disappears being hydrolyzed by protease ClpPA, and a bacterium is killed by MazF [74-76]. In fact, such a suicide mechanism integrates pieces of negative information of very different kinds since the protein synthesis rate depends upon intactness of systems involved in replication, transcription, translation, energy supply, membrane transport, etc. Damage to any of the above system initiates the suicide mechanism purifying bacterial population from damaged bacteria.

In *E. coli*, three suicide mechanisms that are activated by the appearance of a phage in the cell interior have been described. One of them is the formation of ion-permeable channels in the bacterial membrane, whereas the two others are the activation of a protease or ribonuclease specifically cleaving a protein (EF-T$_u$) or RNA (tRNA^{Lys}) respectively, i.e., components required for protein synthesis. Since corresponding death programs are encoded by plasmid or a prophage integrated into the bacterial genome, Lewis suggested that the cell suicide «is a result of competition of parasitic DNAs, with the host as a battlefield» [7]. In any case, it is obvious that the phage-induced suicide is favorable for the bacterial community, purifying it from the phage-infected cells. This may explain why such kind of mechanisms was conserved by evolution (reviewed in refs. [69, 77]).

In all the above cases, the mechanisms of programmed death of bacteria are carried out by proteins that are quite different from those involved in apoptosis of animal cells. On the other hand, some domains inherent in animal pro- and antiapoptotic factors have already been found in bacteria. Perhaps, they are involved in other phenoptotic programs (for reviews, see refs. [7] and [78]).

3.2. PHENOPTOSIS OF UNICELLULAR EUKARYOTES

Phenoptosis has already been documented in unicellular eucaryotes. Fröhlich and coworkers [79] reported that in the yeast *Saccharomyces cerevisiae* small amounts of H$_2$O$_2$ caused cell shrinkage, appearance of phosphatidyl serine in the outer membrane leaflet, chromatin condensation and margination, and cleavage of DNA, all these events resembling the H$_2$O$_2$-induced apoptosis in multicellular organisms. A protein synthesis inhibitor suppressed the death program initiated by H$_2$O$_2$. Similar relationships were revealed by Corte-Real and coworkers who studied acetic acid instead of H$_2$O$_2$ as a toxic agent [80]. Phenoptosis proved to be accompanied by cytochrome *c* release from mitochondria [81].

A programmed death apparently took place in experiments of Longo et al. [82], when the yeast *S. cerevisiae* was kept in expired minimal medium. Expression of the animal antiapoptotic protein Bcl-2 prolonged the life of the yeast cells under these conditions. Death did not occur within the measured period of time (12 days) if pure water was used in place of the expired minimal medium. Similar relationships (except for the Bcl-2 effect) were revealed with some bacteria. It was suggested that such

strategy represent an altruistic suicide of the majority of cells to allow a small number
of cells to survive in the minimal media. This cannot help when nutrition is completely
absent (water instead of a minimal medium), so the cells use a strategy another than
phenoptosis [7].

In April 2002, the yeast programmed death concept has received a direct support.
(1) Madeo et al. [83] disclosed a caspase which is activated by H_2O_2 or aging and is
required for the protein-synthesis-dependent death of yeast. Thus, a specific apoptosis-
mediating protein was identified for the first time in *S. cerevisiae*. (2) Severin and
Hyman [84] discovered that death of yeast, induced by a high level of a pheromone, is
programmed. In particular, the death was prevented by cycloheximide and cyclosporin
A. It required mitochondrial DNA, cytochrome *c* and the pheromone-initiated protein
kinase cascade. When haploids of opposite mating types were mixed, some cells died,
the inhibitory pattern being the same as in the case of the killing by pheromone.
Inhibition of mating was favourable for death.

Thus, pheromone not only activated mating but also eliminated yeast cells failing
to mate. This programmed death switched the yeast population from vegetative to
sexual reproduction and accelerated changing of generations. Hence, the probability for
appearance of new traits, when ambient conditions turned for the worse, could be
enhanced (Fig. 6).

A programmed death mechanism seems to be operative in protozoa. In
Tethahynema, staurosporine was reported to induce morphological changes resembling
those in apoptotic animal cells. The transcription inhibitor actinomycin D prevented the
killing by staurosporine [86].

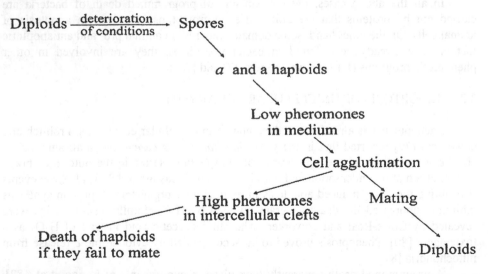

Figure 6. The life cycle of yeast: oscillation between long-lived and short-lived modes. Deterioration of
ambient conditions induces the vegetative-to-sexual reproduction switch mediated by excretion of
pheromones to the medium. Low concentrations of pheromones cause cell agglutination and mating of α and
a haploid cells. Those cells that were agglutinated but failed to mate are killed by high concentrations of
pheromone appearing in narrow clefts between agglutinated cells. This mechanism purifies the yeast
population from such cells and accelerates change of generations in the population. (From Skulachev [85]).

3.3. PHENOPTOSIS OF MULTICELLULAR ORGANISMS THAT REPRODUCE ONLY ONCE

In some of such species, the organism is constructed in a way predetermining death shortly after reproduction. Remember mayflies. Their imagoes die in few days since they cannot eat due to lack of a functional mouth, and their intestines are filled with air [1]. In the mite *Adactylidium*, the young hatch inside the mother's body and eat their way out [87], just as buds formed inside a *Bac. subtilis* cell are released to the outer medium by lysis of the cell (see above). However, much more often a special phenoptotic program is switched on immediately after the act of reproduction. The male of some squids dies just after transferring his spermatophore to a female [88]. The female of some spiders eats the male after copulation. Bamboo can live for 15-20 years reproducing vegetatively, but then, in the year of florescence, dies immediately after the ripening of the seeds. The pacific salmon dies soon after spawning, and this happens due to actuation of a specific corticosteroid-mediated program (rather than to exhaustion of the forces required for life) since dysfunction of the adrenal cortex prevents this kind of sudden death (for discussion, see refs. [1,9,87]).

3.4. AGING OF REPEATEDLY REPRODUCING ORGANISMS: WHY IS IT SLOW?

On the face of it, also in these organisms age-dependent phenoptosis, if exists, might be organized as sudden death. Bowles [89] mentioned a marine bird which suddenly dies at age about 50, without any indications of aging. However, this is certainly an exception rather than a rule.

For evolution of repeatedly reproducing organisms, slow aging should be expected to have a considerable advantage over acute phenoptosis. The function of an age-dependent phenoptosis is to reduce the pollution of the population by long-lived ancestors, thereby stimulating progressive evolution. Slow phenoptosis facilitates performance of this function. In fact, the appearance of a useful trait may compensate for unfavorable effects of aging within certain time limits, thereby giving some reproductive advantage to an individual acquiring such a trait. A strong, large-bodied deer, even after reaching a rather old age, still has a better chance than his younger but weaker rival to win a spring battle for a female or to escape from wolves. If two animals are of the same age but slightly differ in a useful trait, such a difference might become significant for the reproductive success when viability lowers with age. Therefore aging may be regarded, within framework of the phenoptosis concept, as a mechanism allowing useful traits to be revealed and selected during biological evolution [68, 90].

3.5. END-UNDER-REPLICATION OF LINEAR DNA AS A MOLECULAR MECHANISM OF AGING

Bowles [89] suggested that, historically, the living cell invented the first specialized mechanism of aging when linear DNA was used instead of the circular DNA inherent in the majority of bacteria and *Archae*. This event immediately resulted in a specific kind of DNA aging, a process consisting of replication-linked shortening of DNA. Such shortening inevitably accompanies replication of linear DNA, since even now the replication complex operates with linear DNA in the same way as it does with circular DNA. To produce an exact copy of a template, this complex should have some nucleotide residues to the left and to the right from the place where it combines with DNA. This is always the case if it deals with a circular DNA. However, with a linear DNA, the operation of this mechanism results in under-replication of the ends of the DNA molecule, as was first indicated by Olovnikov [91]. The question arises why eukaryotes, during many millions of years of their evolution, failed to improve such an important enzyme as that carrying out DNA replication, to adapt it to linear DNA, while at the same time they solved many much more difficult problems. According to Bowles [89], it happened first of all because under-replication is a mechanism applied by unicellular eukaryotes to accelerate the change of generations by shortening of the lifespan.

Apparently, this mechanism was eventually perfected such that special non-coding sequences (telomeres) were added to the ends of linear DNA. The shortening of the telomere could be used by the cell to count cell divisions without damaging those DNA sequences that encode RNA. Thus, the old (genetic) DNA function was separated from the new one, i.e., the cell division counting [4].

It is not yet clear whether telomere shortening determines the lifespan of multicellular organisms or only of some cells composing these organisms. It has been suggested that the limited number of divisions determined by telomere shortening is a line of anticancer defense of higher organisms, since knockout of a gene required for telomere formation inhibits some kinds of cancerogenesis [7,92]. On the other hand, there are some indications that the intact telomere is unfavorable for the development of other kinds of cancer [93] (for discussion, see ref. [94]). In any case, it is obvious that under-replication of DNA appeared already in unicellular eukaryotes, i.e., long before carcinogenesis, so their application in anticancer defense must represent the modification of another, much older function [4].

3.6. MUTATIONS PROLONGING THE LIFE

A paradoxical prediction of the concept regarding aging as a phenoptotic program consists in that should be mutations damaging this program and, hence, prolonging the life. In the literature, one can already find several cases of this kind. The most demonstrative example was observed by Hekimi and coworkers [95,96] who succeeded in extending by factor 5.5 the life span of nematode *Caenorhabditis elegans* if two genes were knock-outed. One of these genes was shown to encode an enzyme catalyzing final step of the CoQ biosynthesis.

In mice, Pelicci and his colleagues [97] have reported that (i) animals lacking a particular 66-kDa protein (p66^{shc}) lived 30% longer and (ii) fibroblasts derived from these mice did not respond to an H_2O_2 addition by initiating apoptosis.

It is clear from these data that p66^{shc} is involved in ROS-induced apoptosis. Apparently, in young mice it helps to purify the organism from ROS-overproducing cells. This is good for the organism and might, in principle, prolong its life. However, the same function of p66^{shc} appears to be involved in a phenoptotic cascade, and this effect becomes dominating in old mice. As a result, p66[shc] shortens the lifespan [4,98].

The same reasoning can be used to explain an observation of Kirkwood and coworkers [99] who found a positive correlation of the lifespan of various mammals and the *in vitro* resistance of their fibroblasts to oxidative stress.

3.7. PARADOX OF THE DONEHOWER'S MICE

In the first issue of the Nature magazine, 2002, Donehower and coworkers [100] have described mice with a deletion in the *p53* gene that expresses a truncated RNA encoding a carboxyl-terminal 24 kDa fragment of p53. Heterozygote mutant mice (*p53*[+/m]) exhibited, on some reason, enhanced p53 activity and resistance to spontaneous tumors. In fact, non of the thirty-five *p53*[+/m] mice developed overt, life-threatening tumors, whereas over 45% *p53*[+/+] and over 80% *p53*[+/-] mice, respectively, developed such tumors. Surprisingly to the authors, the *p53*[+/m] mice lived *shorter* by about 20%. Shortening of the life span proved to be a result of an early aging. The aging in question included such traits as reduction of body weight; loss of mass of liver, kidney, and spleen; lymphoid and muscle atrophy; osteoporosis; and hunchbacked spine.

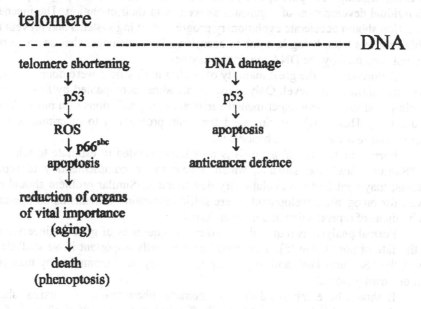

Figure 7. Two p53-dependent mechanisms changing the life span and resistance to cancer. High p53 level is favorable for anticancer defence but stimulates aging. Low p53 level postpones aging but is favorable for cancer. In this way, the Nature prevents immortality which is might slow down progressive development of the living creatures.

These data may be explained (Figure 7) assuming that p53, performing its "guard of genome" function, (1) assures longevity by causing cells that have damaged DNA (and, hence, would become malignant) to commit apoptotic suicide, but (2) decreases life span by causing the apoptosis of cells with shortened telomeres that are still quite functional. As a result, p53 may be involved in optimization of the life span of organisms. Mechanistically, the two p53-linked effects seem to be different, the latter specifically including $p66^{shc}$. Thus one may hope to strongly prolong life span by a concerted activating of p53 and inhibiting $p66^{shc}$ [4].In any case, it seems probable that not only $p66^{shc}$ but also p53 plays a role in controlling the life span of the organism. This is hardly surprizing if we take into account that p53 regulates the cell cycle, synthesis of reparative enzymes and apoptosis. This protein has 13 sites of phosphorylation attacked by various protein kinases, 6 sites of acetylation and a site of modification by another protein (SUMO) affecting the p53 stability [101] (Figure 8). To some degree, p53 resembles a keyboard of a computer controlling the very fate of the organism.

4. Conclusions and Outlook: How Can Formal Analysis Help in Understanding of the Programmed Death Mechanisms?

The present state of the art in the problem of programmed death can be summarized in that such kind of phenomena exist not only in the cells of multicellular organisms (apoptosis) but also at subcellular (mitoptosis) and supracellular (collective apoptosis, organoptosis, phenoptosis) levels. For sure, these events play important role in individual development of organisms as well as in their evolution. The mechanisms in question should accelerate evolutionary progress of living systems and prevent results of this progress from being lost due to occasional DNA damages that on some reasons were not corrected by the DNA repairing enzymes.

Unfortunately, the great majority of studies in this field were done at qualitative rather than quantitative level. Only a few works were accompanied by formal analysis, modeling and subsequent experimental verification of predictions of a model (see, e.g., [5,102,103]). This is why in this field the main problem is to discriminate between *possible* and *really occurring* phenomena.

Formal mathematical analysis is now badly needed to estimate to what degree the "Samurai law" mechanisms, which appear to be complementary to reparation systems, may contribute to evolutionary development. Similar problem should be also solved for ontogenic development where self-elimination mechanisms co-operate with mechanisms of reproduction and differentiation.

Formal analysis is required to predict consequences of gene engineering studies for the fate of population [5]. This would be especially important if we shall decide to cancel the "Samurai law" and to prolong in this way the human life by manipulating with our own genome.

It should be emphasized that, for humans, phenoptosis, if it exists, should be regarded as an animal atavism. Metchnikoff [104] stated in 1907: «Traits of animal origin are inherent in human, who appeared as a result of a long cycle of development.

Achieving a level of mental development, which is unknown among animals, he retains many traits that are not only unnecessary for him, but even obviously harmful». Phenoptosis must be attributed to the most harmful among such traits. In wild nature, phenoptosis is useful first of all for survival and evolution of communities of organisms in aggressive environments. Humans organize their life to minimize its dependence upon environmental conditions. Even such phenoptotic function as defense against asocial monsters (that can appear in the progeny of old parents due to mutations accumulating with age) may be replaced by, say, social regulations forbidding the bearing of children beyond some critical age [4].

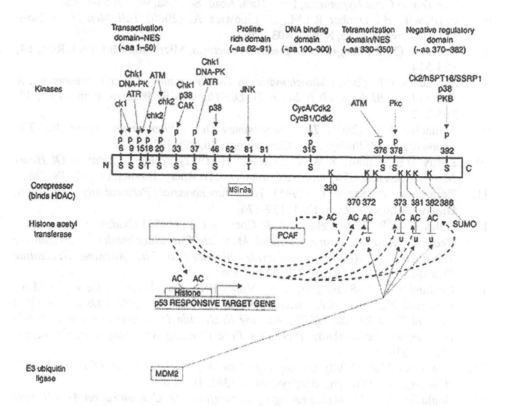

Figure 8. Sites of posttranslational modifications of the p53 protein. Abbreviations: AC, acetyl; MDM2, a ring-finger ubiquitin ligase; NES, nuclear export sequence; PCAF, a histone acetylase; u, uniquitin. (From Wahl and Carr [101]).

5. Acknowledgments

This work was supported by the Ludwig Institute for Cancer Research (Grant BRO 863) and the Russian Foundation for Basic Research (Grant 00-15-97799).

6. References:

1. Weismann A. (1889) *Essays upon heredity and kindred biological problems,* Claderon Press, Oxford.
2. Medawar P.B. (1952) *An Unsolved Problem Of Biology,* Lewis H.K., London
3. Bowles J.T. (2000) Shattered: Medawar's Test Tubes And Their Enduring Legacy Of Chaos, *Med. Hypotheses,* 54, 326-339.
4. Skulachev V.P. (2002) *Programmed Death Phenomena: From Organelle To Organism,* Ann. N.Y. Acad. Sci., 959, 214-237.

5.   Muir W.M. & Howard R.D. (1999) *Possible Ecological Risks Of Transgenic Organism Release When Transgenes Affect Mating Success: Sexual Selection And The Trojan Gene Hypothesis*, Proc. Natl. Acad. Sci. Usa, 96, 13853-13866.

6.   Pawlowski B., Dunber R.I.M., & Lipowicz A. (2000) *Tall Men Have More Reproductive Success*, Nature, 403, 156.

7.   Lewis K. (2000) *Programmed Death In Bacteria*, Microbiol. Mol. Biol. Rev., 64, 503-514

8.   Skulachev V.P. (2000) *Mitochondria In The Programmed Death Phenomena; A Principle Of Biology: «It Is Better To Die Than To Be Wrong»*, Iubmb Life, 49, 365-372

9.   Skulachev V.P. (2001) *The Programmed Death Phenomena, Aging And The Samurai Law Of Biology*, Exp.Gerontology, 36, 995-1024

10. Zorov D.B., Kinally K.W. & Tedeschi H (1992) *Voltage Activation Of Heart Inner Mitochondrial Membrane Channels*, J. Bioenerg. Biomembr. 24, 119-124

11. Zoratti M. & Szabo I. (1995) *The Mitochondrial Permeability Transition*, Biochim. Biophys. Acta, 1241, 139-176

12. Halestrap A.P., Woodfield K.-Y. & Connern C.P. (1997) *Oxidative Stress, Thiol Reagents, And Membrane Potential Modulate The Mitochondrial Permeability Transition By Affecting Nucleotide Binding To The Adenine Nucleotide Translocase*, J.Biol.Chem., 272, 3346-3354

13. Costantini P., A.-S. Belzacq A.-S., Vieira H.L., Larochette N., De Pablo M.A., Zamzami N., Susin S.A., Brenner C. & Kroemer G. (2000) *Oxidation Of A Critical Thiol Residue Of The Adenine Nucleotide Translocator Enforces Bcl-2-Independent Permeability Transition Pore Opening And Apoptosis*, Oncogene, 19, 307-314

14. Skulachev V.P. (1998) *Uncoupling: New Approaches To An Old Problem Of Bioenergetics*. Biochim. Biophys. Acta, 1363, 100-124

15. Skulachev V.P. (1994) *Lowering Of Intracellular O_2 Concentration As A Special Function Of Respiratory Systems Of Cells*, Biochemistry (Moscow), 59, 1433-1434

16. Skulachev V.P. (1996) *Role Of Uncoupled And Non-Coupled Oxidations In Maintenance Of Safely Low Levels Of Oxygen And Its One-Electron Reductants*, Quart. Rev. Biophys., 29, 169-202

17. Skulachev V.P. (1996) *Why Are Mitochondria Involved In Apoptosis? Permeability Transition Pores And Apoptosis As Selective Mechanisms To Eliminate Superoxide-Producing Mitochondria And Cell*, Febs Lett., 397, 7-10

18. Parrish J., Li L., Klotz K., Ledwich D., Wang X. & Xue D. (2001) *Mitochondrial Endonuclease G Is Important For Apoptosis In C. Elegans*, Nature 412, 90-94

19. Li L.Y., Luo X. & Wang X, (2001) *Endonuclease G Is An Apoptotic Dnase When Released From Mitochondria*, Nature 412, 95-99

20. Widlak P., Li L.Y., Wang X. & Garrard W.T. (2001) *Action Of Recombinant Human Apoptotic Endonuclease G On Naked Dna And Chromatin Substrates: Cooperation With Exonuclease And Dnase I*, J. Biol. Chem., 276, 48404-48409

21. Susin S. A., Zamzami N., Castedo M., Hirsch T., Macho A., Daugas E., Geuskens M. & Kroemer G. (1996) *Bcl-2 Inhibits The Mitochondrial Release Of An Apoptogenic Protease*, J. Exp. Med., 184, 1331-1341

22. Susin S.A., Lorenzo H.K., Zamzami N., Marzo I., Snow B.E., Brothers G.M., Mangion J., Jacotot E., Costantini P., Loeffler M., Larochette N., Goodlett D.R., Aebersold R., Siderovski D.P., Penninger J.M. & Kroemer G. (1999) *Molecular Characterization Of Mitochondrial Apoptosis-Inducing Factor*, Nature, 397, 441-446
23. Liu X., Naekyung C., Yang J., Jemmerson R. & Wang X. (1996) *Induction Of Apoptotic Program In Cell-Free Extracts: Requirement For Datp And Cytochrome C*, Cell, 86, 147-157
24. Yamg J., Liu X., Bhalla K., Kim C.N., Ibrado A.M., Cai J., Peng T.-I., Jones S.P. & Wanf X. (1997) *Prevention Of Apoptosis By Bcl-2: Release Of Cytochrome C From Mitochondria Blocked*, Science, 275, 1129-1132
25. Kluck R.M., Bossy-Wetzel E., Green D.R. & Newmeyer D.D. (1997) *The Release Of Cytochrome C From Mitochondria: A Primary Site For Bcl-2 Regulation Of Apoptosis*, Science, 275, 1132-1136
26. Kluck R.M., Martin S.J., Hoffman B.M., Zhou J.S., Green D.R. & Newmeyer D.D. (1997) *Cytochrome C Activation Of Cpp32-Like Proteolysis Plays A Critical Role In Xenopus Cell-Free Apoptosis System*, Embo J., 16, 4639-4649
27. Samali A., Cai J., Zhivotovsky B., Jones D.P. & Orrenius S. (1999) *Presence Of A Pre- Apoptotic Complex Of Pro-Caspase-3, Hsp60 And Hsp10 In The Mitochondrial Fraction Of Jurkat Cells*, Embo J., 18, 2040-2048
28. Du C., Fang M., Li Y., Li L. & Wang X. (2000) *Smac, A Mitochondrial Protein That Promotes Cytochrome C-Dependent Caspase Activation By Eliminating Iap Inhibition*, Cell, 102, 33-42
29. Verhagen A.M., Ekert P.G., Pakusch M., Silke J., Connolly L.M., Reid G.E., Moritz R.L., Simpson R.J. & Vaux D.L. (2000) *Identification Of Diablo, A Mammalian Protein That Promotes Apoptosis By Binding To And Antagonizing Iap Proteins*, Cell, 102, 43-53
30. Newmeyer D.D., Farschon D.M. & Reed J.C. (1994) *Cell-Free Apoptosis In Xenopus Egg Extracts: Inhibition By Bcl-2 And Requirement For An Organelle Fraction Enriched In Mitochodria*, Cell, 79, 353-364
31. Fletcher G.C., Xue L., Passingham S.K. & Tolkovsky A.M. (2000) *Death Commitment Point Is Advanced By Axotomy In Sympathetic Neurons*, J. Cell Biol, 150, 741-754
32. Xue L., Fletcher G.C. & Tolkovsky A.M. (2001) *Mitochondria Are Selectively Eliminated From Eukaryotic Cells After Blockade Of Caspases During Apoptosis*, Current Biol., 11, 361-365
33. Skulachev V.P. (1998) *Cytochrome C In The Apoptotic And Antioxidant Cascades,*. Febs Lett., 423, 275-280
34. Shimizu S., Narita M. & Tsujimoto Y. (1999) *Bcl-2 Family Proteins Regulate The Release Of Apoptogenic Cytochrome C By The Mitochondrial Channel Vdac*, Nature, 399, 483-487
35. Madesh M. & Hajnoczky G. (2001) *Vdac-Dependent Permeabilization Of The Outer Mitochondrial Membrane By Superoxide Induces Rapid And Massive Cytochrome C Release*, J. Cell Biol., 155, 1003-1015

36. Peachman K.K., Lyles D.S. & Bass D.A. (2001) Mitochondria In Eosinophils: Functional Role In Apoptosis But Not Respiration, Proc. Natl. Acad. Sci. Usa, 98, 1717-1722

37. Milan M., Campuzano S. & Garcia-Bellido A. (1997) *Developmental Parameters Of Cell Death In The Wing Disc Of Drosophila*, Proc. Natl. Acad. Sci. Usa, 94, 5691-5696

38. Lang R., Lustig M., Francois F., Sellinger M. & Plesken Y. (1994) *Apoptosis During Macrophage-Dependent Occular Tissue Remodeling*, Development, 120, 3395-3403

39. Ijiri K. & Potten C.S. (1987) *Cell Death In Cell Hierarchies In Adult Mammalian Tissues, In Perspectives On Mammalian Cell Death*, Potten C.S. Ed., Oxford University Press, Oxford, 326-356

40. Reznikov K.Y. (1991) *Cell Proliferation And Cytogenesis In The Mouse Hippocampus*, Adv. Anat. Embryol. Cell Biol., 122, 1-83

41. Hendry J.H. & Potten C.S. (1998) *Letter To The Editor. Comments On The Paper – Cell Survival In Irradiated Microcolonies: How Influential Are Neighbours?*, Int. J. Radiat. Biol., 73, 575-576

42. Wilson J.W., Pritchard D.M., Hickman J.A. & Potten C.S. (1998) *Radiation-Induced P53 And P21$^{waf-1/Cip1}$ Expression In The Murine Intestinal Epithelium*, Am. J. Pathol., 153, 899-909

43. Dilber M.S. & Smith C.E. (1997) *Suicide Genes And Bystander Killing: Local And Distant Effects*, Gene Therapy, 4, 273-274

44. Paillard F. (1997) *Bystander Effects In Enzyme/Prodrug Gene Therapy, Gene Therapy*, 4, 273-274

45. Reznikov K., Kolesnikova L., Pramanik A., Tan-No K., Gileva I., Yakovleva T., Rigler R., Terenius L. & Balaklin G. (2000) *Clustering Of Apoptotic Cells Via Bystander Killing By Peroxides*, Faseb J., 14, 1754-1764

46. Shao R., Xia W. & Hung M.-C. (2000) *Inhibition Of Angiogenesis And Induction Of Apoptosis Are Involved In E1a-Mediated Bystander Effect And Tumor Suppression*, Cancer Res., 60, 3123-3126

47. Mensil M. & Yamasaki H. (2000) *Bystander Effect In Herpes Simplex Virus-Thymidine Kinase/Ganciclovir Cancer Gene Therapy: Role Of Gap-Junctional Intercellular Communication*, Cancer Res., 60, 3989-3999

48. Kagawa S., He C., Gu J., Koch P., Rha S.-J., Roth, J.A., Curley S.A., Stephens L.G. & Fang B. (2001) *Antitumor Activity And Bystander Effects Of The Tumor Necrosis Factor-Related Apoptosis-Inducing Ligand (Trail) Gene*, Cancer Res., 61, 3330-3338

49. Adachi M., Sampath J., Lan L.-B., Sun D., Hargrove P., Flatley R.M., Tatum A., Ziegelmeier M.Z., Wezeman M., Matherly L.H., Drake R.R. & Schuetz J.D. (2002) *Expression Of Mrp4 Confers Resistance To Ganciclovir And Compromises Bystander Cell Killing*, J. Biol. Chem. In Press, Published June 24, 2002 As Ms M203262200

50. Skulachev V.P. (1998) *Possible Role Of Reactive Oxygen Species In Antiviral Defense.* Biochemistry (Moscow), 63, 1438-1440

51. Clark P.J. & Evans P.C. (1954) *Distance To The Nearest Neighbor As A Measure Of Spatial Relationship In Populations*, Ecology, 35, 445-453

52. Simizu S., Takada M., Umezawa K. & Imoto M. (1998) *Requirement Of Caspase-3(-Like) Protease-Mediated Hydrogen Peroxide Production For Apoptosis Induced By Various Anticancer Drugs*, J. Biol. Chem., 273, 26900-26907
53. Johnson T.M., Yu Z.-X., Ferrans V.J., Lowenstein R.A. & Finkel T. (1996) *Reactive Oxygen Species Are Downstream Mediators Of P53-Dependent Apoptosis*, Proc. Natl. Acad. Sci. Usa, 93, 11848-11852
54. Goldkorn T., Balaban N., Shannon M., Chea V., Matsukuma K., Gilchrist D., Wang H. & Chan C. (1998) *H_2O_2 Acts On Cellular Membrane To Generate Ceramide Signalling And Initiate Apoptosis In Tracheobronchial Epithelial Cell*, J. Cell Sci., 111, 3209-3220
55. Hiraoka W., Vazquez N., Nieves-Neira W., Chanock S.J. & Pommier Y. (1998) *Role Of Oxygen Radicals Generated By Nadph Oxidase In Apoptosis Induced In Human Leukemia Cells*, J. Clin. Invest., 102, 1961-1968
56. Dumont A., Hehner S.P., Hofmann T.G., Ueffing M., Droge W. & Schmitz M.L. (1999) *Hydrogen Peroxide-Induced Apoptosis Is Cd95-Independent, Requires The Release Of Mitochondria-Derived Reactive Oxygen Species And The Activation Of Nf-κB*, Oncogene, 18, 754-757
57. Olejnicka B.T., Dalen H. & Brink U.T. (1999) *Minute Oxidative Stress Is Sufficient To Induce Apoptotic Death Of Nit-1 Insulinoma Cells*, Apmis, 107, 747-761
58. Alexeevsky A., Alexeevsky D., Domnina L., Fetisova E., Ivanova O., Pletjushkina O. & Skulachev V. (In Preparation)
59. Neill S.J., Desikan R., Clarke A., Hurst R.D. & Hancock J.T. (2002) *Hydrogen Peroxide And Nitric Oxide As Signalling Molecules In Plants*, J. Exp. Bot., 53, 1237-1247
60. Zottini M., Formentin E., Scattolin M., Carimi F., Lo Schiavo F. & Terzi M. (2002) *Nitric Oxide Affects Plant Mitochondrial Functionality In Vivo*, Febs Lett. 515, 75-78
61. Patel R.P., Mcandrew J., Sellak H., White C.R., Jo H., Freeman B.A. & Darley-Usmar V.M. (1999) *Biological Aspects Of Reactive Nitrogen Species*, Biochim. Biophys. Acta, 1411, 385-400
62. Borutaite V., Morkuniene R. & Brown G.C. (2000) *Nitric Oxide Donors, Nitrosothiols And Mitochondrial Respiration Inhibitors Induce Caspase Activation By Different Mechanisms*, Febs Lett., 467, 155-159
63. Kashigawa A., Hanada H., Yabuki M., Kano T., Ishisaka R., Sasaki J., Inoue M. & Ursumi K. (1999) *Thyroxine Enhancement And The Role Of Reactive Oxygen Species In Tadpole Tail Apoptosis*, Free Radic. Biol. Med., 26, 1001-1009
64. Poderoso J.J., Carreras M.C., Lisdero C., Riobo N., Schöpfer F. & Boveris A. (1996) *Nitric Oxide Inhibits Electron Transfer And Increases Superoxide Radical Production In Rat Heart Mitochondria And Submitochondrial Particles*, Arch. Biochem. Biophys., 328, 85-92
65. Poderoso J.J., Lisdero C., Schöpfer F., Riobo N., Carreras M.C., Cadenas E. & Boveris A. (1999) *The Regulation Of Mitochondrial Oxygen Uptake By Redox Reactions Involving Nitric Oxide And Ubiquinol*, J. Biol. Chem., 274, 37709-37716

84   VLADIMIR SKULACHEV

66.   Poderoso J.J., Peralta J.G., Lisdero C., Carreras M.C., Radisic.M, Schöpfer F.,
      Cadenas E. & Boveris A. (1998) *Nitric Oxide Regulates Oxygen Uptake And
      Hydrogen Peroxide Release By The Isolated Beating Rat Heart*, Am. J. Physiol.,
      274, C112-C119
67.   Poderoso J.J., Carreras M.C., Schöpfer F., Lisdero C.L., Riobo N.A., Giulivi C.,
      Boveris A.D., Boveris A. & Cadenas E. (1999) *The Reaction Of Nitric Oxide With
      Ubiquinol: Kinetic Properties And Biological Significance*, Free Radic. Biol.
      Med., 26, 925-935
68.   Skulachev V.P. (1997) *Aging Is A Specific Biological Function Rather Than The
      Result Of A Disorder In Complex Living Systems: Biochemical Evidence In
      Support Of Weismann's Hypothesis*, Biochemistry (Moscow), 62, 1191-1195
69.   Skulachev V.P. (1999) *Mitochondrial Physiology And Pathology; Concept Of
      Programmed Death Of Organelles*, Cells And Organisms, Mol. Asp. Med., 20,
      139-184
70.   Skulachev V.P. (1999) *Phenoptosis: Programmed Death Of An Organism*,
      Biochemistry (Moscow), 64, 1418-1426
71.   Walker G.C., Neidhard F.C., Curtiss R.I., Ingraham J.L., Lin C.C.L., Low K.B.,
      Magasanik B., Reznikoff W.S., Riley M., Schaechter M. & Umbarger H.E. (1996)
      The Sos Response Of Escherichia Coli, In Escherichia Coli And Salmonella.
      Cellular And Molecular Biology,. Eds.,Washington: Asm Press
72.   Piddock L.J. & Walters R.N. (1992) *Bactericidal Activities Of Five Quinolones
      For Escherichia Coli Strains With Mutations In Genes Encoding The Sos
      Response Or Cell Devision*, Antimicrob. Agents Chemother, 36, 819-825
73.   Novak R., Charpentier E., Braun J.S. & Tuomanen E. (2000) *Signal Transduction
      By A Death Signal Peptide: Uncovering The Mechanism Of Bacterial Killing By
      Penicillin*, Mol. Cell, 5, 49-57
74.   Aizenman E., Engelberg-Kulka H. & Glaser G. (1996) *An Escherichia Coli
      Chromosomal "Addiction Module" Regulated By Guanosine-3'5'-
      Bispyrophosphate: A Model For Programmed Bacterial Cell Death*, Proc. Natl.
      Acad. Sci. Usa, 93, 6059-6063
75.   Engelberg-Kulka H. & Glaser G. (1999) *Addiction Modules And Programmed
      Death And Anti-Death In Bacterial Cultures*, Annu. Rev. Microbiol., 53, 43-70
76.   Engelberg-Kulka H., Sat B. & Hazan R. (2001) *Bacterial Programmed Cell
      Death And Antibiotics*, Asm News, 67, 617-62
77.   Raff M.C. (1998) *Cell Suicide For Beginners*, Nature, 396, 119-122
78.   Bakal C.J. & Davies J.E. (2000) *No Longer An Exclusive Club: Eukaryotic
      Signalling Domains In Bacteria*, Trends Cell Biol., 10, 32-38
79.   Madeo F., Fröhlich E., Ligr M., Grey M., Sigrist S.J., Wolf D.H. & Fröhlich K.-
      U. (1999) *Oxygen Stress: A Regulator Of Apoptosis In Yeast*, J. Cell Biol., 145,
      757-767
80.   Ludovico P., Sousa M.J., Silva M.T., Leao C. & Corte-Real M. (2001)
      *Saccharomyces Cerevisiae Commits To A Programmed Cell Death Process In
      Response To Acetic Acid*, Microbiology, 147, 2409-2415
81.   Ludovico P., Rodrigues F., Almeida A., Silva M.T., Barrientos A. & Corte-Real
      M. (2002) *Cytochrome C Release And Mitochondria Involvement In Programmed*

Cell Death Induced By Acetic Acid In Saccharomyces Cerevisiae, Mol. Biol. Cell 13 (10.1091/Mbc. E01-12-1061)

82. Longo V.D., Ellerby L.M., Bredesen D.E., Valentine J.S. & Gralle E.B. (1997) *Human Bcl-2 Reverses Survival Defects In Yeast Lacking Superoxide Dismutase And Delays Death Of Wild- Type Yeast*, J. Cell. Biol., 137, 1581-1588

83. Madeo F., Herker E., Maldener C., Wissing S., Lächelt S., Herlan M., Fehr M., Laulber K., Sigrist S.J., Wesselborg S. & Fröhlich K.-U. (2002) *A Caspase-Related Protease Regulates Apoptosis In Yeast*, Mol. Cell., 9, 911-917

84. Severin F.F. & Hyman A.A. (2002) *Pheromone Induces Programmed Cell Death In S. Cerevisiae*, Current Biol., 12, R233-R235

85. Skulachev V.P. (2002) *Programmed Death As Adaptation? In Yeast, Probably Yes*, Febs Letters. (Accepted)

86. Christensen S.T., Chemnitz J., Straarup E.M., Kristiansen K., Wheatley D.N. & Rasmussen M. (1998) *Staurosporine-Induced Cell Death In Tetrahymena Thermophila Has Mixed Characteristics Of Both Apoptotic And Autophagic Degeneration*, Cell Biol. Int., 22, 591-598

87. Kirkwood T.B.L. & Cremer T. (1982) *Cytogerontology Since 1881: A Reappraisal Of August Weismann And A Review Of Modern Progress*, Hum. Genet., 60, 101-121

88. Nesis K.N. (1997) *Cruel Love Among The Squids, In Russian Science; Withstand And Revive*, Byalko A.V. Ed., Moscow, Nauka-Physmatlit. (Russ.)

89. Bowles J.T. (1998) *The Evolution Of Aging: A New Approach To An Old Problem Of Biology*, Med. Hypotheses, 51, 179-221

90. Skulachev V.P. (1999) *The Dual Role Of Oxygen In Aerobic Cells, In Biosciences 2000* Pasternak C.A., (Ed.), Imperial College Press, London, 173-193

91. Olovnikov A.M. (1971) *Principles Of Marginotomy In Template Synthesis Of Polynucleotides*, Dokl. Akad. Nauk Sssr, 201, 1496-1498 (Russ)

92. Chin L., Artandi S.E., Shen Q., Tam A., Lee S.L., Gottlieb G.J., Greider C.W. & Depinho R.A. (1999) *P53 Deficiency Rescues The Adverse Effects Of Telomere Loss And Cooperates With Telomere Dysfunction To Accelerate Carcinogenesis*, Cell, 97, 527-538.

93. Rudolph K.L., Chang S., Lee H.W., Blasco M., Gottlieb G.J., Greider C. & Depinho R.A. (1999) *Longevity, Stress Response, And Cancer In Aging Telomerase-Deficient Mice*, Cell, 96, 701-712

94. De Lange T. & Jacks T. (1999) *For Better Or Worse? Telomerase Inhibition And Cancer*, Cell, 98, 273-275

95. Lakowski B. & Hekimi S. (1996) *Determination Of Life-Span In Caenorhabditis Elegans By Four Clock Genes*, Science, 272, 1010-1013

96. Ewbank J.J., Barnes T.M., Lakowski B., Lussier M., Bussey H. & Hekimi S. (1997) *Structural And Functional Conservation Of The Caenorhabditis Elegans Timing Gene Clk-1*, Science, 275, 980-983

97. Migliaccio E., Giorgio M., Mele S., Pelicci G., Revoldi P., Pandolfi P.P., Lanfrancone L. & Pelicci P.G. (1999) *The P66^{shc} Adaptor Protein Controls Oxidative Stress Response And Life Span In Mammals*, Nature, 402, 309-313

98. Skulachev V.P. (2000) *The P66^{shc} Protein: A Mediator Of The Programmed Death Of An Organism?*, Iubmb-Life, 49, 177-180

99. Kapahi P., Boulton M.E. & Kirkwood T.B.L. (1999) *Positive Correlation Between Mammalian Life Span And Cellular Resistance To Stress*, Free Radic. Biol. Med., 26, 495-500
100. Tyner S.D., Venkatachalam S., Choi J., Jones S., Ghebranious N., Igelmann H., Lu X., Soron G., Cooper B., Brayton C., Hee Park S., Thompson T., Karsenty G., Bradley A. & Donehower L.A. (2002) *P53 Mutant Mice That Display Early Ageing-Associated Phenotypes*, Nature, 415, 45-53
101. Wahl G.M. & Carr A.M. (2001) *The Evolution Of Diverse Biological Responses To Dna Damage: Insights From Yeast And P53*, Nature Cell Biol, 3, E277-E286
102. Hardy K & Stark J. (2002) *Matematical Models Of The Balance Between Apoptosis And Proliferation*, Apoptosis, 7, 373-381
103. Akif'ev A.P. & Potapenko A.I. (2001) *Nuclear Genetic Material As An Initial Substrate Of Aging In Animals*, Genetika, 37, 1145-1458 (Russ.)
104. Metchnikoff I. (1907) *The Prolongation Of Life: Optimistic Studies*. Heinemann: London.

DEVELOPMENT OF MOTOR CONTROL IN VERTEBRATES

JESSE STOLLBERG
University of Hawaii at Manoa
Honolulu, HI 96822

1. Abstract

Three general applications of modeling in experimental neurobiology will be presented. In the first instance the focus will be on the development of synapses during embryogenesis. After a presentation of some background material, modeling approaches to 1) synapse formation at the level of light microscopy and 2) the resulting ultra structure at the level of electron microscopy will be discussed. The third application involves a model for muscle cell innervation driven by activity dependent competition between synapses. Rules for this process, based on empirical studies, are incorporated into the presented model, which explains and predicts synapse elimination and the development of the size principle substantially better than previous approaches.

2. Introduction

2.1 DEVELOPMENT AT THE NEUROMUSCULAR JUNCTION

The first problem under consideration is how specialized components become localized at developing synapses - in particular at the developing neuromuscular junction. Because the neuromuscular junction represents an approachable system in which to study the general question of synaptogenesis, much well deserved attention is being given to this problem in many laboratories. The first part of this problem – how the motor neurons communicate to muscle cells that they should concentrate post-synaptic components at a particular location – has been quite well developed in the last 15 years. A molecule termed agrin is synthesized by motor neurons[1], is released by the nerve terminal[2, 3], induces AChR clusters on aneural muscle cultures[4], and its absence in knockout mice causes a failure in formation of the neuromuscular junctions[5].

The second part of the problem – by what molecular mechanism do muscle cells respond to the signal – is still largely unsolved. In this context we have made use of electric fields to induce receptor aggregation in cultured spherical muscle cells[6-9]. Once induced the receptors continue to aggregate after termination of the field, and several lines of evidence suggest that the mechanisms involved are largely the same as those for agrin induced receptor aggregation [10]. Because these electric fields can be used as a powerful experimental system with which to probe particular questions about aggregation mechanisms.

J. Nation et al. (eds.), Formal Descriptions of Developing Systems, 87–98.
© 2003 *Kluwer Academic Publishers. Printed in the Netherlands.*

The initial aggregation of receptors occurs via lateral migration and the rate of aggregate formation is readily measured in electric field experiments. The first part of this section deals with the comparison of this rate to predictions obtained by extending the diffusion limited aggregation model of Witten and Sander[11]. The second part shows the use of another kind of modeling to quantitatively interpret the results from scanning electron microscopy observations of the ultrastructure of receptor aggregates.

2.2 DEVELOPMENT OF SYNAPSE ELIMINATION AND THE SIZE PRINCIPLE

Synapse elimination is a nearly universal phenomenon in developing nerve-target systems in which the target generally loses 50-85% of the synapses initially formed [12, 13]. At the skeletal muscle this is seen as a reduction from an average of 3-4 synapses per muscle cell fiber to one (with few exceptions). This overproduction and subsequent removal of the majority of synapses suggests that some developmental problem is being solved. It had been speculated that it might produce correction of topographical errors in innervation, but extensive studies failed to show significant errors in need of correction.

The size principle[14] is the name given to the observation that the motor neurons in a given motor pool can be ranked both in threshold for firing and in the number of muscle fibers they control, and that these two rankings are correlated. In other words higher threshold motor neurons control more muscle fibers than lower threshold motor neurons.

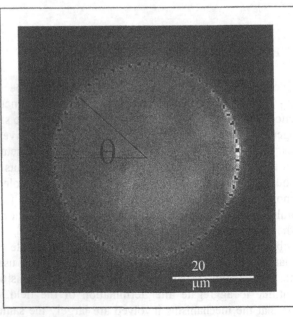

Figure 1. A cultured spherical muscle cell was placed in an electric field and then labeled with fluorescent alpha-Bungarotoxin to reveal acetylcholine receptor distribution. The distribution shows a significant aggregate on the right (cathodal pole) of the cell. Also shown are every fourth of the 256 sectors imposed to analyze quantitatively the receptor density as a function of position (θ). These values are used to construct the data plots shown in Figure 2.

In explaining the development of these two phenomena we turn to the concept of a Hebbian synapse. This is a theoretical construct – for which there is evidence in neuronal systems – in which synapses with activities more highly correlated with target firing grow in strength at the expense of those with less correlated firing patterns. A computer model is presented which demonstrates that under a wide range of conditions this kind of synaptic competition can lead to the development of monosynapticity and the size principle.

3. Results

3.1. THE RATE OF RECEPTOR AGGREGATION

3.1.1 Experimental Results

Electric fields induce receptor aggregation at the cathodal pole of the cell (Figure 1). The response of receptors to the field is unlike the ideal (predicted) behavior both in terms of steady state distribution, and in terms of the behavior after field termination [6-8, 15]. In this latter instance, the results show that the receptor aggregate continues to develop after the field is terminated. As shown the distribution of fluorescently labeled receptors is quantified by the imposition, via software, of 256 sectors representing the cell perimeter.

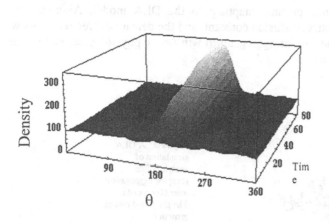

Figure 2. 19 cultured muscle cells were analyzed as in Figure 1 at 5-minute intervals during and after an electric field of 20 minutes duration. The results are presented in terms of the angle θ of Figure 1, where 180 indicates the cathodal cell pole. Receptor density is shown on the vertical axis.

To examine the rate of aggregate growth electrophoresis chambers were assembled on a microscope stage, and a number of cells were followed at 5 minute intervals (Figure 2). The Figure confirms that aggregation continues after the (20') electric field and also shows the depression in receptor density adjacent to the aggregate which is characteristic of a diffusion trap mechanism. Finally, and most pertinent to the present discussion, it reveals the rate of receptor aggregation.

3.1.2. Modeling Results

Diffusion Limited Aggregation (DLA) is a paradigm studied in physics by which aggregation can take place[11]. Generally speaking the model consists of a grid with 1 particle stuck to its center – grids adjacent to the particle are designated sticky, in the sense that if any freely diffusing particle lands there it will become stuck. Now a particle is released at the periphery and allowed to diffuse until it becomes stuck, at which point 1) its adjacent grids become sticky, and 2) a new free particle is released on the periphery. It has been shown that an aggregate derived in such a manner has a fractal dimension of approximately 5/3.

While intriguing, the model as conceived cannot address the question of how fast aggregates go. To incorporate time and space into the model we observe that the mean square displacement for a single step in DLA is

$$\bar{x}^2 = \frac{4g^2}{4} = g^2 \tag{1}$$

where g is the size of the grid and steps available for diffusion are up, down, left, and right. If the diagonal steps are included the mean square displacement becomes 1.5 x^2. Now the mean square displacement is related to the diffusion constant in 2 dimensions by

$$\bar{x}^2 = 4 \bullet D \bullet t \tag{2}$$

which provides an equivalence or time mapping to the DLA model. Assuming the measured values for receptor size, diffusion constant, and the density of free receptors we are in a position to model the rate of aggregate growth under physiological conditions, assuming that it is diffusion limited.

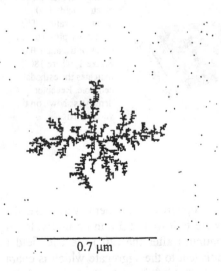

Figure 3. A DLA simulation of acetylcholine receptor aggregation after 60 seconds. The predicted rate of growth is considerably faster than that found in

0.7 μm

The results of such a simulation are shown in Figure 3, and the predicted rate of aggregation is clearly larger than that observed in the experimental system (in which ca. 20 minutes are required to detect aggregation). This rate does of course scale with the diffusion constant, and is nearly proportional to the (measured) density of free receptors at time = 0 [16, 17]. The assumed valence (the number of sites at which each receptor can "bind"; 4 or 8 for rectangular grids, 3 or 6 for hexagonal grids) does not affect the outcome sufficiently to call this outcome into question. Receptor diffusion (i.e. grid size) was scaled down progressively from a step size of 1 to 0.5 to 0.25 particle diameters; the results show rapid progression to asymptotic behavior not significantly different, in this context, to that reported here[17]. Clearly, something other than diffusion is limiting the rate at which receptors aggregate. If we for the moment assume a scaffold of molecules to which AChRs can bind (see next section) the finding supports the notion that receptors do not simply bind passively to this scaffold but are in fact required for its stabilization and growth.

3.2. THE ULTRASTRUCTURE OF DEVELOPING AGGREGATES

3.2.1 Experimental Results

In order to answer questions about the ultrastructural arrangement of receptors within aggregates, sister muscle cell cultures were either stimulated for 2 hours with agrin, or left unstimulated (control). The cultures were then labeled with 12 nanometer gold particles as indicated in Figure 4 in preparation for scanning electron microscopy imaging of receptor locations[18-20]. Membrane regions from both kinds of cells were scanned using a backscattered detector to enhance the contrast between the biological specimens and the gold particles.

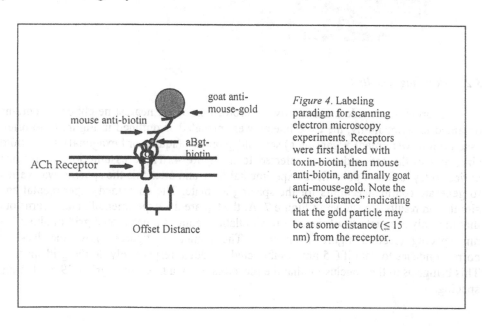

Figure 4. Labeling paradigm for scanning electron microscopy experiments. Receptors were first labeled with toxin-biotin, then mouse anti-biotin, and finally goat anti-mouse-gold. Note the "offset distance" indicating that the gold particle may be at some distance (≤ 15 nm) from the receptor.

A typical image from agrin-stimulated cell membrane is shown in Figure 5. The first step in the analysis is to prepare a list of the XY coordinates of each gold particle; in the case of montage constructions these coordinates are converted to global values using overlapping landmark features[18]. For each XY coordinate in such a list, we then associate a "nearest neighbor distance" – its value is simply the distance in nanometers to the particle closest to the one under examination. In this way the sets of lists (1 set for control, 1 for agrin stimulated membrane) are analyzed to produce nearest-neighbor frequency histograms. The rationale underlying this approach is the question: are receptors bound to a regular, repeating scaffold (in which case some distances are more likely than others) or are they arranged pseudo randomly such that there is no discreet set of favored distances.

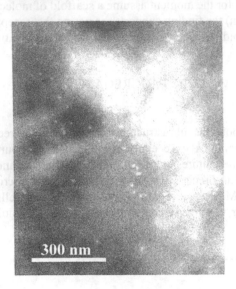

Figure 5. Agrin stimulated membrane. Receptors are labeled with 12 nm gold particles, and viewed with backscattered scanning electron microscopy.

3.2.2 Modeling Results

In order to provide a quantitative interpretation of the nearest neighbor histograms obtained experimentally, the experiments were modeled as outlined in Figure 6. In brief, assumptions were made regarding a) scaffold geometry (square or hexagonal), b) scaffold size, and c) the offset distance referred to in Figure 4. These assumptions were then varied to try to find a good fit to experimental data, and after this the spacing was varied to generate confidence limits for the spacing estimate. The combined experimental and simulation results are shown in Figure 7. At the top are the experimental data. It turns out that the only adequate fit of the data to simulation requires a hexagonal grid of about 9.9 nm spacing (second figure from top). The remaining figures show the lack of correspondence to data if 0.5 nm is subtracted or added, respectively, to the grid spacing. This brings us to the conclusion that the receptors lie on a hexagonal grid of 9.9 ± 0.5 nm spacing.

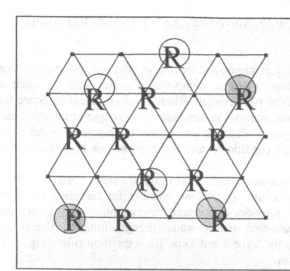

Figure 6. Modeling of nearest neighbor distances. A grid and spacing are assumed, receptors randomly placed (exactly) on the grid, and gold particles are placed on receptors consistent with a steric hindrance of 12 nm (which precludes labeling of adjacent grid points in this case), Finally, the gold particles are randomly displaced by the offset distance. The illustration

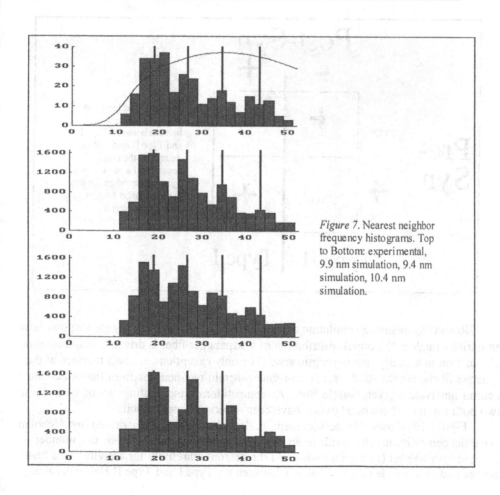

Figure 7. Nearest neighbor frequency histograms. Top to Bottom: experimental, 9.9 nm simulation, 9.4 nm simulation, 10.4 nm simulation.

3.3. DEVELOPMENT OF SYNAPSE ELIMINATION AND THE SIZE PRINCIPLE

3.3.1 Model Results

The model presented here[21] is predicated on the qualitative observation that under the right conditions Hebbian synaptic competition could lead to synapse elimination and the development of the size principle. There is experimental evidence for Hebbian synapses in visual and motor control systems, and both synapse elimination and the size principle are readily observed. The purpose of the quantitative model is to ascertain 1) just how broad the "right conditions" are, and 2) what new predictions arise from a quantitative model.

The model assumes 20 motor neurons each send out 300 neurites onto a field of 2000 muscle fibers at random (these are approximate number for the abductor digiti minimi muscle of the hand). This provides an average innervation of 3 synapses per muscle fiber – all synapses are assumed to have equal strength initially. Values are chosen for the relative strengths of the Type I and Type II competition rules (Figure 8) and the competition simulation is run.

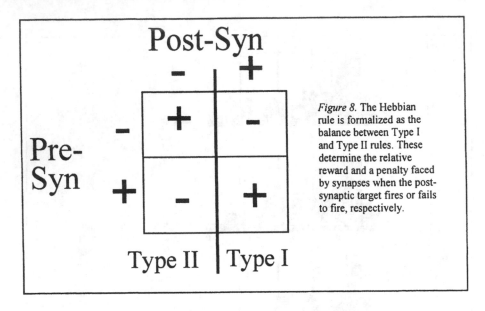

Figure 8. The Hebbian rule is formalized as the balance between Type I and Type II rules. These determine the relative reward and a penalty faced by synapses when the post-synaptic target fires or fails to fire, respectively.

Results from such a simulation are shown in Figures 9 and 10. Figure 9 shows how an initially random (binomial) distribution of synapses per fiber is driven through synapse elimination to a nearly monosynaptic state. The only exceptions, a small number of dual synapse fibers, are due to the rare occurrence wherein two neurites from the same motor neuron innervate a given muscle fiber. As competition cannot distinguish between these two both survive. These rare "twins" have been observed experimentally.

Figure 10 shows the development of the size principle as driven by Hebbian synaptic competition. The result is an exponential relationship between the number of synapses per neuron (motor unit size) and the neuron's threshold for activity. This result is obtained over a wide range of balance between the Type I and Type II Hebbian rules.

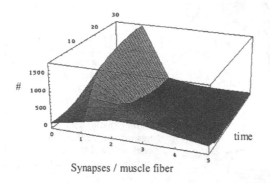

Figure 9. Development of monosynapticity. The distribution of synapses per fiber is random (binomial) initially, but quickly is driven to favor one synapse per fiber.

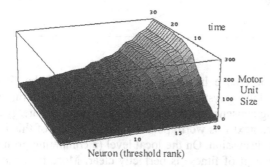

Figure 10. Development of the size principle. The motor unit size (number of muscle fibers controlled by a neuron) is shown against time. Initially all synapses are below threshold, so all neurons control 0 fibers. The distribution becomes exponential.

On the other hand small differences in the balance between the two rules (or in the developmental stage at which manipulation begins) do predict reverse outcomes in experiments separating out a fraction of the motor neuron pool for stimulation or inactivation. Significantly, the opposing results have both been found experimentally[22, 23]. Finally, the model predicts that topographic error correction will take place if and only if such errors are in the minority (Figure 11, top). If erroneous and correct connections are similarly weighted, the model predicts a poor overall innervation pattern and a lack of the size principle. This is a key point, as the existence of the size principle in regenerating projections has been shown to depend on presence or absence of other "correct" innervation[24].

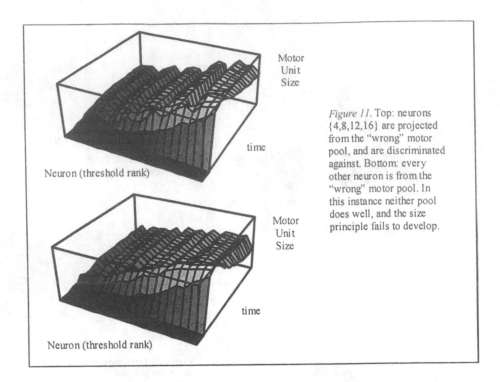

Figure 11. Top: neurons {4,8,12,16} are projected from the "wrong" motor pool, and are discriminated against. Bottom: every other neuron is from the "wrong" motor pool. In this instance neither pool does well, and the size principle fails to develop.

One of the themes of this conference has been an examination of the fitness concept. We shared a sense that developing systems result in *something* being optimized, but were unable to reach a convincing consensus as to what that something is (energy and robustness were discussed). In this context it is worth pointing out an aspect of this issue neatly illustrated in the model under discussion. On the local level (a competing group of synapses on a muscle fiber), the concept of fitness is perfectly clear. More fit synapses are those that fire their target muscle fibers more successfully, and these fitter synapses are the ones that survive the competition. Because of the model assumptions, which also hold in developing vertebrates, this results in an optimization of fine motor control at the more global level (the overall innervation pattern of a muscle). The lesson is clear - optimization of a local fitness metric by a set of constituent parts of a system can result in a quite different kind of optimization system wide.

4. Summary and Conclusions

Three examples of the use of modeling to interpret experimental results have been shown. In the first instance a new model of Diffusion Limited Aggregation, incorporating space and time, was used to rule out such a process in the aggregation of acetylcholine receptors on muscle cell membrane. The second example entailed the use of a simple stochastic model to predict/discover a specific geometry and spacing of receptors within aggregates, which was consistent with experimental observations of nearest neighbor distances. Finally, a model of Hebbian synaptic competition was presented which explains many otherwise difficult observations in the literature, and makes predictions unavailable from the qualitative postulates underlying the model.

Acknowledgements: This research was supported by National Science Foundation Grant IBN97-24035, and American Heart Association - Hawaii Affiliate Grant HIGS-17-95.

5. References

1. Magill-Solc, C. and U.J. McMahan (1988) Motor neurons contain agrin-like molecules, *J Cell Biol* 107, 1825-33.

2. Cohen, M.W., F. Moody-Corbett, and E.W. Godfrey (1994) Neuritic deposition of agrin on culture substrate: implications for nerve-muscle synaptogenesis, *J Neurosci* 14, 3293-303.

3. McMahan, U.J. (1990) The agrin hypothesis, *Cold Spring Harbour Symp Quant Biol* 55, 407-418.

4. Nitkin, R.M., et al. (1987) Identification of agrin, a synaptic organizing protein from Torpedo electric organ, *J Cell Biol* 105, 2471-8.

5. Gautam, M., et al. (1996) Defective Neuromuscular Synaptogenesis in Agrin-Deficient Mutant Mice, *Cell* 85, 525-535.

6. Stollberg, J. and S.E. Fraser (1988) Acetylcholine receptors and concanavalin A-binding sites on cultured Xenopus muscle cells: electrophoresis, diffusion, and aggregation, *J Cell Biol* 107, 1397-408.

7. Stollberg, J. and S.E. Fraser (1990) Acetylcholine receptor clustering is triggered by a change in the density of a nonreceptor molecule, *J Cell Biol* 111, 2029-2039.

8. Stollberg, J. and S.E. Fraser (1990) Local accumulation of acetylcholine receptors is neither necessary nor sufficient to induce cluster formation, *J Neurosci* 10, 247-55.

9. Orida, N. and M.-m. Poo (1978) Electrophoretic movement and localization of acetylcholine receptors in the embryonic muscle cell membrane, *Nature* 275, 31-35.

10. Sabrina, F. and J. Stollberg (1997) Common Molecular Mechanisms in Field- and Agrin-Induced Acetylcholine Receptor Clustering, *Cell Molec Neurobiol* 17, 207-225.

11. Witten, T.A. and L.M. Sander (1981) Diffusion-Limited Aggregation, A Kinetic Critical Phenomenon, *Phys Rev Lett* 47, 1400-1403.

12. Thompson, W.J. (1985) Activity and Synapse Elimination at the Neuromuscular Junction, *Cell. Molec. Neurobiol.* 5, 167-182.

13. Colman, H. and J.W. Lichtman (1993) Interactions between nerve and muscle: Synapse elimination at the developing neuromuscular junction, *Developmental Biology* 156, 1-10.

14. Henneman, E. (1957) Relation between Size of Neurons and Their Susceptibility to Discharge, *Science* 126, 1345-1347.

15. Stollberg, J. and S.E. Fraser (1989) Electric field-induced redistribution of ACh receptors on cultured muscle cells: electromigration, diffusion, and aggregation, *Biol Bull* 176(S), 157-163.

16. Stollberg, J. (1994) A model for Diffusion-Limited Aggregation in Membranes, *Com Mol Cell Biophys* 8, 188-198.

17. Stollberg, J. (1995) Density and diffusion limited aggregation in membranes, *Bull Math Biol* 57, 651-677.

18. Kunkel, D.D. and J. Stollberg (1998) Structural Organization of Developing Acetylcholine Receptor Aggregates, *J Neurobiol* 32, 613-626.

19. Kunkel, D.D., L. Lee, and J. Stollberg (2001) Ultrastructure of Acetylcholine Receptor Aggregates Parallels Mechanisms of Aggregation, *BioMed Central Neuroscience* 2.

20. Stollberg, J. and D.D. Kunkel (2000) Quantitative Assessment of Aggregate Structure, *Comments Mol. Cell. Biophys.* 10, 25-52.

21. Stollberg, J. (1995) Synapse Elimination, the Size Principle, and Hebbian Synapses, *J Neurobiol* 26, 273-282.

22. Callaway, E.M., J.M. Soha, and D.C. Van Essen (1987) Competition favouring inactive over active motor neurons during synapse elimination, *Nature* 328, 422-6.

23. Ridge, R.M. and W.J. Betz (1984) The effect of selective, chronic stimulation on motor unit size in developing rat muscle, *J Neurosci* 4, 2614-20.

24. Thomas, C.K., et al. (1987) Patterns of reinnervation and motor unit recruitment in human hand muscles after complete ulnar and median nerve section and resuture, *J Neurol Neurosurg Psychiatry* 50, 259-68.

DISLOCATIONS IN THE REPETITIVE UNIT PATTERNS OF BIOLOGICAL SYSTEMS

BEATA ZAGÓRSKA-MAREK and DANUTA WISS
Institute of Plant Biology, Wrocław University,
Kanonia 6/8, 50-328 Wrocław, Poland

1. Repeat patterns

Periodicity, one of the most simple and frequent components of the process of pattern formation, observed in nonliving matter such as crystals or physical waves, is also present in living organisms. The architecture of an early animal or plant embryo depends upon the action of some master genes that section the 3-D space occupied by the growing population of embryonic cells into segments, organized along all three directions of growth (Alberts *et al.* [1], Howell [14]). The segments may be identical, as metamers emerging along the longitudinal axis of round or flat worms or muscles in a fish body, or plant modules such as nodes and internodes of the shoot. They may also assume a different identity due to the secondary processes of differentiation. Among numerous examples let us name just a few: the anterior and posterior segments of *Drosophila* larva, the concentric layers of dermal, ground and vascular tissues in plant cylindrical vegetative organs, the consecutive whorls of floral organs such as petals, stamens and carpels. These are the best known illustrations of the phenomenon in discussion. However there are many other examples of periodic patterns in biological systems. Some of them are rigid and represent a final product of certain developmental processes. Such for instance are the striped patterns of plant or animal body pigmentation (Fig. 1) or rippled patterns of its superficial sculpture (Fig. 2). Others, the most interesting, are dynamic, forming and frequently transforming in the course of an individual's development. In our article we will present the latter as special cases being of special interest to our research group. They represent the greatest challenge in our attempts to produce their adequate description, which is possible only when the laws of their development are traced down.

Figure 1. Repeat patterns of pigmentation: in a zebra coat (left) and in a pansy flower (right).

J. Nation et al. (eds.), Formal Descriptions of Developing Systems, 99–117.

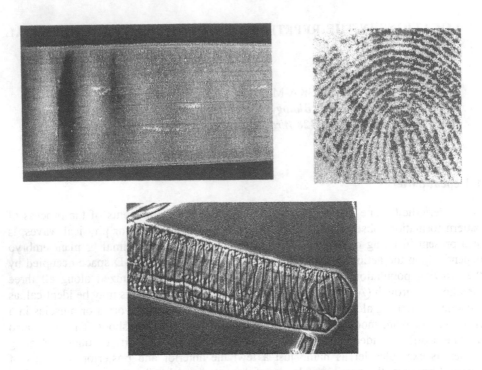

Figure 2. Ripple-mark patterns in living organisms: an undulated leaf of grass (left), the author's fingerprint (right), thickenings of the secondary cell wall in a conducting plant cell (below).

2. Dislocations as faults in repeat patterns

Periodic patterns are not always perfect. Their disturbances are very interesting for the reason that aberrations, deviations from the rule, are often more informative than the principle itself. In biology this is proven by the analysis of mutant phenotypes, the conclusions of which are important for better understanding of morphogenetic pathways and evolutionary trends. The disturbances of patterns either provide an insight to the mechanism governing pattern formation or change pattern properties. In the *Nautilus* shell (Fig. 3) one may readily notice some imperfections in an otherwise regular pattern of evenly spaced pigmented stripes. These are bifurcations of the stripes, which multiply towards the external edge of the shell coil. Such bifurcations have been defined in crystallography as dislocations – the defects that are often present in a regular crystalline lattice (Fig. 3). In a planar lattice, planar defects mean that an additional rows of atoms ions or molecules are being added to the existing pattern. For ions, the size of the ion is the determining factor in the crystal lattice arrangement. What causes then the appearance of dislocations in the striped pattern of a *Nautilus* shell?

 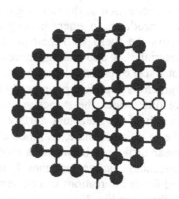

Figure 3. Nautilus pompilius (left) – bifurcating of stripes in pigmentation pattern of the shell is more intense in an older (initial) part of the shell, where each stripe bifurcates twice or even three times, whereas in the youngest part stripes almost do not divide. Photograph (modified) from David Attenborough's Life on Earth. Scheme of a crystalline planar lattice with dislocation (right). White circles show an additional row of elements.

3. The causes of defects

In some molluscs, shells are more helical than spiral – the radius of consecutive coils of the shell tube increases slowly from the apex and a shell growing forward becomes elongated, having clear longitudinal axis. In *Nautilus* the shell spiral is almost ideally equiangular, constructed on the principle of the Golden Ratio (Fig. 4).

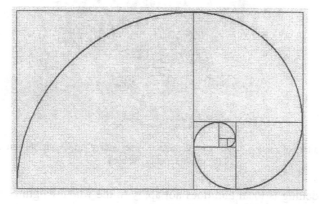

Figure 4. The construction of planar, equiangular spiral. The length of the side of each consecutive square belongs to the Fibonacci sequence 1,1,2,3,5,8....

It is a logarithmic spiral in its simplest planar form – the shell tube's revolutions occur in one plane – the radius increases rapidly and both opposite edges of the tube become extremely unequal in length. The pigmented band's width slightly increases outward. As already pointed by Meinhardt [16] the presence of dislocations, transverse to the direction of the tube's growth, probably means that the rhythm of the *Nautilus'* cells producing pigment is coupled to some extent (the width is not exactly the same across the shell) with their productivity. When they produce a lot of shell building material, they also undergo more cycles of pigment production. In this case it seems that not a size of pattern element, a stripe itself, but the increase of the surface, on which the pigment is deposited periodically is responsible for dislocation effect. Even though they are little wider, the stripes are more numerous on the outside, where the longitudinal growth of the shell tube surface, on which they appear, is more intensive. We would like to point out here, that theoretically in the "logarithmic shell" growing forward the linear difference between its inner and outer edge lengths (both being in a Fibonacci sequence) changes continuously in such a manner, that with time there should be fewer and fewer bifurcations in a pattern. Indeed, this is exactly what we see in real nautiluses, as shown in Fig. 3.

A similar explanation can be applied to the striped pattern of zebra or coral fish – the dislocations most often form where the surface of the developing animal body is shaped by differential growth, and thus the pattern is disturbed by bumps or bending. Specific patterns characteristic for different zebra species develop due to genetically controlled differences in timing of the pattern origin and location in space of this differential growth in the developing embryo (Bard [3, 4]). Initially all the stripes are of the same width. According to Gilbert [8] "... when the region between the pigmented bands becomes too wide, secondary stripes emerge, as if suppression (of pigment formation) was weakening" (Fig. 5).

Figure 5. Primary pattern of zebra is clearly complemented by secondary stripes, visible here on the hind legs of the mother and of the baby.

If this widening is local and does not apply to the whole length of the stripes or ripples, dislocation emergence is inevitable. It is generally accepted that the patterns with alternative pigmented and non-pigmented bands, exemplified by zebra stripes develop in a way predicted by Turing reaction-diffusion models. Some effects are explainable by assuming that two or more Turing-type reaction-diffusion systems interact (Murray [17]), although in most of the cases neither activators nor inhibitors have been defined.

There is an important difference between a *Nautilus* shell and a zebra embryo. In the mollusc the coloring pattern results from the condition, changing in time and in space, of the cells continuously extending the shell tube. In a zebra the whole surface of an embryo body is organized at a certain moment of development into the stripes, with this initial pattern being subsequently modified by differential growth and secondary pigmentation. Fascinating analogies to both cases exist in one unique plant tissue, in cambium. This lateral, cylindrical sheet of plant embryonic cells is often beautifully organized spatially (zebra analogy) into a structural wave - a periodic pattern, which, in addition, is not stationary but travels along cambial surface (Hejnowicz [10, 12], Zagórska-Marek [21, 22]). Thanks to the special activity of cambial cells producing consecutive annual rings of a tree, this spatial organization is continuously recorded (*Nautilus* analogy) in figured wavy wood (Fig. 6).

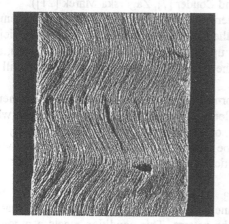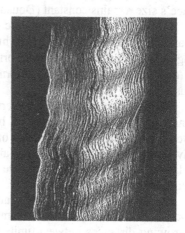

Figure 6. Wavy wood as seen from the side (left), which is attached to cambium (tangential face), and in 3-D view (right). The consecutive annual rings contain records of cambial wavy structure. This way the radial face of wood becomes rippled. (The pictures mysteriously changed sides– left has became right and vice versa, thus the explanations below are totally misleading)

Nautilus and zebra cases show that on one hand the same phenotypic effects may be accomplished in different ways, with the engagement of different developmental strategies. On the other hand, many various phenotypic effects observed in unrelated organisms may have at their bottom the same theme – for instance periodicity and dislocations. This applies not only to the macro-scale but also to the micro-scale of development. The same laws are observed in spatial organization of cellular processes, as exemplified by the structure of the plant cell wall shown in Fig. 2. When simulating pattern development in living things, we should always be aware however of biological limitations and restraints to the assumptions of the proposed models.

In the cases discussed, it seems that the size of the repeat pattern element remains more or less constant during development, and the factor introducing dislocations is a localized change in size of the surface on which the pattern appears. However, this is not always the case. In a repeat pattern of fingerprint both negative (of trough line) and positive (of crest line) dislocations are present simultaneously (Fig. 2). They are noticeably numerous in the lower right quadrant of a print where they bifurcate in one direction. Sometimes two dislocations of different type, when they bifurcate in opposite directions, seem to neutralize each other's effect. This happens just above the "loop" of the fingerprint shown in Fig. 2. Such cases are harder to explain, because there is no evident disturbance in surface growth that would justify their formation. Another case of "neutral" dislocations has been reported for cambium, where in principle there is no longitudinal expansion of the tissue (Zagórska-Marek [21]). This will be shown and discussed later in the text. There is not known to us a satisfactory explanation of the appearance of dislocations there, where neither pattern element nor the surface itself seems to change in size. The local deviation from the main direction of propagation of the diffusion-reaction system should be considered in possible models of this effect.

In computer simulations of phyllotactic pattern growth, the size of pattern elements, changing systematically and continuously, may cause dislocation, while the surface's size remains constant (Douady and Couder [7], Zagórska-Marek [21]).

For quality of the phyllotactic pattern, the proportion between pattern element size and the size of the surface on which the pattern is formed is decisive. A sufficient change of this during development brings up at first the disturbance – dislocation, and then new organization of the pattern. The case of phyllotactic dislocations will be discussed later.

In the sandy ripplemark pattern formed by oscillating water, the phenomenon ingeniously studied and interpreted by Hertha Ayrton [2], dislocations appear when ripples are moving, consuming each other, or when there is a superficial inequality of a sandy bottom. They probably may also be caused by a change in a period of water oscillation, altering size of formative vortices, and brushes moving the sand particles under the shelter of existing crests.

It seems that the system causing stationary, spatial periodic patterns with dislocations must be dynamic for some time. Otherwise it would not be able to react to the growing distances between units of the pattern. This condition meets a model of morphogenetic waves with constant period (Hejnowicz [11, 12]). Superposition of such waves propagated continuously produces a stationary beat-pattern.

Dislocations emerge wherever there is a sufficient local deviation from the established parameters of the system responsible for the appearance of periodic pattern.

4. The morphology of dislocations

In the same repeat pattern, as we already mentioned, dislocations may appear as negative and positive, i.e., the potential of bifurcating is a property of both elements of the pattern: of pigmented and non-pigmented areas in striped patterns, of ridges and furrows in undulated surfaces.

Dislocations in the repeat pattern may be divided into horizontal and vertical, relative to the principal direction of growth of the surface on which the pattern is forming. Classifying them in this way is not always easy though, because in a developing organism there are two such directions for surfaces and three for volumetric growth of the organs. Fortunately, both animals and plants exhibit clear polarity in their development. Establishing the main axis in a developing embryo with its apical and basal end is a fundamental event of great importance for further appropriate growth and differentiation. The same applies to developing, axial in nature, organs. Thus the reference point for definition of dislocations may be the longitudinal axis of an organ or of the whole body. In this sense, dislocations in the zebra and *Nautilus* belong to a horizontal type.

Local undulation of a leaf blade in monocots, once it becomes visible, is composed of ripples perpendicular to the direction of longitudinal expansion of the blade (Figs. 2, 7, 8). This shows that the expansion of the leaf surface changes regularly its rate along the longitudinal axis of the organ. Interestingly enough this undulation is not stationary but initially it is moving along the blade's longitudinal axis (Hejnowicz [13]). Dislocations in the rippled leaf are also horizontal. Wherever they develop, they indicate that not only along the leaf axis but

Figure 7. Rippled surfaces of iris leaves. Undulations show in the middle of the blade leaving the edges free (upper left), on its very edge (upper right) or across the whole blade (bottom). Crests are sometimes bifurcated (upper right and bottom).

also in the parallel areas of the leaf surface, positioned side by side, the rate of expansion is uneven. One example of this possibility is shown on the Fig. 7, where undulations on the lateral edge of the iris leaf are clearly arrested laterally by the presence of a longitudinal vein, and on the very edge there is one small dislocation. In most of the leaves of monocots the ripples appear less or more seldom, rather accidentally and unpredictably (Fig. 7). There is however one special plant, *Carex plantaginea* Lam., whose leaves are always undulated (Fig. 8). In these leaves the arrest takes place on both left and right sides of an area with the excessive expansion. Each strip of the lamina, bordered by the major veins acting as the strings, is undulated except for upper, lateral strips. In neighboring strips the undulation is shifted by half of the wave's length. This produces a peculiar chequered effect on the lamina surface. The length of superficial waves in these leaves varies (Fig. 8). It is usually in a range of 6-10 mm, sometimes, locally, in a lesser range of 3-6 mm.

106   BEATA ZAGÓRSKA-MAREK and DANUTA WISS

Waves of different length may occur in the same leaf. Under high magnification the very minute local undulations invisible to the naked eye disclose themselves as also present in this peculiar system (Fig. 8 – right photograph). They occur in the stripes of tissue bordered by the smallest veins still parallel one to another. This is a unique case of such strongly expressed tendency for differential growth of the leaf blade in a periodic manner. We have just initiated this research on *Carex*; thus the results of these preliminary observations do not give an answer to many basic questions. From Hejnowicz's study on tulip leaves, we have learned that spatial and temporal oscillations in pH of the epidermal cell walls facilitate this undulation (Hejnowicz [13]). At least in this case we know the chemical factor responsible directly for the repeat pattern development.

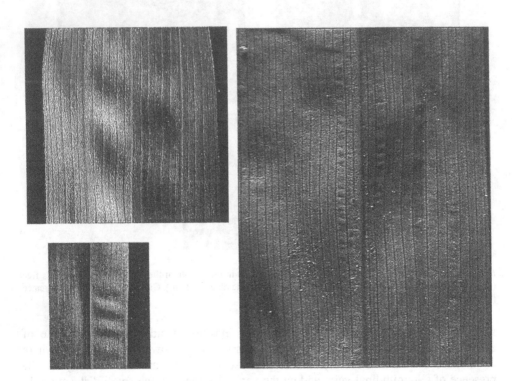

Figure 8. Superficial undulations of leaves in *Carex plantaginea* have different lengths even in the same leaf. The lateral veins restrict undulated areas. In stripes of undulated leaf, positioned side by side, the crests alter horizontally with troughs. Right photograph was taken at higher magnification than other two.

In plant leaves or in their modifications such as flower petals, the venation pattern exhibits vertical dislocations. Bifurcations of veins are of V-type i.e. they are oriented towards the outside edge of the leaf or petal and this means, that the number of units (vascular bundles) in a repeat pattern of venation developmentally increases (Fig. 9).

Figure 9. Bifurcating outwards pigmented veins of an iris (left), and geranium (right) petals, indicate the sites of differential expansion. (the same as in Fig.6)

They are good indicators of the intensity of superficial growth in various directions. Sometimes during longitudinal and lateral expansion of these planar structures, the mid-veins do not divide or divide rarely whereas lateral veins bifurcate many times. This shows that the lateral parts expand intensively along their edges whereas terminal edge at the same time does not expand as much laterally. In case of not pointed, but two-lobed, structures it is the central vein that splits first (Fig. 9).

Differential growth of the surface is clearly responsible for numerous vertical bifurcations of vascular strands in the fan-shaped, two-lobed leaf of *Ginkgo* (Fig. 10). The blade of the leaf is expanding but not all strands bifurcate on the same radius from the starting point, i.e., the base of the leaf where the stalk changes into the blade. This phenomenon is worth of pointing out. There is no ideal regularity in natural, not simulated repeat patterns. Bifurcations in *Ginkgo*, appearing at first sight randomly, may result not only from uneven expansion of elongating and expanding laterally fan-shaped blade. Also the distances between repetitive units of the pattern – parallel vascular strands, if initially unequal - may be held responsible for this effect. In the case of *Adianthum*, a fern shown on Fig. 11, the leaves positioned closely on the same plant evidently differ in symmetry of blade expansion.

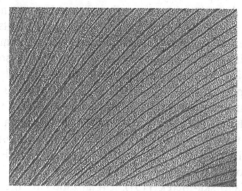

Figure 10. Bilobed leaf of *Ginkgo* is served by intensively bifurcating veins.

Figure 11. Two leaves of this fern are asymmetrical as far as their venation pattern tells.

Multiplication of repeat pattern elements through vertical dislocations is natural during superficial expansion. The reversal of this trend brings up a developmental decrease in a number of pattern elements. Vertical dislocation of V-type in this case stands "upside down" with respect to the direction of growth. To illustrate this case we shall turn to the plant SAM – shoot apical meristem, responsible for regular initiation of lateral organs. Their organographic order is known as phyllotactic pattern.

5. Special cases

5.1. PHYLLOTACTIC LATTICES

Indeterminate growth of plants, especially perennial ones, requires new modules being added to the plant body constantly in a course of continuous organogenesis. This is exemplified by either consecutive annual rings of secondary vascular tissue produced by lateral embryonic cells of cambium, or new growth increments, that every year elongate axial, cylindrical plant organs (shoots or roots) due to activity of their apical meristems. The latter, besides elongating the main organs, also produce periodically the primordia of lateral structures. Packing of these primordia on the lateral surface of shoot apical meristem (SAM), in its almost absolute regularity, resembles crystal lattice – a repeat pattern. All lateral primordia can be connected with imaginary spiral lines, running parallel to each other within either of two opposite sets. Two most conspicuous lines selected each from another set are mutually orthogonal. All intersecting lines form a phyllotactic lattice. The building block of this lattice is a triad of neighboring primordia, called phyllotaxis triangular unit (PTU). It can be compared to the repetitive module – crystal unit cell present in the atomic or molecular, crystal networks (Zagórska-Marek [19]). Organization of the lattice is usually very regular, but not always the same (Fig. 12). It varies depending upon the number of lines present in the lattice and their inclination to the direction of system growth. The pair of numbers referring to the lines in each set defines the lattice organization – i.e., phyllotactic pattern. It is sometimes convenient to provide these numbers with indices of line orientation: z (to the right) and s (to the left).

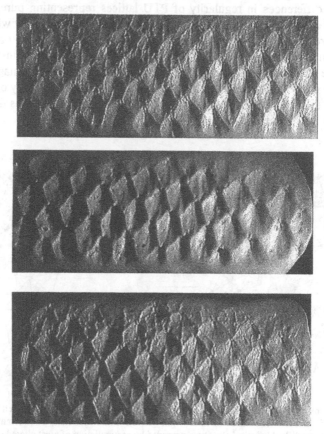

Figure 12. Phyllotactic lattices of magnolia obtained by rolling generative shoot over the surface of modeling clay (Zagórska-Marek [20]). In each lattice the same pattern element, replicated twice and seen on the opposite edges, has been labeled. This facilitates counting the lines connecting pattern elements and identifying organization of the lattice, different in each case: 3s:7z (upper), 4z:8s (middle) and 6z:8s (lower). Note that neither of these represents the most common case in plants, the main Fibonacci organization, in which the numbers of lines belong to the mathematical sequence 1,1,2,3,5,8....

In some cases, the regularity of PTU pattern is broken by the appearance of dislocations (Fig.13), i.e., bifurcations of lines connecting pattern elements in the phyllotactic lattice (Zagórska-Marek [19], [20]). Dislocations are **vertical** of two types. The less frequent **V-type** means the developmental addition of a new line, similar to vein bifurcations in leaves. The rather common **Λ-type** means termination of an existing line. The difference in frequency is understandable in light of the fact that SAMs are paraboloid in shape, so that the size of organogenic surface on which primordia are initiated rather decreases. Even a single dislocation means that the given system of primordia packing, utilized by SAM for some period of time, alters – becomes transformed ontogenetically into another. The cause of such ontogenetic transformations remains unknown.

The question we have asked recently is whether there are any signs within the system preceding dislocation before it shows up in the lattice. The next question is: are there any differences in regularity of PTU lattices representing pure, but at the same time different, packing systems? In another words we want to know whether all packing systems are equal in their regularity thus, perhaps in stability. If not, the dislocation could be interpreted as an unavoidable effect of the initial pattern's instability.

The model system we used in our investigations was the magnolia embryonic, generative shoot. It has been selected because of the great variability of packing systems present in this structure (Zagórska-Marek [20]). Another reason was that dislocations in magnolias are quite frequent.

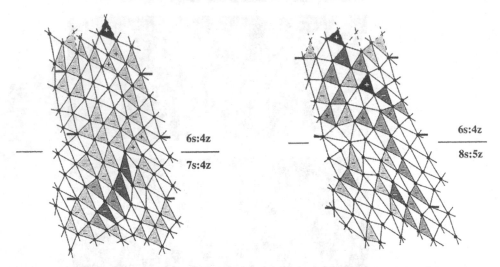

Figure 13. Phyllotactic lattices with dislocations of Λ-type: one in a set of 7s lines (left lattice) and three – one in a set of 5z and two in a set of 8s lines (right lattice). In the first case, transformation of the phyllotactic pattern is accomplished by one dislocation, preceded by accumulation of most altered (in -) PTUs below (dark grey triangles). In the resulting pattern deviations of PTUs from expected values are moderate in + and in - (light grey triangles). The second transformation of phyllotaxis is much more dramatic. It is also preceded by a kind of superficial deficit in a lattice (all triangles are below expected value) but three dislocations result in a formation of a not relaxed, but very unstable, bijugate pattern.

A comparison of magnolia lattices mapped until now showed that in the lattices with dislocations, the deviations of PTU's geometry from parameters predicted theoretically are much more dramatic (Fig. 13) than in lattices representing pure patterns (Fig. 14). In one case (left on Fig. 13), in the pattern preceding transformation, highly altered PTUs were grouped below the site of later appearance of dislocation, in exactly the same (circumferentially) sector of the lattice. Dislocation evidently caused a decrease of fluctuations in PTU geometry, as they were minute in the lattice above the level of imperfection. However, in another case (right on Fig. 13), fluctuations persisted and were distributed evenly within the lattice, suggesting great instability of the pattern resulting from transformation. Is that new packing system unstable *per se*? Or it is rather that one dislocation was insufficient to relax accumulated earlier disparities of SAM growth parameters? Classical Fibonacci lattices contained the least disturbed PTUs, while the purely bijugate lattice (right on Fig. 14) was more unstable. Does it mean that magnolia does not like making bijugate phyllotaxis? We know that there are some plants such as *Torreya* that exclusively make bijugy as their most favored pattern. This research is still in progress; its results are still inconclusive, although very promising.

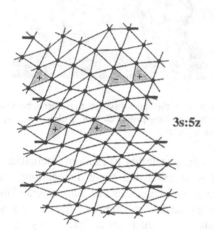
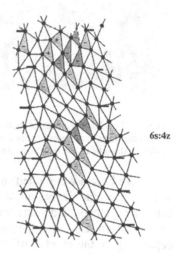

Figure 14. Phyllotactic lattices without dislocations. Pure patterns differ slightly in stability of PTUs. The most regular are lattices with the main Fibonacci organization, where deviations are below the threshold or as small as these shown in the left lattice. In other patterns exemplified by bijugate lattice on the right deviations are greater.

The available models of phyllotaxis allow the development of dislocations leading to qualitative change in lattice organization. This has been accomplished in Battjes' computer simulations (Zagórska-Marek [22]) under following assumptions:

- every newly initiated primordium fills the first available space (Hofmeister rule),
- the size of organ primordia or the size of organogenic surface of SAM change in time; the change is either abrupt and strictly localized, or systematic and continuous.

The results of modeling have to be confronted constantly with the empirical data. Otherwise, we will never understand correctly the mechanisms that govern morphogenetic processes engaged in phyllotactic pattern formation. In our lattices, the measured parameters of PTU geometry were angular. We do not know then what really changes within SAM facilitating dislocation – the size of pattern element, the size of organogenic surface, or both? Two striking examples of most regular lattices with the most common, main Fibonacci organization, where either one element of the pattern is missing or two elements act as one without disturbing the lattice's order (Fig. 15), suggest that there is yet a lot of conceptual thinking ahead. It is possible that the phyllotactic pattern generated by SAM has its blueprint elsewhere. In the near future, we want to compare the results of our empirical studies on phyllotactic lattice stability and PTU geometry with the results of computer simulations.

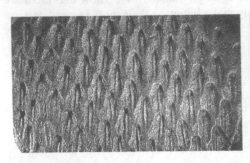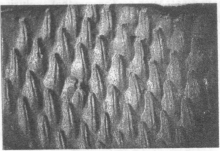

Figure 15. Two Fibonacci lattices, both with 3z:5s:8z organization. In the upper, two pattern elements became so close that they were treated in the development as one. In the lower, one pattern element is clearly missing, yet the phyllotactic lattice developed further without even noticing that!

We have not exploited so far in our models the possibility of random fluctuations in the primordia size within a certain range. It is possible that some dislocations are better than others, and some transformations are more likely to occur than others. The same pattern can also be transformed ontogenetically in many different ways, depending on the number of dislocations and where (in which set of lines) they show up.

5.2. HORIZONTAL DISLOCATIONS IN CAMBIAL PATTERNS

Another plant meristem where dislocations can be encountered is the cambium. In storied cambium, the arrangement of elongated axially fusiform cells, in their positional order, resembles smectic liquid crystals (Zagórska-Marek [22]). Repeat units of the pattern are horizontal tiers of cells, called storeys. They frequently bifurcate to the right or to the left on the cambial surface. Dislocations originate from initially nonstoried (nematic) arrangement of cells. The surface of cambium increases in a tree ontogeny due to multiplication of dividing cells. The cambial cylinder has at the beginning of each tree ontogeny just about 1mm in diameter, reaching later a size measured in meters. When all cells divide only longitudinally, as they do in storied cambium, the horizontal clones of cells of the same lineage unfold on the cambial surface with time. Dislocations occur where three clones of different origin meet. Their ancestors were three neighboring, nonstoried cells. As the fusiform cells have the potential of regulating their length, the dynamic rearrangement of bifurcations cannot be excluded. Nevertheless, these dislocations seem to be rather static.

The second pattern, in which dislocations are present in cambium, is the striped domain pattern discovered by Hejnowicz [9]. Here the repetitive units of the pattern are the domains. In each domain, elongated fusiform cambial cells change their orientation in one direction – to the left in the S-domain or to the right in the Z-domain. The domain pattern is propagated vertically. Thus it can be interpreted as the effect of a vertical shift in a phase of the cycle among oscillating cells. Hejnowicz hypothesises that the domains result from a travelling cambial wave, whose morphogenetic potential causes the cells to deviate from a vertical position (Hejnowicz [10, 11,12]). The domain pattern is translated into a visible tangential (flat) undulation of the cambial surface. Dislocations occur where the areas of uniform cell inclination bifurcate horizontally. Wavy structure of cambium with all these defects is recorded, as has already been mentioned earlier in this article, in wavy wood (Figs. 6, 16).

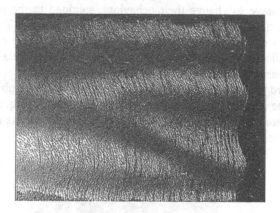

Figure 16. The tangential section of ash wavy wood, with two opposing dislocations. Differential light reflection from the surface gives the impression of white stripes, where grain orientation is to the left, and dark one where it is opposite.

There is no clue as to how the dislocations develop in the domain pattern. Initially, we speculated that because cambial waves may differ in length and velocity, the defects result from local differences in these parameters.

Figure 17. A model of temporal changes in a shape of dislocation appearing in the domain pattern when in the local area (left strip of each diagram) the wave-length and velocity are smaller than elsewhere in cambium. The area is moving upward with the main pattern.

Our first attempt of modelling this showed that under these assumptions dislocations should not be stable. They should transform in shape changing connections with neighboring domains with time (Fig. 17). It appeared however that, through the period of 18 years, neither the shape of the two natural dislocations studied altered nor horizontal shift of their forking points took place (Zagórska-Marek [21]).

The forks in the domain pattern are puzzling. They contain this weird point – a meeting place for the cells from the left and from the right, changing inclination in opposite directions. There must be a kind of silencing taking place there with regard to the intensity of cell inclination change. Another reflection is that in one particular site of cambium where cells were oscillating properly until the dislocation arrived, the same cells suddenly have to change their behaviour, subdued to the positional information brought by the deformation of travelling morphogenetic field. This field conveys also the information on the size of vertical phase shift for cycling cells and the angle of deviation. In all this, it is a beautiful example of the supracellular organization present in a multicellular organism.

The third and least known repeat pattern present in cambium is similar to that described in the leaf blades. Ripples of the cambial surface, frequent in beeches, appear locally. They are usually horizontal, sometimes oblique to the stem axis (Fig. 18). As in all other systems, the repeat units are able to bifurcate.

Figure 18. Smooth bark of the beech tree discloses pronounced ripple-mark pattern of cambial surface hidden beneath. Ripples are oblique and one of the crests bifurcates upward.

6. Evolutionary background and conclusion

The strategy of conquering as much of 2-D and later 3-D space through bifurcations of existing units of the repetitive unit patterns as possible is as ancient as multicellular organisms are. The simplest repetitive units of living organisms are cells. At first they formed linear 1-D structures – filaments. The transition from 1-D to 2-D growth through branching (Fig. 19) or by forming flat structures was possible when the terminal cell in a filament acquired the ability to change temporarily the orientation of its divisions with respect to the main direction of growth. This resulted in cellular dislocations such as those that can be observed on the edge of growing fern gametophyte (Fig. 20). The strategy of dilating structures through addition of pattern repetitive units was one of very early accomplishments in evolution. Once perfected, it allowed skipping the stage of the filament and forming disk-shaped flat structure right from the first one cell, as exemplified by development of green algae *Coleochaete orbicularis* (Fig. 20).

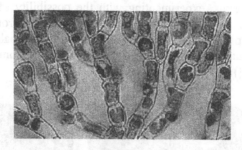

Figure 19. Branching filaments of a green algae *Coleochaete pulvinata*. Photo (modified) courtesy Charles Delwiche.

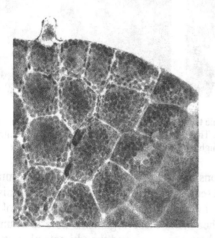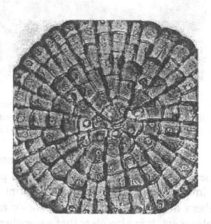

Figure 20. Bifurcations of cellular radial filaments in fern gametophyte (left) and in a thallus of green algae *Coleochaete orbicularis* increase the size of these planar structures. They develop because marginal cells of each radial file, sensing the increase in circumferential tension, from time to time undergo the division parallel instead of perpendicular to the direction of file growth. The same strategy is employed by a higher plant's cambium increasing in diameter with time.

Simple and ancient, because rooted in a history of nonliving matter, the rule of adding a new unit to the existing set of units in a repeat pattern is still a powerful tool in evolutionary as well as in ontogenetic development of living organisms. The perianth of some flowering plants, typically tetramerous, may occasionally, on the same individual, be pentamerous. Sometimes, for unknown reasons, the successful addition of one discrete element, having the same identity, apparently fails. The resulting flower's petal in such case is bifurcated (Fig. 21). Is this the same as in magnolia phyllotactic lattices, where sometimes we do not know exactly whether we are dealing with one or two pattern elements in the lattice (see Fig. 15)?

In *Arabidopsis*, a typical crucifer having four petals in a wild form, perianth may become pentamerous in *pan*-1 mutant (Running, Mayerowitz [18]). In another model plant *Antirrhinum*, the mutations of some genes such as *cycloidea* and *radialis* change the symmetry of the flower from dorsiventral to radial, with its merosity increasing the number of corolla parts from 5 to 6 (Luo *et al.* [15], Clark *et al.* [5], Coen [6]). In evolution of Angiosperms, experimenting with the possibility of adding new organs along the circumference of the floral apex led to transition from initially trimerous flowers, today still typical for monocots, magnolias and some paleoherbs (*Asarum*), to tetramerous and pentamerous flowers of eudicots, evolving from a common ancestor (Zagórska-Marek [23]).

Figure 21. The corollas of *Forsythia* (left) are in principle tetramerous. However, among many flowers on the same plant there are some that have different merosity. They are penta- or even hexamerous. Forget-me-not flower (right) is usually pentamerous; climbing up to a higher solution apparently was unsuccessful here.

Not one universal mechanism responsible for dislocation effect, a common theme, a motif of all biological patterns presented in this article has been recognised. The involvement of the genetic blueprint is indisputable. Genes definitely control the numbers of parts in living organisms or rather make these organisms able to count in development. It is also clear that somewhere between genes and the ultimate phenotypic result of this countdown, besides the undefined activators and inhibitors also other forces than chemical, biophysical for instance, are engaged. On Hugo Steinhaus' gravestone in a quiet small cemetery on the outskirts of our town there is a sentence to be read and appreciated: "Between the spirit and the matter mathematics mediates".

7. References:

1. Alberts B., Johnson A., Lewis J., Raff M., Roberts K., Walter P. (2002) *Molecular Biology of the Cell*, Fourth Edition, Garland Science, New York.
2. Ayrton H. (1910) The Origin and Growth of Ripple-mark, *Proceedings of the Royal Society of London* A84, 285-304.
3. Bard J. B. L. (1977) A unity underlying the different zebra striping patterns, *J. Zool.* (London) 183, 527-539.

4. Bard J. B. L. (1981) A model for generating aspects of zebra and other mammalian coat patterns, *J. Theor. Biol.* 19, 363-385.
5. Clark J.I., Carpenter R., Luo D., Coen E.S. (1999) Regulation of *cycloidea*, a gene controlling floral asymmetry in *Antirrhinum majus.*
6. Coen E.S. (1999) *The Art of Genes. How Organisms Make Themselves,* Oxford University Press, New York.
7. Douady S., Couder Y. (1996) Phyllotaxis as a self-organizing iterative process. I, II, III, *J. Theor. Biol.* 178, 255- 312.
8. Gilbert, S. F. (1997) *Developmental Biology,* Sixth Edition, Sinauer Associates, Inc. Publ., Sunderland, Massachusetts. plus http://www.devbio.com/
9. Hejnowicz Z. (1971) Upward movement of the domain pattern in the cambium producing wavy grain in *Picea excelsa, Acta Soc.Bot.Poloniae* 40, 499-512.
10. Hejnowicz Z. (1973) Morphogenetic waves in cambia of trees, *Plant Sci. Lett.* 1, 359-366.
11. Hejnowicz Z. (1974) Pulsations of domain length as support for the hypothesis of morphogenetic waves in the cambium, *Acta Soc. Bot. Poloniae* 43, 261-271.
12. Hejnowicz Z. (1975) A model of morphogenetic map and clock, *J. Theor. Biol.* 54, 345-362.
13. Hejnowicz Z. (1992) Travelling Pattern of Acidity in the Epidermis of Tulip leaves, *Bot. Acta* 105, 266 – 272.
14. Howell S.H. (1998) *Molecular Genetics of Plant Development,* Cambridge University Press, Cambridge, New York, Melbourne.
15. Luo D., Carpenter R., Vincent C., Copsey L., Coen E.S. (1996) Origin of floral asymmetry in *Antirrhinum, Nature* 383, 794-799.
16. Meinhardt H. (1995) *The Algorithmic Beauty of Sea Shells,* Springer-Verlag, Berlin, Heidelberg, New York.
17. Murray, J. D. (1981) A pre-pattern formation mechanism for animal coat markings, *J. Theor. Biol.* 88, 161-199.
18. Running M.P., Mayerowitz E.M. (1996) Mutations in the *perianthia* gene of *Arabidopsis* specifically alter floral organ number and initiation pattern, *Development* 122, 1261-1269.
19. Zagórska-Marek B. (1987) Phyllotaxis triangular unit; phyllotactic transitions as the consequences of the apical wedge disclinations in a crystal-like pattern of the units, *Acta Soc. Bot. Poloniae* 56, 229-255.
20. Zagórska-Marek B. (1994) Phyllotaxic diversity in *Magnolia* flowers, *Acta Soc.Bot. Poloniae* 63,117-137
21. Zagórska-Marek B. (1995) Morphogenetic waves in cambium and figured wood formation, in: *The Cambial Derivatives,* M.Iqbal (Ed.), *Encyclopedia of Plant Anatomy,* Gebrüder Borntraeger, Berlin, Stuttgart.
22. Zagórska-Marek B. (2000) Plant meristems and their patterns, in: *Pattern formation in biology, vision and dynamics,* A. Carbone, M. Gromov, P. Prusinkiewicz (Eds.) World Scientific Publishing Company, Singapore, pp. 217-239.
23. Zagórska-Marek B. (2000) Mechanisms of patterning in flowers, *Advances in Cell Biology* 27, 97-109 (in Polish).

THE DESCRIPTION OF GROWTH OF PLANT ORGANS:

A CONTINUOUS APPROACH BASED ON THE GROWTH TENSOR

JERZY NAKIELSKI AND ZYGMUNT HEJNOWICZ
Department of Biophysics and Cell Biology
University of Silesia
Jagiellonska 28, 40-032 Katowice, Poland

Abstract. Our *developing systems* are growing plant organs. Plant organs grow symplastically, i.e., in a co-ordinated way. In our approach we treat the symplastic growth of a plant organ as an example of irreversible deformation (plastic strain). Specificity of the organ in comparison to elasto-plastic or elasto-viscous solids is expressed mainly in cell divisions. Symplastic growth leads to the concept of a growth tensor. Cell divisions occur in the principal planes of this tensor.

1. Symplastic growth

In our approach we treat the *growth* of a plant organ as irreversible deformation (plastic strain) of the mass of cells which, in contrast to animals, are surrounded with relatively *rigid walls*. These walls determine the shapes of plant cells, but the development of these shapes is subordinated to the growth of the organ; the walls of neighboring cells are joined by the middle lamellae composed usually of a solid pectic material. Moreover, the neighboring protoplasts are joined by plasmodesmata, which are cytoplasmic connections lined by membrane. Such a coordinated growth of cells within an organ, during which contiguous walls do not slide or slip with respect to each other, has been called symplastic growth (Priestley [35], Erickson [6], Romberger *et al.* [39]). The symplastic growth, in two dimensions, can be visualized as deformation of a pattern on rubber sheet. Allowing puncturing of the sheet would account for the formation of intercellular spaces.

2. Growth tensor and principal directions

The growth of a plant organ when considered more closely, is an irreversible deformation of a cell wall network (its plastic strain). If ds is a line element in the cell wall oriented along s, the *local rate* of this deformation (*strain rate*), is redefined as the *relative elemental rate of growth* of the element in the direction s ($RERG_{l(s)}$).

$$RERG_{l(s)} = \frac{dV_s}{ds} \tag{1}$$

(Richards and Kavanagh [38], Hejnowicz [12]) where V_s is a displacement velocity of the element.

J. Nation et al. (eds.), Formal Descriptions of Developing Systems, 119–136.

If the velocity vector field, $\mathbf{V}$, is known, $\frac{d\mathbf{V}}{ds}$ can be expressed as the absolute derivative of $\mathbf{V}$ along the curve s (Hejnowicz and Romberger [19]):

$$\frac{d\mathbf{V}}{ds} = \frac{D\mathbf{V}^k}{ds} \qquad (2)$$

The absolute derivative can, in turn, be expressed by the covariant derivative

$$\frac{D\mathbf{V}^k}{ds} = \mathbf{V}^k_{,n} \frac{dy^n}{ds} \qquad (3)$$

in which the n factors of $\frac{dy^n}{ds}$ are contravariant components of the unit vector representing the direction s:

$$\frac{dy^n}{ds} = e^n \qquad (4)$$

Thus:

$$\frac{D\mathbf{V}^k}{ds} = \mathbf{V}^k_{,n} e^n \qquad (5)$$

To evaluate $RERG_{l(s)}$ which is a scalar, we must take the inner product of $\mathbf{V}^k_{,n} e^n$ and the vector $\mathbf{e}_s$. Because $\mathbf{V}^k_{,n} e^n$ is a contravariant vector, to find the inner product we must act on it with the vector $\mathbf{e}_s$ represented by its covariant components e_k. Thus

$$RERG_{l(s)} = \mathbf{V}^k_{,n} e^n e_k = T^k_n e^n e_k \qquad (6)$$

We call the tensor T^k_n the growth tensor (or more correctly growth rate tensor). It is the covariant derivative of $\mathbf{V}$ (Hejnowicz and Romberger [19]).

Specification of the growth tensor requires the choice of a coordinate system and knowledge of the displacement velocity ($\mathbf{V}$) of points (such as cell-wall intersections) in growing organ as a function of their positions.

Concerning displacement velocities, experimental studies provide two types of data. One comes from studies of the living growing organ, for example, displacements of markers on the surface of the organ (Erickson and Sax [7], Ishikawa *et al.* [22]), and deformation of the organ surface by use of the replica method (Williams and Green [43]). The other, basing on the property that growth manifests itself in cell pattern, comes from an analysis of cell packets as groups of cells with common origin (Nakielski [27], Hejnowicz et al. [18]).

If $RERG_l$ is not the same in all directions, then there must be three mutually orthogonal directions along which $RERG_l$ attains extreme values (Fig. 1). These are the *principal directions of growth* (PDG). Between successive points of the organ along a coordinate line these directions change in a continuous way. The lines tangent to the principal directions are the *trajectories of the principal directions*. Three such mutually orthogonal trajectories pass through every point in the organ. The surfaces tangent to pairs of principal trajectories are the *principal surfaces*. An empirical statement is that the *surface* of each organ coincides with the principal surface of one orientational type. This principal surface is called *periclinal*. The principal surfaces of the two remaining orientational types are *normal* to the organ surface. In an elongated organ one of them is *anticlinal longitudinal*, the other *anticlinal transverse*.

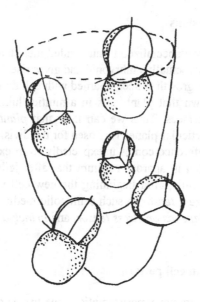

Figure 1.   3D-plots showing directional variation of $RERG_l$ in some points of a root apex. Three principal directions of growth, p, a, and l, are indicated.

The principal directions can be recognized by finite analysis of growing surface as crosses of maximal and minimal growth rates (Goodall and Green [11], Dumais and Kwiatkowska [5]) or by analysis of the cell wall arrangements at the surface and in the section of the organ (Hejnowicz [14], Hejnowicz *et al.* [18]).

Based on the knowledge of principal directions, a curvilinear coordinate system having lines that coincide with trajectories of these directions can be found. In such a system, called the *natural coordinate system* for a particular organ (Hejnowicz [13]), specification of the field **V** is simplified. Assuming a rotational symmetry of the system (u, v, φ), the matrix of the growth tensor has the form (Nakielski [26]):

$$(7)$$

$$\begin{bmatrix} \frac{1}{h_u}\left(\frac{\partial V_u}{\partial u} + \frac{1}{h_v}\frac{\partial h_u}{\partial v}V_v\right) & \frac{1}{h_u}\left(\frac{\partial V_v}{\partial u} - \frac{1}{h_v}\frac{\partial h_u}{\partial v}V_u\right) & \frac{1}{h_u}\frac{\partial V_\varphi}{\partial u} \\ \frac{1}{h_v}\left(\frac{\partial V_u}{\partial v} - \frac{1}{h_u}\frac{\partial h_v}{\partial u}V_v\right) & \frac{1}{h_v}\left(\frac{\partial V_v}{\partial v} + \frac{1}{h_u}\frac{\partial h_v}{\partial u}V_u\right) & \frac{1}{h_v}\frac{\partial V_\varphi}{\partial v} \\ \frac{1}{h_\varphi}\left(\frac{\partial V_u}{\partial \varphi} - \frac{1}{h_u}\frac{\partial h_\varphi}{\partial u}V_\varphi\right) & \frac{1}{h_\varphi}\left(\frac{\partial V_v}{\partial \varphi} - \frac{1}{h_v}\frac{\partial h_\varphi}{\partial v}V_\varphi\right) & \frac{1}{h_\varphi}\left(\frac{\partial V_\varphi}{\partial \varphi} + \frac{1}{h_u}\frac{\partial h_\varphi}{\partial u}V_u + \frac{1}{h_v}\frac{\partial h_\varphi}{\partial v}V_v\right) \end{bmatrix}$$

where V_u, V_v, V_φ are components of **V** and h_u, h_u, h_φ are scale factors of the coordinate system.

The distribution of $RERG_{l(s)}$ within and at the surface of shoot and root apices was described (Hejnowicz *et al.* [16], [17]; Nakielski [26], [28]; Hejnowicz and Karczewski [15]).

3. Directions of cell divisions

Plants, unlike animals, have specialized tissue, called meristem, dedicated to growth in the sense that it remains perpetually embryonic and gives rise to new cells throughout the plant's life. In such tissue, growth is accompanied with cell divisions.

It has long been known that plant cells in a multicellular organ, divide mainly in periclinal or anticlinal surfaces. Thus we can infer that *plant cells divide in principal planes* (which of three particular planes is chosen for the division is a separate problem). Strong support for our inference comes unexpectedly from experiments in which plant protoplasts were embedded in agarose medium that after jellying was stressed (Lynch and Lintilhac [25]). The cells, after beginning the new cell wall formation, divided in the *principal planes of stress tensor* in such semi-solid medium. It will be substantiated later that *growth tensor depends on stress tensor*, and *principal planes of the two tensors coincide*.

4. Principal directions in cell pattern

It was mentioned that plants have meristematic tissue that is devoted to cell multiplication. Such tissue, responsible for the formation of above- and below- ground portions of the plant, is located at the tips of root and shoot, in dome-like shaped organs - the root and shoot apices. Although these two apices differ, the rules governing their cellular organization are remarkably similar. Looking at their cell pattern in the axial section, Figure 2 shows an example of the root apex. We can see that cell walls are arranged into two families of mutually orthogonal lines, *periclines* and *anticlines* (Sachs [40]). The lines, considered smooth regardless of the zigzag resulting from the shape of individual cells, preserve their orthogonal intersections during growth, and the walls tangent to them, though displaced with respect to the tip, retain a recognizable periclinal and anticlinal orientation. Because also newly formed partition walls are either anticlinal or periclinal, the cell pattern is regular and may remain steady as the individual cells move down through it. Why is it so?

The answer relates to the tensor properties of growth. If PDGs for the considered organ are known:

(i) periclines and anticlines represent lines formed by two types PDGs;

(ii) two mutually orthogonal elements of the cell wall network preserve their orthogonal intersection during growth only if they are oriented along PDGs, otherwise the angle between them changes into acute or obtuse;

(iii) partition walls in cell divisions are formed in planes defined by two PDGs.

Feature (iii) indicates that cells are able to detect directional cues included in the principal directions and obey them during growth.

5. Growth rate and tensile stress in cell wall

The $RERG_{l(s)}$ is a function of *tensile stress* (σ) of the cell wall.

The tensile stress of cell walls results directly from *turgor pressure*, as in isolated cells. However, in cells of the organ there are also *tissue stresses* which give an indirect component. In a particular wall the two components are superimposed.

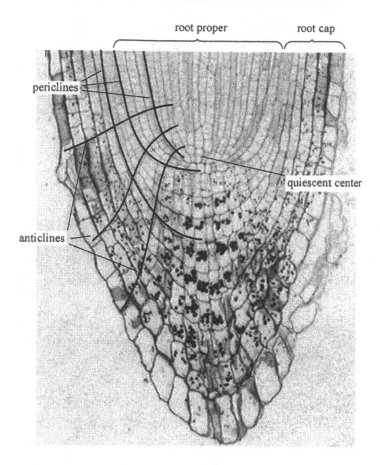

Figure 2. Root apex of radish (axial section). Its cells are arranged in files radiating from a topographic center (Quiescent Center). Cell walls are arranged into the pattern of periclines and anticlines (unpublished photograph provided by I. Potocka).

The **turgor pressure** (P) is hydrostatic pressure that arises as an *osmotic phenomenon* in a cell. In steady state P equals to the difference between *osmotic pressure* inside the cell and in the wall (outside the membrane). The growing cell is pressurized by its turgor pressure which causes an equivalent tension in the cell wall. To induce cell enlargement, it is necessary to relax the cell wall tension by some chemical wall-loosening reaction. This results in a decrease of the turgor and thus of the water potential of the protoplast, which consequently provide a driving force for the passive influx of water and a corresponding increase of cell volume. At the same time the original turgor pressure is restored. Steady-state growth results when all these processes take place simultaneously.

Turgor pressure acts uniformly in all directions. However, *tensile stress* in the cell wall is usually *anisotropic*. The magnitude of this tensile stress in the wall varies with direction *depending on cell shape*. *Tensile stress* in a given direction is thus a unique *physical quantity* that *depends* on *cell shape*. An example is shown in Figure 3. In a spherical cell, tensile stress in the cell wall does not depend on a direction, whereas in a

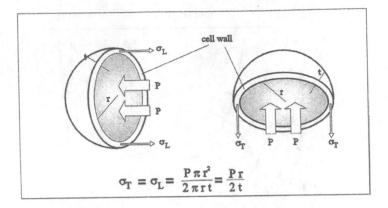

$$\sigma_T = \sigma_L = \frac{P\,\pi\,r^2}{2\,\pi\,r\,t} = \frac{P\,r}{2\,t}$$

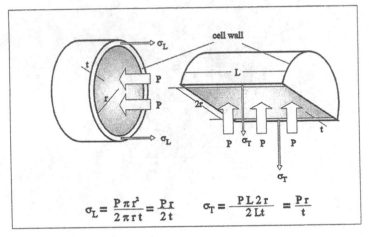

$$\sigma_L = \frac{P\,\pi\,r^2}{2\,\pi\,r\,t} = \frac{P\,r}{2\,t} \qquad \sigma_T = \frac{P\,L\,2\,r}{2\,L\,t} = \frac{P\,r}{t}$$

Figure 3. Tensile stresses (σ) in the wall of the spherical (top) and the cylindrical (bottom) cell under turgor pressure (P). (r – radius, t – wall thickness, σ_T – transversal stress, σ_L – longitudinal stress).

cylindrical cell, the transversal stress is twice the longitudinal one.

As to the inter-dependence of RERG and stress, empirical relationship is: for *volumetric growth* of cells (Lockhart [24]):

$$RERG_{vol} = \varepsilon \cdot (P - P_0) \tag{8}$$

or for *elongation* of cylindrical cells (Ortega [33]):

$$RERG_l = \varepsilon \cdot (\sigma - \sigma_0) \tag{9}$$

where ε-extensibility, P_0-threshold pressure.

A turgorized cell is a *prestressed (tensegral) structure*: the protoplast is compressed the cell wall is stretched. **Tissue stresses** (TSs) in a cell wall are the stresses (tensile and compressive) which occur in tissues of an intact turgid stem and which are not due to external forces; thus they represent *prestresses* in the organ (Hejnowicz *et al.* [20]).

In static equilibrium, the sum of forces involved in the generation of TSs is zero. This means that the tensile TS in a particular direction is accompanied by compressive stress in this direction. However, because stress is force per unit cross-sectional area (we consider here normal stresses only), the magnitudes of the tensile and compressive TSs may differ depending on the cross-sectional areas of the layers or strands upon which the forces act. The plural form (TSs) must be used when speaking of an organ. The singular form (TS) is used for a particular tissue in an organ.

If the organ is composed of *turgid tissues* that *differ in mechanical parameters* (cell wall thickness, cell diameter, elastic moduli of cell wall, P), then an unavoidable physical consequence of turgor pressure is occurrence of TSs. Another origin of TSs may be *differential growth of tissues* (Peter and Tomos [34]), so we need to distinguish between *structure-based TSs* and *differential growth-based TSs*.

The structure-based TSs are a consequence of turgor pressure in cells which differ mechanically in an organ. We have thus two stress-effects of P: tensile stress in cell wall as the *primary effect of P*, always occurring in the wall irrespective whether the cell is in an organ or is isolated, and TSs as the *secondary effect of P*, occurring only when tissues are confined within an organ.

Measurements of TS in stems and comparison of the results with values calculated from the structural differences in the tissues of the stem allow one to conclude that the structure-based TSs readily explain the TSs occurring in stems (Hejnowicz and Sievers [21]).

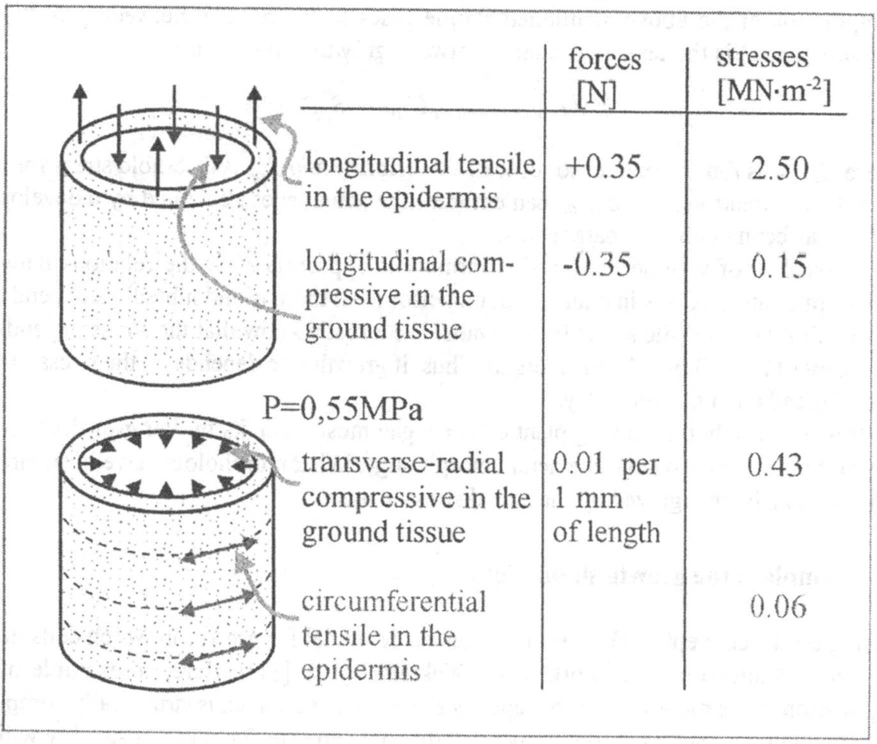

	forces [N]	stresses [MN·m^{-2}]
longitudinal tensile in the epidermis	+0.35	2.50
longitudinal compressive in the ground tissue	-0.35	0.15
P=0,55MPa		
transverse-radial compressive in the ground tissue	0.01 per 1 mm of length	0.43
circumferential tensile in the epidermis		0.06

Figure 4. Scheme explaining the tissue stresses in a sunflower hypocotyl.

The TSs (at a particular point) strongly depend on the direction (Fig. 4). In sunflower hypocotyl two groups of tissues can be distinguished: the peripheral tissue (epidermis and subepidermis) and the inner tissue (ground tissue). The former is composed of cells characterized by thicker wall and bigger cell diameter. It is under tensile TS in contrast to the inner tissue which is under the compressive TS. The forces involved in the generation of TSs are equal (in magnitudes). However, the sectional area differ and therefore the tensile TS in the peripheral tissue is relatively large (approximately 5 times higher then turgor pressure). Observe that longitudinal stresses are higher then transversal ones.

It should be noted that the TS in a particular cell wall is due to the properties of all other cells in the organ and depends on the geometry of the organ. We have thus two important statements: (i) the tensile stress in the cell wall due to the primary effect of P depends on the geometry of a particular cell, and (ii) TSs depend on the geometry of all cells in the organ. In general, therefore, the tensile stresses in cell walls of a turgid organ depend on the actual geometry of the organ.

It should be mentioned that usually the tensile TS in the longitudinal direction in turgid stems occurs in peripheral tissues (epidermis, collenchyma) and its value may be much higher than that due to the primary effect of P (as shown in Figure 4).

6. Growth and stress on tensor level

As known, stress is a tensor quantity S_{pr} (notation in Cartesian coordinates). An obvious extrapolation of the above mentioned simple cases of the relation between growth and pressure (stress) is the tensorial relation between growth and stresses:

$$T_{mn} = E_{mnpr}(S_{pr} - S_{pr}^0) \qquad (10)$$

where E_{nmpr} is fourth-rank tensor of wall extensibility and S_{pr}^0 is threshold stress for irreversible deformation. The E_{nmpr} can embrace the actual cues generated by a developing cell and affecting cell wall parameters.

Regardless of whether or not this formula is appropriate for the relations between growth rates and stresses in plant organ on tensor level there is undoubtedly a dependence of growth rate on tensile stress in cell walls. We already know that the stress depends on geometry of the cell or cells in an organ. Thus, if growth rate depends on the stress, it also must depend upon the geometry.

It is obvious that a growing plant cell or organ must relate its further growth to the result of its previous growth – the actual morphology. What morphology have been already achieved can be recognized by the actual stress status.

7. Example of the growth simulation

Basing on the concept of GT, the computer-made model for growth in which cells divide in principal directions was worked-out (Nakielski [30], [31]). Here, an example of the application of the model to the root apex is shown. The root apex is structurally composed of two integrated parts: the root proper and the root cap (Fig. 2). It grows steadily without a rotation around its axis, but formally it is a figure of revolution. At the tip of the root proper there is a zone of mitotically less active cells (all neighbors are more active), called the Quiescent Center, QC (Clowes [4]), which is of paramount importance for growth organization and cellular organization of the root apex (Feldman [8]; Barlow [1]; Webster and MacLeod [42]). Simulations will show how an idealized cell pattern of the apex with a QC develops during growth.

The model is two-dimensional and, because of the rotational symmetry of the root apex, refers to the axial section. Besides being based on GT, the model includes an algorithm according to which cells divide in principal planes. The application of the model needs two types of data, one about the field of $RERG_l$ and the other about some details of cell pattern.

7.1. GROWTH TENSOR

GT and the Root Natural Coordinate System (R-NS) dedicated for roots with a the QC were described (Hejnowicz and Karczewski [15]). In the R-NS(u, v, φ) in which the axial plane corresponds to $\varphi = const$, PDG trajectories are conveniently described by two families of mutually orthogonal lines u and v (Fig. 5). Two lines, u_0 and v_0 divide the field into four zones: 1 and 2 – the root proper, 3 and 4 – the root cap. The zone 1 represents the QC, the zone 2 - the remaining part of the root proper whereas the zones 3 and 4 correspond to central and lateral parts of the root cap respectively. $RERG_l$ in the QC are equal to zero, whereas they attain relatively high values outside. The equations for V_u and V_v ($V_\varphi = 0$) are the following:

zone 1: $V_u = 0, \quad V_v = 0$

zone 2: $V_u = c(u - u_0), \quad V_v = 0$

zone 3: $V_u = 0, \quad V_v = -k \cdot sin(qv)$

zone 4: $V_u = c \cdot (u - u_0), \quad V_v = -k \cdot sin(qv)$

where $V_u = \frac{du}{dt}$ $V_v = \frac{dv}{dt}$ and $q = \pi/v_0$, c, d are constants. $c = 0.3$, $k = 1.0$, $u_0 = 0.15$, $v_0 = 1.0$.

7.2. CELL PATTERN

The cell pattern is visualized as a two-dimensional cell-wall network, in which cells are represented by polygons. Two neighboring polygons have a common side (adjacent walls of two cells), three such polygons have a common three-way junction (as a place in which three cells meet).

7.3. RULES FOR GROWTH AND CELL DIVISIONS IN THE APEX

GT acting on the cell wall network results in growth. During growth, the network is extended, deformed and new cells, formed by cell divisions. Rules for growth and cell division are the following:

I. Cells are given their vertices in R-NS at $t = 0$. During growth, locations of the vertices change. New locations are calculated from old ones by integration of V_u and V_v with respect to time.

II. The growing cell increases in area. If the value of the cell area assumed to be critical, A_{cr}, is exceeded the cell divides (Fig. 6). It is then replaced by two daughter cells. The location and the orientation of the division wall are considered separately.

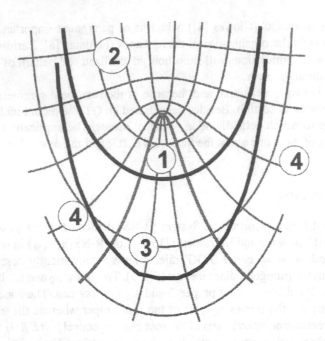

Figure 5. Tensor field for the root apex with the Quiescent Center represented by the pattern of PDG trajectories in the axial plane; the root proper and the root cap are outlined. PDG trajectories are described by lines of the Natural System (light gray). Two lines (boldly drawn) divide the field into four zones assumed to generate the root proper (1, 2) and the root cap (3, 4); zone 1 represents the Quiescent Center.

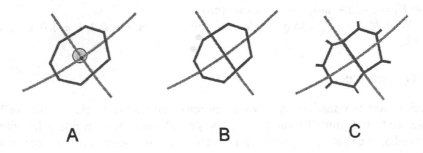

A **B** **C**

Figure 6. Cell Division. Two mutually orthogonal lines (gray) represent PDG trajectories. A The division wall will pass through the point found randomly inside the circular region including the cell center. B At this point two PDGs are considered (the third one is perpendicular to the figure plane). The cell divides along the PDG that gives the shorter wall. C The cell wall network is locally rearranged by shortening of the newly formed wall. For details see text.

III. The location of the cell division is governed by the following procedure. The cell center, as the center of area, and the distance from the center to the nearest wall, d, are calculated. The circular region with the radius $r = q \cdot d$ were $0 \square q \ll 1$ around the center is defined and, inside it the point M, is found in a random way (Fig. 6A). Through this point the division wall will pass.

IV. The orientation of the cell division depends on PDGs. Two PDGs at the point M are found (the third is perpendicular to the axial plane). The length of a potential wall along each of them is calculated and the PDG for which the wall will be shorter is chosen (Fig. 6B). Along this direction the division wall is built.

V. In the model, the division wall is a segment of a straight line. Points of attachment of the wall form new vertices, i.e. new three-way junctions. At the last step of cell division, three-way junctions are re-arranged (Fig. 6C), as a result of the shortening of the newly formed wall by $1/m$ of its former length.

VI. The division wall avoids attachment to an existing vertex (four-way junction). If this happens (very seldom), the point of attachment of the wall is displaced by a segment ε (very short).

7.3.1. Parameters

For a given cell, the occurrence of the cell division depends on A_{cr} (the rule II). It is assumed that $A_{cr} = 156\% \, A_{av}$ where A_{av} is an average cell area which changes, in relative values, from 1, for the smallest cells (cortex), to 1.9, for the bigger cells (peripheral parts of the root cap). The remaining parameters are constant for all cells: $q = 0.25$, $\varepsilon = 0.09z$ (where z is the mean length of the smallest cells), $m = 14$ (all found heuristically).

7.4. RESULTS

7.4.1. Individual simplified cells

Fates of individual cells during growth are determined by the position of these cells in the GT field. To show how the algorithm works consider growth of a group of originally rectangular cells located within a growth field specified for a particular root apex (Fig. 7). Figure 7A shows only growth. The cells enlarge, deform and displace into new position. The pattern of their expansion depends on location: cells located in the zones 1 and 2 growth relatively slowly and mainly proximally, cells located in the zones 3 and 4 grow faster and mainly distally. Notice deformation of cell walls. Mutually orthogonal walls preserve their orthogonal alignment only if they are tangent to PDGs. Figure 7B shows growth with cell divisions. All considered cells divided (without the application of rule V), then some their derivatives divided forming the cell tetrads. The division walls are oriented in PDGs, they are tangent either to the longitudinal or transversal PDG trajectories.

7.4.2. Root apex

The developmental sequence of the formation of the root apex (timesteps 0, 1−3) and then its further steady state growth (timesteps 4 − 7) is shown in Figures 8 and 9. The incoming cell pattern consists of cells located in the distal portion of the root apex including the QC. Notice the orientation of the cells with respect to PDG; cell walls coincide with PDG trajectories. At every time-step of the application of GT field to the cell-wall network, the PDG trajectories coincide with periclines and antilines. One of these trajectories, the one denoted by v_0 , generates the root/cap border.

The cell pattern extends in such a way that each of the four zones of GT field generates growth exactly in the region for which it was defined. Cells localized in the QC do not grow at all, whereas cells located in the other regions of the root apex grow and divide

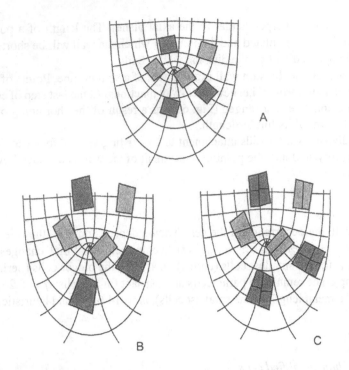

Figure 7. Growth and divisions of simplified cells located originally in different parts of the root apex. In the background PDG trajectories are shown, the indicated trajectory separates the zones 1,2 and 3,4. A t=0. Cells have similar, rectangular shape; notice their different orientation with respect to the PDGs. B Growth without cell division – cells enlarge and displace according to their location. C Growth including cell division – cells divided in PDGs, division walls are tangent to either periclinal or anticlinal PDG trajectories.

moving away from the QC. As assumed, the cells located in the zone 2 form the root proper, and the cells located in the zones 3 and 4 form the root cap.

The dividing cells obey PDGs and new walls are tangent to PDG trajectories oriented either periclinally or anticlinally. At successive times, zigzag wall arrangements similar to those observed in real root apices are formed. Cell divisions are either transversal, multiplying the number of cells in the existing cell file, or longitudinal, generating new cell files. The transversal divisions are in the great majority and they have an anticlinal orientation in the root proper and a periclinal orientation in the root cap.

All newly formed walls, as lying along principal directions, preserve their orientation with respect to these directions. Because of this, the cell pattern is steady over time and, as shown in Figure 10, can be considered self-perpetuating.

8. Plant organ as a living developmental system

An organism during its lifetime can be likened to a cybernetically directed machine running through a sequence of programs: embryonic development, juvenile stage, mature reproductive stage, senescence, and death (Romberger *et al.* [39]). Every program re-

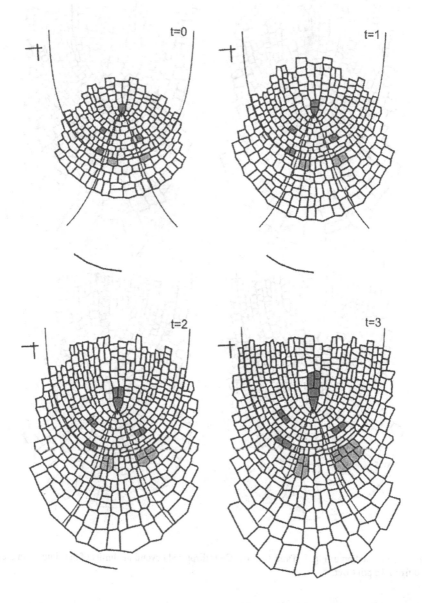

Figure 8. Sequence of growth of the idealized root apex with Quiescent Center. Timesteps 0, 1 − 3. In simulation the tensor model for growth was used. Different zones of GT field are delineated by two coordinate lines, $u = u_0$ and $v = v_0$ (boldly drawn). At $t = 0$, some cells are shaded to show cell packets originated from them at successive times. Short curvilinear segments on the left side of the root apex represent limits of the simulation area. Cells divide along PDGs hence the root apex grows steadily maintaining its cell pattern in time.

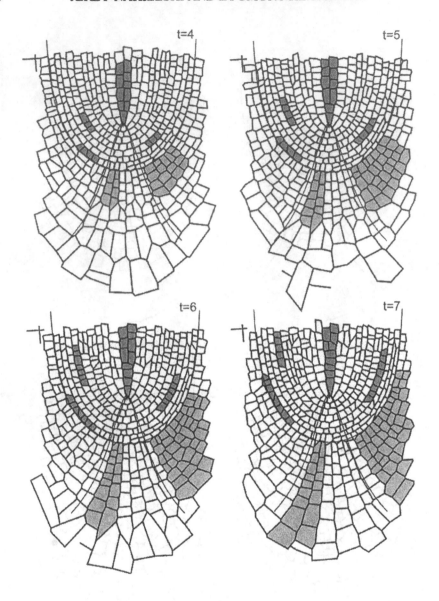

Figure 9.   Continuation of Fig. 8. Timesteps 4-7. Cells displaced beyond the limits of the simulation area were omitted from the print-out.

quires structural and functional integration and coordination among the cells similar to that which occur during symplastic growth.

The present approach appears useful for the formal description of cell development. This refers not only to the steady-state case of growth, as shown in the previous section, but also the developmental changes occurring in the root apex during the regeneration of a root cap (Nakielski [29]) and after hormone treatment (Nakielski and Barlow [32]) have been satisfactorily described.

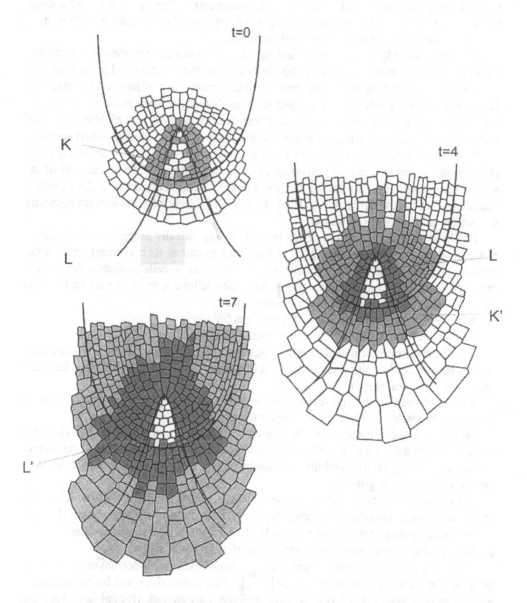

Figure 10. The stability of the cell pattern of the simulated root apex. Three timesteps from Figs. 8 and 9 are shown. In $t = 0$, the group of cells surrounding QC is denoted by K, at $t = 4$, the group K' is formed from it. In $t = 4$, in turn, the group of cells L similar to K is denoted and again, at $t = 7$, the group L' is formed from it. The groups K and L are similar, hence the groups K' and L' are similar too. In this sense, the cell pattern can be considered self-perpetuating.

We realize that in our approach, the growth of the plant organ has been treated principally as the deformation of a solid, or more precisely as the deformation of a cellular solid (Gibson and Ashby [10]). However, in comparison to the deformation of a solid, some departures due to the fact that the organ is composed of living cells, are allowed:

(i) The cells can divide i.e. new walls are built in;.

(ii) Cell tips or edges can grow intrusively, i.e., the tip or edge dissolves enzymatically the middle lamella between two neighboring cells in front of the tip or edge, and intrudes between the interrupting the plasmodesmata. The narrow band of new wall continually being formed by the advancing edge almost immediately becomes glued to the wall of the neighboring cell by pectic materials. These substances form new middle lamella of ordinary appearance. Secondary plasmodesmatal connection can be established across it.

(iii) Between cells, usually along an edge (where three or more cells meet) intercellular gas spaces may arise by dissolution of middle lamellae. Certain patterns of the symplastic growth allow for increase of the spaces to create a big intercellular chamber, producing honeycomb aerenchyma. The latter provides the greatest possible strength with the least possible amount of tissue.

(iv) Individual cells or group of cell may die in a genetically programmed way either leaving cell walls appropriately developed to fulfill functions such as transport of water (xylem), mechanical support (sclerenchyma), or whole cell or cells (including their walls) undergo lysis. The lysed cells collapse or leave intercellular space that may be filled by air or secretes. All this may affect the organ geometry.

We see the strength of our tensor approach as follows:

(i) Physics (mechanics) of a growing organ is followed.

(ii) The concept of principal directions appears. The directions can be recognized not only as the crosses in finite element analysis of growing surface but also as periclines and anticlines of the cell pattern.

(iii) Inference that the cell divides in one of the principal planes of growth or stress tensor can be made. This inference is supported experimentally.

(iv) Known dependence of growth rate on tensile stress in cell wall on a scalar level can be extrapolated to the tensor level in which each tensor depends on the actual geometry of the growing system, and the extensibility tensor can effectively embrace developmental cues for organ growth.

(v) Growth of a plant organ can be simulated for a particular pattern of periclines and anticlines (natural coordinate system), for a particular field of displacement velocities, and on using appropriate algorithms details of cell partitioning can be specified, like the avoidance of four-way junctions, maintenance of mean cell dimensions etc.

Referring to growth of plant organs, the approach based on Lindenmayer systems can be used (reviewed by Prusinkiewicz [36], [37]). This approach cannot be considered as an alternative because it is based on quite different backgrounds (formal languages and theory of automata) and physics of growth is not represented in it. Nevertheless, some algorithms of Lindenmayer systems, especially those used in "cell systems" (De Boer et al. [2], [3], Fracchia et al. [9]) can be applied in our approach.

Our mathematical approach concerns the tensor analysis - a field unfamiliar for biologists. However, if a biologist has appropriate computer programs, all operational problems disappear. It is a matter of the concept of *tensor quantity* that should be understood by biologists by analogy to the concept of vector quantity. If R denotes a region in space, and if for every point of this region a vector in each direction can be defined, then we may have a tensor point function. This is an intuitive explanation of a tensor understandable by a typical biologist. If such vectors in different directions are defined by Nature, then they represent a true tensor for sure. Such vectors cannot depend on the coordinate system applied, and the transformation requirements for the tensor are automatically obeyed. Vector analysis is also difficult for biologists, but they do not have problems in understanding what a vector quantity is. With the concept of *vector*, connections are made between notions such as *magnitude* and *direction*. An important notion that originates from the concept of *tensor* is that of *principal directions*. A plant anatomist sees periclines and anticlines in the cell wall network but usually does not understand that they represent the principal direction of a growth tensor, and presumably also those of a stress tensor.

References

1. Barlow, P.W. (2002) Cellular pattering in root meristems: Its origins and significance, in Y. Waisel, A. Eshel and U. Kafkafi (eds.), *Plant Roots. The Hidden Half* (3rd Edition), Marcel Dekker Inc., pp. 49–82.
2. De Boer, M.J.M. and De Does, M. (1990) Relationship between cell division pattern and global shape of young fern gametophytes. I. A model study, *Bot. Gaz.* 151, 423–434.
3. De Boer, M.J.M., Fracchia, F.D. and Prusinkiewicz, P. (1992) A model for cellular development in morphogenetic fields, in G. Rosenberg and A. Salomaa (eds), *Lindenmayer Systems: Impact on Theoretical Computer Science, Computer Graphics, and Developmental Biology*, Springer-Verlag, Berlin, pp. 351–370.
4. Clowes, F.A.L. (1956) Localization of nucleic acids synthesis in root meristems, *J. Exp. Bot.* 7, 397–312.
5. Dumais, J. and Kwiatkowska, D. (2002) Analysis of surface growth in shoot apices, *Plant J.* 31, 229–241.
6. Erickson, R.O. (1986) Symplastic growth and symplasmic transport, *Plant Physiol.* 82, 1153.
7. Erickson, R.O. and Sax, K.B. (1956) Rates of cell division and cell elongation in the growth of the primary root of Zea mays, *Proc. Amer. Phil. Soc.* 100, 499–514.
8. Feldman, L.J. (1984) The development and dynamics of the root apical meristem, *Amer. J. Bot.* 71, 1308–1314.
9. Fracchia, F.D., Prusinkiewicz, P. and De Boer, M.J.M. (1990) Animation of the development of multicellular structures, in *Computer Animation '90*, N. Magnenat-Thalmann and D. Thalmann (eds.), Springer-Verlag, Tokyo, pp. 1–19.
10. Gibson, L.J. and Ashby, M.F. (1988) *Cellular Solids. Structure and Properties*, Pergamon Press, Oxford.
11. Goodall, C.R. and Green, P.B. (1986) Quantitative analysis of surface growth, *Bot. Gaz.* 147, 1–15.
12. Hejnowicz, Z. (1982) Vector and Scalar Fields in Modeling of Spatial Variations of Growth Rates within Plant Organs, *J. Theor. Biol.* 96, 161–173.
13. Hejnowicz, Z. (1984) Trajectories of principal growth directions. Natural coordinate system in plant growth, *Acta Soc. Bot. Pol.* 53, 29–42.
14. Hejnowicz, Z. (1989) Differential growth resulting in the specification of different types of cellular architecture in root meristems, *Environ. Exp. Bot.* 29, 85–93.
15. Hejnowicz, Z. and Karczewski, J. (1993) Modeling of meristematic growth of root apices in a natural coordinate system, *Amer. J. Bot.* 80, 309–315.
16. Hejnowicz, Z., Nakielski, J. and Hejnowicz, K. (1984) Modeling of spatial variations of growth within apical domes by means of the growth tensor. I. Growth specified on dome axis, *Acta Soc. Bot. Pol.* 53, 17–28.
17. Hejnowicz, Z., Nakielski, J. and Hejnowicz, K. (1984) Modeling of spatial variations of growth within apical domes by means of the growth tensor. II. Growth specified on dome surface, *Acta Soc. Bot. Pol.* 53, 301–316.
18. Hejnowicz, Z., Nakielski, J., Włoch, W. and Bełtowski, M. (1988) Growth and development of shoot apex in barley. III. Study of growth rate variation by means of the growth tensor, *Acta Soc. Bot. Pol.* 57, 31–50.
19. Hejnowicz, Z. and Romberger, J.A. (1984) Growth tensor of plant organs, *J.Theor. Bot.* 110, 93–114.
20. Hejnowicz Z., Rusin A. and Rusin, T. (2000) Tensile tissue stress effects the orientation of cortical microtubules in the epidermis of sunflower hypocotyl, *J. Plant Growth Reg.* 19, 31–44.

21. Hejnowicz, Z. and Sievers, A. (1996) Tissue stresses in organs of herbaceous plants. III. Elastic properties of the tissues of sunflower hypocotyl and origin of tissue stresses, *J. Exp. Bot.* **47**, 519–528.
22. Ishikawa, H., Hasenstein, K.H. and Evans, M.L. (1991) Computer-based video digitizer analysis of surface extension in maize roots, *Planta* **183**, 381–390.
23. Lindenmayer, A. (1968) Mathematical models for cellular interaction in development, Parts I and II, *J. Theor. Biol.* **18**, 280–315.
24. Lockhart, J.A. (1965) An analysis of irreversible plant cell elongation, *J. Theor. Biol.* **8**, 264–275.
25. Lynch, T.M and Lintilhac, P.M. (1997) Mechanical signals in plant development: a new method for single cell studies, *Devel. Biol.* **191**, 246–256.
26. Nakielski, J. (1987) Spatial variations of growth within domes having different patterns of principal growth directions, *Acta Soc. Bot. Pol.* **56**, 611–623.
27. Nakielski, J. (1987) Variations of growth in shoot apical dome of spruce seedlings: A study using the growth tensor, *Acta Soc. Bot. Pol.* **56**, 625–643.
28. Nakielski, J. (1991) Distribution of linear growth rates in different directions in root apical meristems, *Acta Soc. Bot. Pol.* **60**, 77–86.
29. Nakielski, J. (1992) Regeneration in the root apex. Modelling study by means of the growth tensor, in *Mechanics of Swelling*, T.K. Karalis (ed.), NATO ASI Series, Springer-Verlag, Berlin Heidelberg, Vol. H 64, pp. 179–191.
30. Nakielski, J. (1997) Growth field and cell displacements within the root apex, in *Dynamics of Cell and Tissue Motion*, W. Alt, A. Deutsch and G. Dunn (eds.), Birkhauser, Basel, pp. 267–724.
31. Nakielski, J.(2000) Tensorial model for growth and cell division in the shoot apex, in *Pattern Formation in Biology, Vision and Dynamics*, A.Carbone, M.Gromov and P. Prusinkiewicz (eds.), World Scientific, Singapore, 252–286.
32. Nakielski, J. and Barlow, P.W. (1995) Principal direction of growth and the generation of cell patterns in wild-type and gib-1 mutant roots of tomato (Lycopersicon esculentum Mill.) grown in vitro, *Planta* **196**, 30–39.
33. Ortega, J.K.E. (1985) Augmented growth equation for cell wall extension, *Plant Physiol.*, **79**, 318–320.
34. Peters, W.S. and Tomos, A.D (1996) The history of tissue tension, *Ann. Bot.* **77**, 657–665.
35. Priestley, J. H. (1930) Studies in the physiology of cambial activity. II. The concept of sliding growth, *New Phytol.* **29**, 96–140.
36. Prusinkiewicz, P. (1994) Visual models of morphogenesis, *Artificial Life* **1**, 61–74.
37. Prusinkiewicz, P. (1998) Modeling of spatial structure and development of plants: a review, *Scientia Horticulturae* **74**, 113–149.
38. Richards, O.W. and Kavanagh, A.J. (1943) The analysis of the relative growth gradients and changing form of growing organisms: Illustrated by Tobacco leaf, *Amer. Nat.* **77**, 385–399.
39. Romberger, J.A., Hejnowicz, Z. and Hill, J.F. (1993) *Plant Structure: Function and Development*, Springer-Verlag, New York.
40. Sachs, J. (1887) Lecture XXVII. Relations between growth and cell-division in the embryonic tissues, in *Lectures in Plant Physiology* (translated by H.M. Ward), Clarendon Press, Oxford, 431–459.
41. Silk, W.K. and Erickson, R.O. (1979) Kinematics of plant growth, *J.Theor. Biol.* **76**, 481–501.
42. Webster, P.L. and Macleod, R.D. (1996) The root apical meristems and its margins, in *Plant Roots. The Hidden Half*, 2nd edition, Y.Waisel, A. Eshel and U. Kafkafi (eds.), Marcel Dekker Inc., New York, 51–76.
43. Williams, M.H. and Green, P.B. (1988) Sequential scanning electron microscopy of a growing plant meristem, *Protoplasma* **147**, 77–79.

USING DETERMINISTIC CHAOS THEORY FOR THE ANALYSIS OF SLEEP EEG

JOHN D. RAND[1], HERVÉ P. COLLIN[2], LINDA E. KAPUNIAI[3], DAVID H. CROWELL[3], JAMES PEARCE[3]

[1]Department of Physiology, University of Hawaii, Honolulu Hawaii;
[2]Department of Physics, University of Hawaii, Honolulu Hawaii;
[3]Sleep Disorders Center, Straub Clinic & Hospital, Honolulu Hawaii.

1. Abstract

Dynamic system characterization involves the identification and formal description of the quantitative relations between inputs and outputs from the system being explored. In most nonlinear systems inputs are unknown and measurements of the dynamic properties of the system are sought from the measured output. The neural networks in the brain, evaluated during sleep, constitute one such dynamic system. The electroencephalogram (EEG) is the principle output time series that is used to characterize one feature of the underlying dynamics of the neural system recorded from sleep. The EEG patterns are the result of extracellular electrical current flow generated in the brain, which underlies the recording electrode. The EEG is the primary tool used in the qualification and classification of sleep. The aim of this study was to develop a set of tests that examine the nature of a time series, such as the EEG, to establish the presence of nonlinear dynamics. There is no single test that unambiguously identifies nonlinear dynamics in a time series but there are a number of tests that taken together are a powerful predictor of deterministic chaos. A singular stage 2 EEG sleep epoch is put through the full set of tests to 1) describe the test applications to "real data" and to introduce some of the tools of deterministic chaos theory; 2) identify nonlinear dynamics in sleep EEG that warrant further investigation using a nonlinear system model. The theory of each test is described and its proper application is exemplified using the sleep EEG data. The consequences of nonlinear test applications and their inappropriate use are discussed. These formal descriptions indicate the presence of deterministic chaos in EEG time series recorded during sleep and suggest that the tools for deterministic chaos are appropriate for system characterization of sleep EEG.

2. Introduction

Dynamic system characterization involves the identification and formal description of the quantitative relations between inputs and outputs from the system being explored. In most nonlinear systems, inputs are unknown and measurements of the dynamic properties of the system are sought from the measured output called a time

J. Nation et al. (eds.), Formal Descriptions of Developing Systems, 137–149.

series. Thus, it is impossible to know the quantitative relationship between inputs and outputs. To characterize these nonlinear systems a formal relationship or model must be imposed to analyze its dynamics. Finding the appropiate model is often difficult but can be classified according to one of the following five types:

- Equilibrium: An initial disturbance of a mechanical or electrical oscillator decreases to zero amplitude as a result of damping.
- Periodic: The steady-state oscillation of linear electrical or mechanical systems caused by sinusoidal input or limit cycle motion in nonlinear systems.
- Quasiperiodic: Linear oscillations in systems with more than one degree of freedom - behavior of some nonlinear systems.
- Chaotic: Complex oscillations in nonlinear systems that exhibit relatively broad frequency spectra and sensitivity to initial conditions and system parameters.
- Stochastic: Systems in which the output parameters cannot be predicted from inputs.

Early studies of the recorded time series associated with the human brain, such as the EEG, speculated that these output signals were the result of a vast number of conservative, linear oscillators operating in concert with one another, but which could be considered independent from one another [1]. The output time series looks complex because it contains numerous frequencies associated with independent linear oscillators. The problem with treating the EEG as a periodic or stochastic process is the implication that the neural ensembles that govern the EEG have either a fixed relationship or are purely random and this is inconsistent with biophysical structure. In recent years, physiologists have been combating this problem by applying other mathematical tools to the analysis of the EEG and other time series. In essence, nonlinear dynamic analysis assumes that a biophysical model, such as a neural network, is what generates the EEG signals. This model displays chaotic, rather than stochastic, behavior and therefore can be developed within the active field of mathematical research called "deterministic chaos". In this model the human brain is a collection of nonlinear processors that are varying in time and the output of which, as measured by various invasive and noninvasive techniques, displays chaotic behavior. The appropriate tools to analyze the resulting temporal data recorded from the brain are the tools of chaos.

The EEG is the primary time series used in the qualification of sleep. It is a recording of the electrical potential difference between two points on the surface of the brain or from the outer surface of the scalp. The EEG patterns are the result of extracellular electrical current flow generated in the brain, which underlies the recording electrode. The electric current flow reflects activity of the individual cortical neurons [2]. It is believed that the EEG is largely due to the summed postsynaptic potentials that can be either excitatory or inhibitory, near the surface of the cortex close to the recording electrode or deeper in the brain [3]. The general pattern is most often irregular, but at times distinct periodicities do appear which allow classification into categories, some of which correlate with wakefulness and the different stages of sleep.

In order to justify the application of deterministic chaos, as the proper model for representing the quantitative relationships between inputs and outputs recorded during sleep, a number of tests should be performed to establish chaotic dynamics. No individual test can absolutely establish the presence of chaos in the time series, but in concert, these tests are an excellent indicator of nonlinear dynamics and the proper application of the chaos model. Hence, establishing deterministic chaos involves running some preliminary tests, which are indicative and consistent with chaos. If deterministic chaos is indicated, then a set of nonlinear correlates such as those described below is used to quantify the time series.

2.1. VISUAL TIME SERIES INSPECTION

Visual patterns are often seen in an otherwise random looking time series. These visual patterns are clearly not periodic and have proven impossible to detect systematically with current analytical techniques [3], but are recognizable patterns nonetheless. Sometimes this is the first indication of the presence of chaos [4]. The clinical definitions for sleep stage and transient arousal scoring are based on the visual identification of such a pattern [14].

2.2. PHASE SPACE ATTRACTOR

Viewing the phase space attractor is a powerful way of determining the nature of the time series. Essentially, one looks for a bounded, strange attractor. If random processes govern the system, then the trajectory will eventually visit all points in the space. Hence the stochastic attractor is unbounded (has no attractor region) in phase space. Periodic systems display a local attractor in phase space called a limit cycle. Chaotic systems, however, have the characteristic of settling into a form (often strange looking) within an invariant subspace of the larger dynamical space called an "attractor" region. Instead of wandering randomly they are attracted into a distinct, describable form. The phase space region is generated by plotting the original time series against the same series but lagged by an optimal value.

The optimal lag reveals the most favorable attractor and is determined by finding the first minimum of the autocorrelation between the plotted time series. Phase space is not limited to two-dimensions but can include multi-dimensional space where each orthogonal axis in the plot is a lagged time series. Each sequentially lagged time series is plotted against one another and represents an embedding dimension. For example, the average mutual information (AMI) algorithm reveals an optimal lag of the original time series of three. Then a new lagged time series is generated from the original that contains the same time series values but the starting point of this new lagged time series is at the 3rd point. An embedding dimension of three would require three such orthogonal time series plotted against each other (the original time series, the first lagged times, and the second lagged time series which begins at the 6th point). It is imperative that the embedding dimension be chosen large enough so that it reveals the complete attractor phase space (see below).

2.3. FAST FOURIER TRANSFORM

The FFT breaks the time series into a sum of sinusoidal functions to transform time domain data into frequency domain counterparts. The FFT is a qualitative measure for chaos. Chaotic time series show a broad distribution in their power spectrum unlike that of most other and in particular periodic time series. Random data also show the same broad distribution and cannot be distinguished from chaos by visual inspection alone. For a more detailed description of Fourier transforms see [5].

2.4. CORRELATION DIMENSION

The correlation dimension (D_2) is the most popular quantitative nonlinear parameter that is used to characterize a chaotic time series. D2 is generated using an iterative process that captures the spatial geometry of the phase space attractor [6, 7]. The process involves choosing a starting point somewhere in the two-dimensional (embedding dimension of two) attractor and determining the number of plotted points that fall within a given radius. The smallest radius is chosen using the AMI procedure [13]. The AMI is determined by finding a minimum autocorrelation in the given time series and is used to determine the optimum time series lag. Once the number of points are determined within a radius (R) for an embedding dimension of two, a new radius is chosen and the process is reiterated until a radius is chosen that is large enough to encompass all the points in phase space and no new numbers are gained from larger radii. Thus, the correlation integral (C_R) for a specific embedding dimension is determined using the total number of points within radius R divided by the largest number of mathematically possible points:

$$C_R = \lim_{N \to \infty} \frac{1}{N^2} \sum_{i=1}^{N} \sum_{j=1}^{N} G(\varepsilon - | x_i - x_j |)$$

Once the number of points is determined a sequentially larger embedding dimension is chosen and a new correlation integral is generated. A typical plot of correlation function like Figure 4b) includes 10 curves corresponding to C_R calculated for larger and larger embedding dimension. In summary, the computations provide a batch of radii (R) and their associated correlation sums (C_R) for each of several embedding dimensions. Thus, the correlation is a Log-Log plot of C_R versus R. The D2 value is then determined from a plot of the correlation integrals called the correlation function. D2 is the average slope over the middle 1/3 (linear region) of the correlation function (see Figure 4b). The singular value used to express the D_2 of a chaotic time series is not a whole number as in the case of linear conservative systems. Rather, the chaotic D_2 is fractal in nature and always a fraction rather than a whole number. Again, this does not distinguish it from random time series, but a low D_2 that is fractal is considered a good indicator of deterministic chaos. An important simplification recognized by Grassberger and Proccacia [10], is that the distribution of these curves correspond to a power law. That is:

$$N_R \sim R^{(x)}$$

where x is defined as D2. This suggests that, as in the attractor, the correlation integral of a scattering of points throughout a D_2-dimensional volume will be proportional to R^{D2}, that is

$$C_{(R)} = A\ R^{D2}$$

where A is a constant of proportionality. Taking the logarithm of both sides gives,

$$\log C_{(R)} = D_2 \log R + \log A$$

Thus, the slope of a plot of $\log C_{(R)}$ versus $\log R$ reveals D_2. An average value for D_2 is determined by taking the average slope over the middle-1/4 of the vertical scale of the correlation function plot. Figure 4b) shows the calculation of D_2 from the correlation function plot. The value of C_R at any particular single value of R is not important, but how C_R changes with R is critical.

2.5. CORRELATION DIMENSION VERSUS EMBEDDING DIMENSION

A common method used to verify the use of the proper embedding dimension is to plot the correlation dimension versus the embedding dimension (D_g). The correct D_g is the lowest approximate asymptotic value of the plateau. It can also be used to establish the range of D_2 for the study and to investigate problems that arise due to small sample size. In theory, the highest D_2 value is associated with the most complex time series. Thus, a stochastic time series should have the highest measure of D_2. As the D_2 increases so will the D_g required to properly view the entire attractor. Because the stochastic time series attractor eventually, if given a large enough time series, will visit all of phase space available and it will, in theory, take an infinitely large D_g to properly embed the attractor. This means that the complexity will also rise to infinity. Since both D_2 and D_g rise to infinity together, no asymptote is reached. This is not the case for the chaotic, periodic or quasi-periodic time series. In each of these cases the D_2 value is asymptotic with the D_g. As the proper complexity value is reached the embedding dimension remains constant, requiring no additional information to properly describe the time series.

2.6. SURROGATE STUDY

A surrogate study essentially re-shuffles the actual values in the time series while maintaining its frequency content and autocorrelation function. Then these new synthetic time series, which have the same amplitudes as the original time series but different phases, are analyzed using the tools of deterministic chaos. This synthetic time series has been randomized, and thus should have a higher complexity than the original. If many surrogates are generated in this way, a range of complexity can be determined. If the original time series complexity falls within the range of the surrogated time series it cannot be distinguished from randomness. The following null hypothesis is then tested. *Ho: Linear Dynamics Null Hypothesis:* The underlying dynamics that generate the EEG at C_4 referenced to A_1 during sleep are linear, with Gaussian white noise random inputs. This null hypothesis is inconsistent with the possibility of chaos, since linear dynamics cannot produce chaos [2]. This means that if a time series is chaotic, the null hypothesis can be rejected. The method for generating the surrogate data consistent with the linear dynamic null hypothesis is to:

1. Compute the Fourier Transform of the original time series. The transform includes both an amplitude and phase at each frequency.
2. Replace the phases with random numbers, ranging between 0 and 2π. This has no effect on the transform amplitudes.
3. Compute the inverse Fourier transform consisting of the amplitude and randomized phases. This produces a new time series we are calling the surrogate. 10 surrogates are generated in this way for each sampled time series, or epoch, in the study.

The surrogate data has the same amplitudes as the original time series. Since the power spectrum is proportional to the amplitudes squared, the surrogate data time series has the exact power spectrum as the original. The autocorrelation is the Fourier transform of the power spectrum. Thus, the autocorrelation function is also the same between the surrogate time series and the original.

3. Methods

A singular stage 2 EEG sleep epoch is put through the full set of tests to 1) describe the test applications to "real data" and to introduce some of the tools of deterministic chaos theory; 2) identify nonlinear dynamics in sleep EEG that warrant further investigation using a nonlinear system model. To validate the chaos model these tests were also run on well-established periodic and random output time series to illustrate the differences between these models.

The following tests were run to establish the presence of chaos in the EEG of sleep stage and arousal:

- Visual Time Series Inspection
- Phase Space Attractor
- Fourier Transform
- Correlation Dimension
- Correlation Dimension vs. Embedding Dimension Plot
- Surrogate Study

The stage 2, 30-second EEG sleep epoch used in this study was recorded during an all-night PSG on a 37-year-old female subject. The EEG (C4-A1) for this subject was recorded during a larger study of 20 healthy male (N=13) and female (N=7) subjects, between 25 and 46 years of age [9]. The subject was free of known neurologic, medical conditions or medications that could alter EEG characteristics, such as hypertensives and/or stimulants. The subject had an apnea/hypopnea index (AHI) <= 5.0, a periodic limb in sleep (PLMS) Index <= 0.2, and the required total sleep time (TST) of 300 minutes. The EEG was recorded on the SensorMedics 4100 SomnoStar system, which was interfaced with a Grass 8-20D or Grass 78E Polygraph and included data storage on an optical disk. The signal was digitized at 100Hz and displayed on a computer monitor while the data were simultaneously recorded.

The software that was used for the nonlinear study was the Contemporary Signal Processor for Windows95 (CspW) program [8]. This commercially available program is a collection of algorithms for the analysis of signals from nonlinear sources. The algorithms are specifically designed for the time series analysis of nonlinear systems and are accessed through a Graphical User Interface (GUI) that prompts the user for parameters and then calls the appropriate executable file to run the chosen algorithm. Three time series, each made up of ~3000 data points, were analyzed using the tests outlined in the methods section [9].

4. Results

In the plots comparing the diagnotic test results, the first time series is a noisy periodic time series (Figure 1a) that was generated at 15 cycles per second, the second time series is a 30-second, stage 2 sleep EEG epoch recorded as part of an all-night sleep study (Figure 1b) and the third time series represents white noise (Figure 1c).

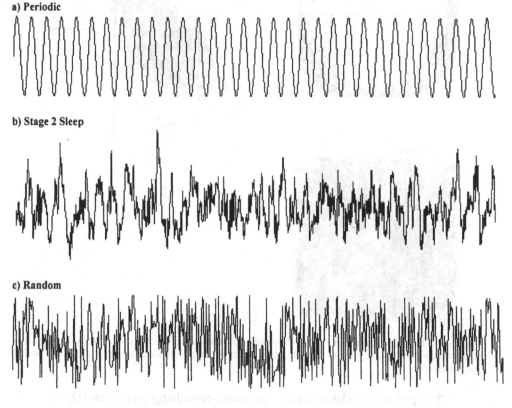

a) Periodic

b) Stage 2 Sleep

c) Random

Figure 1. a. The first time series is a noisy periodic time series that was generated at 15 cycles per second. b. The second time series is a 30-second, stage 2 sleep EEG epoch recorded as part of an all-night sleep study and c. the third time series represents white noise.

Figures 2 and 3 are plots of the phase space attractor and fast Fourier transform respectively. Note that the spectral content of the plot does not display any noticeable peaks. This broad distribution of the power spectrum is a characteristic of chaotic as well as stochastic time series.

The correlation function plots for the three exemplary time series. The D2 for the S2 sleep is clear fractal 3.61 and is determined to be between the values of the calculated D2 for the 15 Hz time series 1.48 and the random data 5.30.

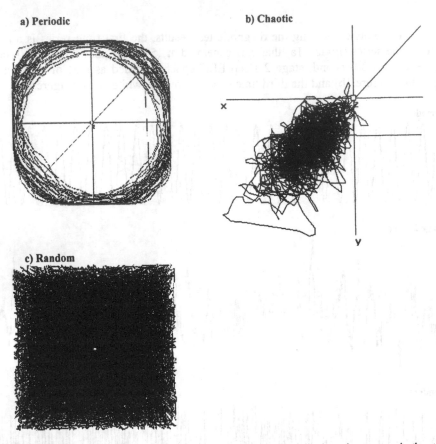

Figure 2. a. The phase space plots of the 15 Hz time series revealing a limit cycle attractor. b. the strange attractor seen in the stage 2 EEG sleep epoch and c. no attractor region in the phase space plot of the random time series.

The plot of correlation dimension versus embedding dimension (D_2 *vs.* D_g) is shown in Figure 5 for the three analyzed time series. Notice the asymptote that appears in Figure 5 b. at a correlation dimension of ~4.0. As the embedding dimension increases the correlation dimension is unchanged. The asymptotic value is sometimes used as a crude measure of D_2 as an alternative to the method illustrated in Figure 4 to characterize the strange attractor. The random data in Figure 5 c. reveals no discrenable asymptote. This is typical of stochastic systems. Both the stage 2 sleep EEG in Figure 5 b. and the sine wave data in Figure 5a. exhibit an asymptote at 2.0 and 4.0 respectively.

a) Periodic

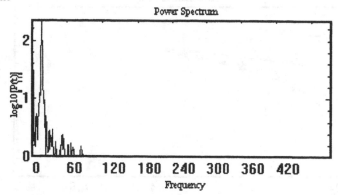

b) Stage 2 Sleep

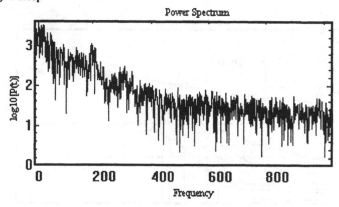

c) Random

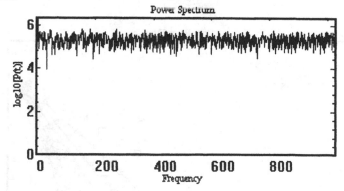

Figure 3. The FFT for three time series are shown here. In a. note the peak power at 15 Hz. Both the FFT for the stage 2 sleep 30 second epoch b. and random time series c. have no discernable peak power in the frequency spectrum.

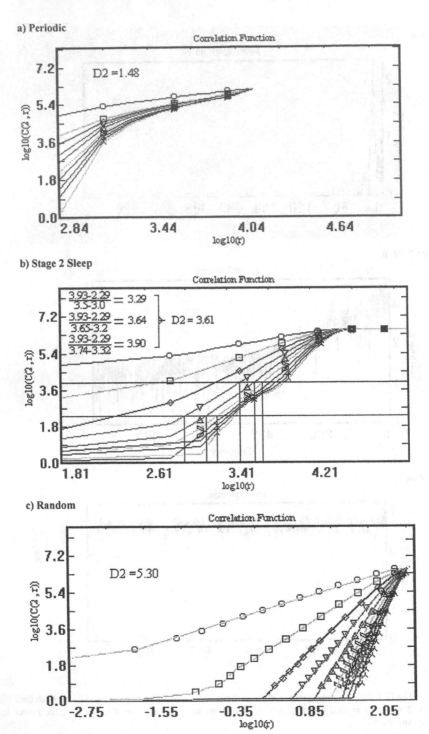

Figure 4. The correlation functions of three time series: a. periodic b. a 30 second epoch of EEG in stage 2 sleep smapled at 100 HZ and c. random or white noise.

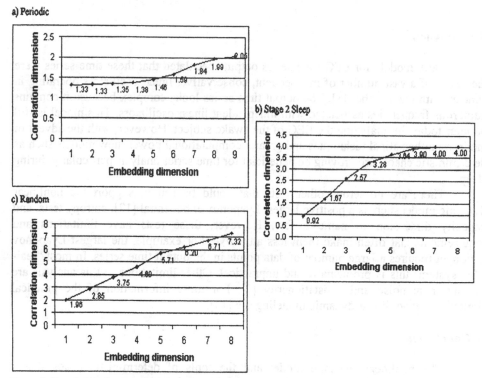

a) Periodic

c) Random

b) Stage 2 Sleep

Figure 5. Correlation dimension plotted against embedding dimension show the varying results for a. periodic sine wave, b. stage 2 sleep EEG and c. random time series.

A surrogate study of S2 sleep indicates that the D2 valve (3.61) was significantly lower (p< 0.001) than the surrogate value for D2; this suggests that the D2 value calculated for stage 2 sleep EEG (3.61) is outside this range indicating that it is inconsistent with a linear stochastic model.

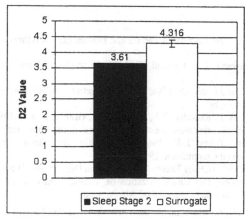

Figure 6. The stage 2 sleep EEG D2 value and the surrogate data generated from the original time series. Note that the surrogate average for D2 is 4.316 +/- 0.15.

5. Discussion

Early models for EEG time series output speculated that these time series were the result of a vast number of independent, conservative, linear oscillators operating in concert with one another [1]. The output time series looks complex because it contains numerous frequencies associated with independent linear oscillators. This model is still popular today for analyzing the EEG of the awake subject. However, with the advent of modern mathematical techniques and greater computational power, new models such as deterministic chaos, are proving be beneficial for time series analysis particularly during sleep.

There are numerous other tests that could be run to support the nonlinear dynamic model such as a positive largest Lyaponov exponential [12], entropy tests, and capacity dimension to name a few. However, these tests have limitations and requirements that often make them less attractive. For example, the largest Lyaponov exponent requires a large number of data points in the output time series. In most "real-life" systems this is problematic and unpractical. Other limitations in this analysis are measurement noise and nonstationarity [7]. For more information of the practical limitations of nonlinear dynamic modeling see [13].

6. Conclusion

The nonlinear dynamic model and the tools of deterministic chaos theory provide additional information that is not redundant and that can compliment the more traditional analysis techniques such as FFT [11]. No individual test can absolutely establish the presence of chaos in the time series, but in concert, these tests are an excellent indicator of nonlinear dynamics and the suitability of the chaos model. These formal descriptions indicate the presence of deterministic chaos in EEG time series recorded during sleep and suggest that the tools for deterministic chaos are appropriate for system characterization of sleep EEG.

7. References

1.   Nunez, P.L. (1995) Neocortical Dynamics and Human EEG Rhythms, Oxford University Press, New York.
2.   Kandel, E., Schwartz, J. and Jessell, T. (1991) Principles of Neural Science 3rd ed., Elsevier, New York.
3.   Niedermeyer, E. and Lopes De Silva, F. (1993) Electroencephalography 4th ed., Lippinscott Williams & Wilkins, Philadelphia.
4.   Elbert, T., Ray, W.J., Kowalik, Z.J., Skinner, J.E., Graf, K.E. and Birbaumer, N. (1994) Chaos and physiology: deterministic chaos in excitable cell assemblies, Physiological Review 74(1), 1- 47.
5.   Cooley, J.W. and Tukey, J.W. (1965) An algorithm for machine calculation of complex Fourier series, Mathematical Computation 19, 297
6.   Williams, G.P. (1997) Chaos Theory Tamed, Joseph Henry Press, Washington DC.
7.   Rand, J. D., (1999) The Characterization of Transient Arousal in Sleep EEG Using Nonlinear Dynamics, UMI Publishers, Ann Arbor, MI, Appendix C.
8.   Applied Chaos LCC and Randle, Inc., (1996) Tools for Dynamics: A Users Manual, CSP for Windows95, Version 1.
9.   Rand J, Kapuniai L, Crowell D, Pearce J., (2001) Nonlinear analysis of 30-second sleep stage epochs,

in: W. Sulis, I. Trofimova (eds), Nonlinear Dynamics in the Life and Social Sciences, IOS Press, Ohmsha, pp. 195-203.

10. Grassberger, P. and Procaccia, I. (1993) Measuring the strangeness of a strange attractors, Physica 9D, 189-208.
11. Fell, J., Roschke, J., Mann, K. and Schaffner, C. (1996) Discrimination of sleep stages: a comparison between spectral and nonlinear EEG measures, Electroencephaloghaphy and Clinical Neurophysiology 98, 401-410.
12. Wolf, A., Swift, J.B., Swinney, H.L. and Vastano, J.A. (1985) Determining Lyapanov exponents from a time series, Physica 19D, 285-317.
13. Abarbanel, H. (1996) Analysis of Observed Chaotic Data, Springer –Verlag, New York.
14. Rechtschaffen, A. and Kales, A. (1968) Manual of standardized terminology, techniques and scoring system for sleep stages of human sleep, Brain Information Services/Brain Research Institute, UCLA.

Part 3.

Emergence

HOW DOES COMPLEXITY DEVELOP?

JACK COHEN D.SC., F.I.BIOL.
Institute of Mathematics,
University of Warwick,
CV4 7AL, UK
and Assisted Conception Unit,
Women's Hospital,
Birmingham B15 2TG, UK
drjackcohen@aol.com

1. Meta-rules and Myths

How Does Complexity Develop?

A great variety of answers springs to mind in response to this question. They range from the simple idea that multiplying simple processes results in more complex causality through the "chaotic" divergence of interactive systems, so that the space between the piano keys comes to include new keys, to recursive systems that change their parameters each time around and indeed to recursive systems that change their rules as they evolve. This is not only a philosophical question, interesting for itself but without application to the everyday world. We are surrounded by systems becoming more complex, from embryos and ecosystems to industrial processes and international law.

There has arisen a very potent myth in the Complexity community that suggests that complex systems have rules, or perhaps meta-rules, in common. So, by understanding embryology we might be able to regulate housing development, or by understanding Artificial Life systems running as cellular automata we might learn to manage businesses, or to persuade people that we can do so or teach them to do so. Although this is lucrative on the retail level of individual consultancies, I am not persuaded that those wholesale applications of Complexity Rules could have served as a secure foundation for a new scientific world-view. Maturana and Varela [20], (but also see Varela, Maturana and Uribe [32]) made a very worthy attempt at such a synoptic position, but this was considerably marred by their misunderstandings of organic reproduction: they believed it to be romantically effective, even efficient. They were wrong, notably in that Nature uses high redundancy and the avoidance of hereditary error by wastage rather than repair (e.g. Cohen [3, 5]).

Comparisons of natural processes with the works of man have many pitfalls: Rapport [26] showed that ecologists accept some economic metaphors very uncritically and that, reciprocally, economic theory has assumed some obvious - but mistaken - tenets of folk ecology (such as restricted input, "cost" of offspring, and "balance"). This contributes greatly to the fragility of the larger models in both sciences. Boisot and Cohen [2] called

J. Nation et al. (eds.), Formal Descriptions of Developing Systems, 153–164.
© 2003 *Kluwer Academic Publishers. Printed in the Netherlands.*

for more disciplined use of simile and metaphor across this bridge but, while widely quoted, have not in general been heeded. Mathematical ecologists and economists have too often posited, as a tractable starting assumption, that equilibrium or "balance" conditions can be modelled, if not actually attained. And the usage CAS, for "complex adaptive system", has usually been assumed to have the property so beloved of 19th century physiologists, homeostasis. The metaphor here is "a ball in a cup", that returns to its lowest-energy position after displacement - or for a one-dimensional case, a thermostatic heating system.

This is a quite superfluous assumption. In principle, very nearly all complex systems are not adaptive in that sense; there is no reason they should be. We are seduced by the vision of organic development, where most trajectories seem to be convergent: adaptive in that CAS sense. More properly, they may exhibit homeorhesis, maintenance of a path rather than of a position. This is a presently fashionable position, but is false-to-fact (Cohen, [5]): very nearly all sexually produced organisms die without breeding; only a tiny minority contributes heredity to the future (Cohen [3, 5]). Of course, complex systems that last long enough to be investigated, or even to be chosen as examples, are like species in ecosystems: they are mostly maintained by feedback systems in this way. That is why Gaia seems to be composed of very many of such perpetuating systems (Lovelock [19]). But in prospect rather than in retrospect, complexifying systems must very nearly all fail soon after their inception, just like organisms.

How is it, then, that life can arise and multiply, evolve and burgeon as it has so clearly done on this planet? How is it that it seems to be worthwhile to start a business? Can there be some meta-rule that underpins ecosystems and economics, replacing the romantic but mistaken notion that most attempts succeed? Is there a universal rule that promotes complexity itself? Can lifting oneself by one's bootstraps into a complex mode bring systems into a more secure domain?

There is one metaphor that puts these questions into a new frame. Stewart and I have used it in several books (e.g. Pratchett, Stewart and Cohen [24]). Life, as a complex system, has lifted itself out of the simpler constraints of chemistry and physics; the mitochondrial proton pump works as if heat engines, and entropy, were irrelevant to it. Complex systems can evolve through successive mechanisms, just as a bridge can retain its function through changes of the material of which it is constructed (even, perhaps, surviving being "made of nothing" - a tunnel). Consider the problem of getting a 100 kg man into synchronous orbit, a Clarke orbit 22,000 miles up over the equator, where the communications satellites are. From the 1960's to middle '70's, there seemed to be an absolute, Newtonian lower limit to the energy you must expend to lift 100 kg to that higher potential energy: two-or three-stage rockets, different fuels and tricks, gave an envelope of possibilities, but there was a minimum that seemed irreducible.

Then the "space bolas" was conceived, and the rules changed. Like the South American weapon, three long "ropes" are tethered at a point, but the rocks are replaced by cabins on the ends of cables hundreds of miles long, orbiting like a giant Ferris Wheel in space with its centre of gravity, say, in 200-mile orbit. An astronaut enters a passing cabin just above the atmosphere and is carried up nearly to the furthest point; if he gets out there, he will be thrown up to catch the next cabin of a higher, larger space bolas, and a third can deposit him in Clarke orbit. The cost can be a twentieth of the energy needed for the rocket ship. Further, once up there so economically, he can do better still. By "letting down a rope" to a friend on the equator, he can begin the construction of a Space Elevator, with cabins going up and down the 22,000 miles at the cost only of friction. We can't build these yet, but they're good physics. They show that investment in innovation can free later technology from earlier constraints. I am using such a "space elevator" to type this paper - it depends upon enormously expensive investment in chip and printer technology (perhaps enough to build a Space Elevator, actually), and is now free of nearly all of the constraints of pen-and-ink or mechanical typewriters. Life has very many such "Space Elevators" - the mitochondrial energetics has left its "rocket fuel" requirements aeons ago. By investing in "higher orbits" by providing yolk or milk or complex developmental homeorhesis, organisms can start off high on the complexity ladder, because their ancestors left rocket fuel behind and invented new, high-tech tricks.

So life gives the impression of convergence, efficiency, slickness - we call this adaptedness. It looks to be a model for other complex systems to emulate; but it works by selection and wastage (Cohen [5]), and it employs many cryptic high-tech "magic" tricks (Cohen and Stewart [10]), as impossible to analyse as a modern mobile telephone with that message on the case: "No User-Serviceable Parts Inside."!

2. Para-questions and anthropic principles

I want to distinguish a particular kind of "why?" question, that seems to be addressing important issues but is in principle unanswerable except in verbal terms, from other "why?" questions that can usefully advance our thinking. All "why?" questions are in a sense "para-questions", addressing a level of organisation above their explicit subject matter. "Because I say so!" is the unsatisfying parental approach too often given. "Because God made it like that!" is the hardly more satisfying deist answer to questions like "Why are there protons and electrons?"; note how different that question is from "Why is the mass of the proton 1820 times that of the electron?" - I discuss those in the context of anthropic answers below. Planetary orbits were satisfactorily described by Newton's Laws, but Newton himself was very unhappy with the transmission of "gravity" across space. "Why" the Moon's orbit was like that, just there, was answered by a higher-level verbal invention, gravity: the Moon was continually falling to Earth, but its velocity just kept it from getting closer, like the cannon-ball fired around the Earth simile. Only when the higher-level model was replaced by a more rational (but equally metaphorical) space-time Einsteinian model did the Moon's orbit become part of a space-curvature picture in which it was seen as part of a geometric many-dimensional model of the Solar System that included all the astronomical bodies and their mutually interactive paths. Then the answers to astronomical "why" questions could relate Solar System geometry to space-time manifolds with dents. And the mathematician can now answer the question of the Solar System's long-term stability with a definitive "Perhaps!"

Are we, then, limited in our answers to "Why is there life despite the tendency of nearly all complex systems to come to pieces?" with "The Universe is just like that!" with some appeal to Space Elevators and the robustness of higher technology? I think we are not, and there are three steps to showing why. The first is Ian Stewart's very elegant demolition of anthropic answers to such questions. The second is his replacement of the "standard thermodynamic model" with Boltzmann entropy, with one that includes gravity and shows an increase of order rather than a decrease. The third employs the Complexity stance and uses concepts like symmetry-breaking and bifurcation theory to show that, as a general principle, the Universe must - and does -"make it up as it goes along.". And "along" therefore means onwards and upwards, evolving systems everywhere and everywhen.

Firstly, this demolition of anthropics, detailed at greater length in Cohen and Stewart [10]. The standard anthropic answers to questions of the form "How is it that the Universe is such a friendly place for people?" come in two forms of different strengths. Both agree that only in such a Universe would there be people to ask the question. All of the elements must be produced, in the interiors of stars, in just the way they are for life's machinery to work, Planck's Constant can't be more than 5

But Stewart points out that his Toyota automobile is just such a mechanism, too: change the threads on the bolts, compared with the nuts, just a tiny bit and it won't work and could fall to pieces, change the size of the wheels by a mere 5% and the tyres won't fit, reduce the tyre size to fit and the wheels won't touch the ground and the car will lose all tractionĔ The anthropic argument shows that there can only be one kind of car, beautifully designed - or Designed - for Ian's locomotory needs. However, one can observe very many perfectly adequate auto designs, all effective, rendering this argument totally silly. Surely any complex interactive system, from autos to universes, that "works" can be credited with this anthropically unique status. And to claim it for our set of working-universe models is naïve in the extreme (Cohen and Stewart [10]).

Secondly, the entropy question and life (again see Cohen and Stewart [10]). Physicists mostly believe that our physics is the physics, not a physics: see the anthropic argument above, but also realise that our set of "fundamental particles" and "universal constants" could have been very different if we had started from other than Rutherford's electron. Physicists have been particularly fond of a thermodynamic model for the universe that sees it, basically, as a Boltzmann-ruled heat engine. They model this as a collection of perfectly elastic balls in a confined space, and perform mathematically trivial or very difficult operations to show that such a system inevitably decreases in order, increases randomness, as time passes. The sacred Second Law of Thermodynamics, that has its roots in this model, models such decrease/increase as an increase of a dimensionless property, entropy. As entropy increases, temperature falls (actually, differences of temperature reduce) and eventually, this model predicts, we will see the Heat Death of the Universe.

This physical-universe-must-run-down stance has immense difficulties with the existence of life, in just the form posed by our Big Question here. The great physicist Schrodinger wrote a delightful little book called "What is Life?" [27] in which he concluded that life was a very aberrant organisation of physical matter/energy that incorporated negative entropy, so that it kept its own entropy high at the expense of its surroundings. He was quite clear, as nearly all physicists still are today, that life can only persist by increasing the entropy of its surroundings more than it decreases its own entropy, by processes like photosynthesis or feeding, and so temporarily postponing its inevitable demise (by a minor Heat Death). Stewart points out ([28], in press, and in Cohen and Stewart [10]) that this model isn't sacred or unique. A very similar, slightly more complicated model, that incorporates gravity affecting the elastic balls in the system, shows a precisely opposite trajectory: it increases in order as time passes! In the real universe, this means that an initially hydrogen/helium filled cosmos amplifies any tiny heterogeneity into a gravitational anomaly around which grows a star, a galaxy, a cosmic order. This, needless to say, agrees with observation. And proponents of such a model are not surprised to find local increases of order, living things, not only fighting increasing entropy individually but also creating a vast tree of increasing complexity on a little planet.

So, just as Einstein's new model put "why" questions about orbits into new, more answerable frames, so Stewart's new thermodynamic model puts "How can life evolve?" into a more rational, answerable frame. Instead of the underlying dynamic of our universe being the dissolution of order towards a Heat Death, it is instead committed to progressive increase of order, at least at the physical level. Instead of "negative entropy" and other contortions of our understanding of physical systems, the dynamic of a progressively ordering cosmos has a place for chemistry as corrupt physics, biology as convoluted chemistry, and so onwards and upwards (Stewart and Cohen [29], Cohen and Stewart [10]). Kauffman [16] has a different take on this, but arrives at much the same conclusions: the universe is inevitably progressive, constantly invading its adjacent possibilities. His evidence is that both biology and technology are continually expanding their phase spaces into "the adjacent possible"; results exploit more possibilities than initials. He warns physicists and chemists that they may be too conservative, by restricting the possible outcomes of their experiments to the same restricted phase space as their initiations.

The problem of a conservative/entropic universe versus a progressive one is most acute in my own field, embryology. I take the central question of embryology to be of the form "How can the complexity of each growing feather, by any measure I can devise, be much more than the complexity of the earliest stages of the whole chicken embryo?". (I did my Ph.D. on feather development - see Cohen and 'Espinasse [6].) Clearly, there is not a Law of Conservation of Complexity. But the progressive increase of ordered structure, so precisely repeating the development of previous generations, is a wonderful process whose provenance must fit into a rational word-view (Keller [17]). It fits Stewart's model much less uncomfortably than Schrodinger's. But can we argue that individual development, and its evolution, is a necessary, or a predictable, part of such a physical universe? I think we can, using modern biological principles and understanding, not the simplified folk biology of Maturana and Varela, and combining it with some insights of complexity theory. (But I don't see any more hope for a rational, honest Management Theory coming from this more enlightened approach).

3. Rules, corruption and speciation

Classical cosmologists took physical rules for granted. A few scientists cast doubt on this procedure. For me the most dramatic was J B S Haldane, who in a story in the "Rationalist Annual" about 1935 (it was in my parents' library, and I later discussed it with JBS himself) suggested that "time" was only characteristic of our familiar universe of mountains, birds flying over them dropping a feather every century, Neapolitan streetsweepers - and perhaps angels - and the rules (he said Laws) by which they operated. Before this, and after it, were "eternities" in which different rules applied. Our time was "eternity", viewed from long before, or after, with rules incomprehensible to creatures much earlier or later. Stewart and I, in a science-fiction story [30] had a little joke that the Universe was more than 40 billion years old (in contrast to the current guess, about 16 billion), but had endured many civilisations whose technology, or even biology, had corrupted the evidence. Big Bang models have problems with the first microseconds, and have invented "inflation" as a solution to link the unimaginable to the more familiar. Stewart and I have faced the problem [29], but so far have not proposed a solution.

The problem is this: when the first chemical elements were synthesised in stellar interiors, was there a notional "chemistry" sitting out in the wings, waiting to be actualised as the first elements tentatively combined to make molecules? When the first organism(s) began to multiply (perhaps it was little more than recursive chemistry, but that doesn't ease the problem much) did God open the Rule Cupboard and search for the Natural Selection booklet? The cosmologies we are presented with by astrophysicists are coy about this problem; they tell us what happened, more-or-less, but not whether/why/how the rules were actualised. Some, including the first Hawking book, are nearly deistic.

The cop-out is easy: the Universe doesn't have rules, they are invented by observing humans to make events easier to classify. Stewart and I [29] have shown this proposal to be inadequate, not least because we're made of universe-stuff and have been selected to have brains that operate (on several levels) to mediate survival by congruence with real-world events. Unsuccessful determination of nature's regularities and rules was a losing option, and our ancestors became ancestors by getting it right, whatever that means. And the reason mathematical science seems congruent with the way the universe works is not, we think, that we have latched on to Universal Mathematical Truths, but that we've invented an Earth-primate maths-and-rule-book that sits between us and the universe's causality (Cohen and Stewart [10]). And it works. Our minds may cheat, and then get confused, by separating events from the rules that govern their behaviours, just as they seem happy to separate structures from processes (for example Newark Airport is definitely a process, different structures each time I go through!). But overall, this is successful - Boeing 747's fly (Cohen and Stewart [7]). So, when we've worked out a picture of causality that incorporates how rules appeared, what will this do to the status of notional Complexity meta-rules?

Let's tell the causal Universe's story first, and then see where we might need The Rule Book. We'll get to the meta-rule book for some of the para-questions.

Our cosmology goes something like this. A tiny speck of unimaginably concentrated energy appeared out of nothing. It rapidly passed through several cooling/expansion phases, then inflated to a reasonable size, of some light-years across. Its temperature had fallen far enough that protons, neutrons, electrons could begin to have existence (was it only quarks before?) and soon associated to make atoms. There was a slight imbalance of normal matter over anti-matter because of some odd muon asymmetry, so it didn't all turn back into nothing but left us the difference, a normal-matter universe. This necessarily has a built-in highest velocity to which even electro-magnetic radiation must conform. Nearly all the matter was hydrogen, some helium, and this began to collapse (Stewart's model) into stars, galaxies, Great Walls. In the stars excruciating physics forced hydrogen atoms to fuse, making higher elements and thereby inventing chemistry.

Why is c 300, 000 m/sec? If we did it again, could it be different each time? Or would it, and Planck's constant and all the other numbers in physics, come out the same again? A few physicists, like Lee Smolin, think that it could be different, that each time a new universe is budded off from ours - from what looks like a Black Hole from this side - its constants may be a little bit different. Why this universe has those good Black Hole constants is that these are the best ones for generating Black Holes. So our progenitor universes were the most prolific, and it's not surprising that we're one of those, because nearly all universes are.

Now we're back with the anthropists and para-questions. Why do just those reactions occur in stars, making for example carbon by the triple-alpha process in this universe? Well, that reaction is fudged by having it occur in stars whose temperature is just that required for the reaction (no anthropics here: your coal fire is at just that temperature required to burn coal). If the physical constants were a little different we might get carbon by warmer or cooler processes, or maybe not get carbon at all but get a rather different element at position 12 which wasn't any good for complicated chemistry. In such a universe, Ian and I believe, another element altogether might have the linkage properties needed. Such a universe would have anthropists arguing that their universe was uniquely suitable for life. Remember the automobile analogy: our system might be a Toyota, theirs a Jaguar. Many ways, then, to make working automobiles or working universes. Again, then, "Why are there electrons?" That's a para-question, and until we know more at higher physical levels (for example "What else could quark combinations do?") it's a useless question. It is unlike the triple-alpha question above, where we do know much of the phase space of possibilities surrounding the actual nuclear reactions in stars. We know enough to see that there are many ways to make chemically-based planets, even if the constants were different - and probably equally interesting universes without chemistry but with pqrsz or xabq. The automobile anthropics there, far from our own universe's phase-space, would provide bicycles and helicopters instead of our familiar automobile metaphors, and probably anthropic life forms confident that those are the only way to exist.

Am I saying that anything could happen as a universe complicates? Perhaps. We don't know enough to rule that out. What we do know, because we have imaginations that can work out the geography of notional phase spaces of the possible around the actual, is that ours is far from the only way to make an interesting universe. And there are reasons for us to believe that most such universes will complicate, make it up as they go along like ours does. Kauffman argues this in the frame of "the adjacent possible", that enables any behaviour or process to generate a more complicated future. And recursive complication will make it more interesting still. The image is of Pooh Bear and his friend Eeyore following the Heffalump's footsteps in the snow and going around the little bush; they then follow the progressively more complicated footsteps round and round, convinced they're on the trail of a whole herdĚ. We argue to the same result, but by a different route (Cohen [5], Pratchett, Stewart and Cohen [24, 25], Cohen and Stewart [10]).

What we would really like now is an example, a case study where the universe shows us its motives as it "makes things up", inventing chemistry in the excruciating physics of star cores, or exploding biology out of recursive chemistry. We can describe what happens easily enough: the nuclear reactions in stars, the many suggestions for the origins of life on clay surfaces or in mineral-oil coacervates are not illuminating about the new rules. What we need to know is how these new systems take on a life of their own, becoming chemistry with peculiar electron-orbital rules that make DNA a cryptic possibility, or biology with natural selection, potentially generating monkeys, people and astronauts way down the line.

The initial difficulty is that people really do like preformation as the explanation of pattern (Keller [17]): we like to claim that pattern happened because what was there beforehand cryptically laid down the pre-pattern. In classical embryology this was the homunculus in the sperm, or the eggs as Russian dolls with all their descendants (down to the Day of Judgement, see Cohen [4]). Too often in modern biology, even embryology, it is the "DNA blueprint" whose replicated (pre-formed) pattern is supposed to form the basis of the embryo's structure, then to complicate this into the adult (Cohen and Medley [8]). No explanation of increasing complexity is required, because it doesn't increase. I demonstrate the Belousov-Zabotinski chemical reaction, which produces very overt bulls-eye patterns very reliably on an overhead projector, then re-exhibits new patterns three or four times after thorough mixing, to show that pattern can indeed arise de novo (e.g. Janke et al [13]) and references therein, Cohen and Stewart [7, 29]. Duane Gish, Vice-President of the Creation Research Society, accused me of conjuring, of cheating, when we debated whether a Creator/Designer was necessary while the B-Z was repeating on an OHP (at Birmingham University in 1988). At one level he was right. The B-Z reaction was invented to demonstrate this very point, and its robustness was improved by Art Winfree and me in 1972. So we want a different example, not a tuned-up toy teaching example, that models real increase of complexity in the natural world, preferably counter-intuitively so that the preformationist explanations are rejected.

The rabbit, of course, is now to be exhibited out of the hat. Not surprisingly, it has the signature of Ian Stewart across its rump. It is the very persuasive demonstration that - even in the complete absence of heterogeneity - populations of identical organisms will become two species as a parameter changes. Panmictic (random mating) - or asexual - organisms all living in the same environment, eating the same food and subject to the same predators and parasites will speciate, even if the only parameter that changes is the predictable increase of density as they multiply, resulting in more stringent competition. Such sympatric speciation, we are just discovering, is very common in nature.

Biologists (e.g. Mayr [22]) would much rather see speciation as the result of an environmental heterogeneity, allopatric speciation: on this side of the mountain/river/desert, population 1 evolves into species 1, while on the other side population 2 becomes species 2 as it adapts to different problems. There is no problem of initiation of a new pattern; the contingent fragmentation of the environment can be blamed. Darwin's finches, on all those little Galapagos islands, were an ideal example. Except that, like many clearly allopatric classical scenarios, the Natural History now looks more sympatric. Much of the speciation happened on an island, not divergence between different island populations (Grant [12]). And most of the species not only can interbreed still, but do. An even better sympatric example is the explosive cichlid fish evolution in the African Rift Valley Lakes: in each lake an explosion of hundreds of species diverged into a rather similar spectrum of types (Fryer and Iles [11], Kornfield and Smith [18]). In Lakes Victoria, Malawi, and Tanganyika (0.5 million, 2 million, 7 million years old) there are fish that live by eating other fishes eyes, or their babies, or the scales near their tails (they nearly catch the prey species!). In the lakes their stomachs have mostly the preferred diet; in aquaria they all eat proprietary fish food. Those from Lake Victoria are 5x different in linear size; some are plankton feeders, some grub algal communities off rocks, some eat insects - as well as those specialist feeders. Yet their genetic divergence is less than that of Homo sapiens, our own one species (Mayer et al [21])!

It is always possible to tell an allopatric story, if the sympatric one seems impossible as Natural History: even when there were only two fish starting the Victoria cichlids, one is upstream of, or deeper than, the other. Their progeny will find and exploit differences. As soon as there's ten species, they partition the environment for each other, and a cascade of increasing heterogeneity ensues - a herd of allopatric Heffalumps. That is why it is that so very many biologists believe that speciation is effectively contingent on environmental divergence, and why it is so important that Stewart has proven mathematically that sympatric speciation is normal, ordinary, just what the universe would do once it has species.

Sympatric speciation is the perfect model. It exposes the rules. Kauffman, in The Origins of Order [14], nearly addressed this issue full-square. He showed how many simple systems became complex, from linked threads-and-buttons to interacting light bulb circuits to (theoretical) gene arrays. He showed that a hundred items (N) should be linked by about ten interactive paths (K) for complexity to appear as emergent properties. K < 10, the system freezes or oscillates; K > 10, order breaks down. At about 10, the system is interesting. He did some maths, but was not convincing about the genesis of the rules. About his clever phenomena, there's little doubt: the universe likes to generate complexity "at the Edge of Chaos", as the media put it. In his later books [15, 16] he builds a worldview on this discovery, that systems bumble into the adjacent possibilities as often as they can, complicating themselves and anything they're context for. He likes African Lake cichlids. But he doesn't know why they speciate, only how.

4. Why speciation?

The new look concerns the graining of models of evolution, not the question of what gets selected. This is not a question of levels, of Dawkins' genes being selected, where other biologists have species-selection, and most biologists think the environment "sees" organisms. Population geneticists pretend that the environment sees proportions of different gene alleles (versis) in a population, and that each version has a "fitness value" S, making its carrier more or less likely to be represented in future generations. Reproductive biologists think that the environment mostly culls babies, so that early life-history stages may be better adapted - or not.

Stewart sees what he calls "pods"; these each contain (refer to) those members of a population that are, as far as selection is concerned, identical. So a population consists of a number of pods, and all individuals are competing for food, living space and so on; but each pod has a slightly different "style". This reflects the Lewontin-school observation that species are rather heterogeneous (about 10% heterozygosity in wild populations; that contrasts with the classical population-genetics view that a population has very-nearly-identical members except for those carrying a few mutations), and does not matter for the mathematics - all pods start out equal. Because each individual (as a member of its pod) "sees" all the other pods, there is initially a high degree of symmetry. But as the population density, or some other parameter, increases there is a sudden bifurcation, and two or three groups of pods diverge. Details of the divergence differ from occasion to occasion, but the result is two species where one was before, usually exploiting the new environment slightly more efficiently. In general, what we see is two (occasionally three, temporarily) normal (or more likely, otherĚ.) distributions where there was one before. Characters retain their average value, surprisingly; and we imagine that speciation may often happen (at least in the computer) without overall change of allele frequencies - only in what allele combinations survive in organisms to form part of the ecosystem (Cohen and Stewart [9]).

It is very tempting to see this pattern as much more general: the atoms at a star's core must be in different states, must form pod-like sub-populations that are competing, at least for space. Wachterhauser's recursively-chemical earliest life forms were competing for catalytic substrate and for chemical feedstock and energy [33]. Stewart would surely have those systems become different pre-organisms, not identical versions. They would exploit the substrate more effectively than one "species", and this primitive ecosystem would begin to complexify. As soon as one realises that there is a universal tendency towards divergence, via symmetry-breaking and evolution of systems toward bifurcations as parameters change, an evolving universe becomes inevitable.

And the rules? They are, surely, the axes of difference along which the pods differentiate. Darwin's finches mostly differ effectively in beak size and shape. So seed is to beak as. . . becomes a rule for evolving finches (Stewart, Elmhirst, and Cohen [31]). Whatever the constraints on the (electron-depleted) nuclei that began to have sexual congress in the cores of stars (instead of polite bouncing encounters), these must, in this view, have begun to determine what flavour of chemistry would result. Similarly, the kinds of human/pre-human encounters and cultures that interacted to begin Homo sapiens determined our subsequent cultural trajectories by building into our adapting brains (Barkow, Tooby and Cosmides [1]) all those prejudices - and abilities - that we exploit daily.

This combination of historical, contextual and interactional constraints is very familiar to the biologist working with phylogenies, and trying to understand present organisms by putting them in phylogenetic context. As a last biological example that should illustrate just how many Heffalumps there are in a usual biological herd - how many recursions have knitted rules into the biology of contemporary organisms - here is a thought about heteronomous (autoparasitoid) hymenopterans. These are tiny wasps whose larvae grow in the bodies of other larval insects, but with complex life history strategies. Remember the complicity (Cohen and Stewart [7]) that linked thirsty early insects to the wet eyes of early mammals. Then the development of piercing/sucking mouthparts, so that blood-sucking became a viable mode, opening up a whole new world of parasitic possibilities to viruses, and to protozoans like malaria and sleeping-sickness. Then bear in mind that some of these insects had piercing ovipositors too (perhaps using the same developmental mechanisms as sharp mouthparts - arthropods are very like extended Swiss-Army knives), and they used these to lay their eggs in other insects. Then put yourself in the position of a female wasp, with control over the future sex of the eggs she lays (she doesn't let sperms fertilise the male-producing eggs), coming upon a new host: a scale-insect feeding on a leaf. Her options in her rule-based world are different, depending on whether another wasp of the same or another species, her close kin or not, lays eggs in the same host. She must control her offspring sex ratios for optimum representation in the future, according to a very convoluted set of constraints and opportunities that have knotted up the originally simple natural-selection-for-parasites rule-book. Some of these wasps (heteronomous species, see above) have male-producing eggs laid in different insect hosts from the female-producing eggs, sometimes parasitising not the host insect larva but the parasite larvae inside it (hyper-parasitism). One more twist: some of these (autoparasitoids, see above) have female larvae parasitising host insects, but male larvae hyper-parasitising female larvae laid by other (?) females of their own species (Ode and Hunter [23])! And they navigate the sex ratios appropriate to these different strategies.

That is a recursive complicity of the kind that some biologists really enjoy (not the reductionist biotechnologists). These scientists really know that the universe loves complexity, and fosters it. The parasitic hymenopterans are very successful: more than 10,000 species is a conservative estimate. This shows us, if not a family of meta-rules, at least a hint of why complex knitted-up ecosystems evolve and maintain themselves. Mostly, we leave out the parasites, and their parasites, when we use them as examples to show how complex systems can work, just as we leave out all those baby-deaths when we model evolution by natural selection. Perhaps the universe is telling us that we should follow our own Complexity-philosophy precepts: don't simplify the system to learn from it, accept it in its full exuberant state. Only then will we learn to draw the lessons from it, as Stewart has done, that illuminate its meta-rules.

References

1. Barkow, J.H., Cosmides, L. & Tooby, J. (Eds.). The Adapted Mind: Evolutionary Psychology and the Generation of Culture. New York: Oxford University Press
2. Boisot M and Cohen J (2000) "Shall I Compare Thee to... an Organization?" Emergence 2 (4) 113-135
3. Cohen J (1977) Reproduction. London: Butterworths
4. Cohen, J (1988) Review of The Ovary of Eve: Egg and Sperm and Preformation by C. Pinto-Correia. Endeavour 22 83-4
5. Cohen J (2001) Concepts: Knife-edge of design. Nature 411 529
6. Cohen J and 'Espinasse P G (1961) On the normal and abnormal development of the feather. J. Embryol. exp. Morph. 9, 223-251
7. Cohen, J. and Stewart, I. (1994) The Collapse of Chaos; simple laws in a complex world New York: Penguin, Viking
8. Cohen, J and Medley, G (2000) Stop working and Start Thinking: a guide to becoming a scientist Cheltenham: Nelson-Thornes
9. Cohen, J and Stewart, I N (2000) Polymorphism viewed as phenotypic symmetry-breaking. In Non-Linear Phenomena in Biological and Physical Sciences (Ed S K Malik, M K Chandrashekaran, and N Pradhan). New Delhi: Indian Nat Sci Acad. 1-64
10. Cohen J. and Stewart I (2002) Evolving the Alien: the science of extra-terrestrial life London:Ebury Press
11. Fryer, G., and T.D. Iles. (1972) The cichlid fishes of the Great Lakes of Africa. Edinburgh: Oliver & Boyd; Neptune City, New Jersey: TFH Publications
12. Grant P R (1999) The Ecology and Evolution of Darwin's Finches. Princeton Univ. Press
13. Janke W, Henze C and Winfree A T (1988) Chemical vortex dynamics in three-dimensional excitable media. Nature 336 662-665 and references therein
14. Kauffman, S (1993).The origins of order. Oxford: Oxford University Press.
15. Kauffman, S (1994) At Home in the Universe. New York: Viking,
16. Kauffman, S (2000) Investigations Oxford: Oxford University Press
17. Keller E F (2002) Making Sense of Life; Explaining Biological Development With Models, Metaphors And Machines. Cambridge, MA: Harvard University Press
18. Kornfield, I., and P.F. Smith. 2000. African cichlid fishes: model systems for evolutionary biology. Annual Review of Ecology and Systematics 31: 163-196
19. Lovelock, J (1988) The Ages of Gaia: a biography of our living earth. Oxford: OUP
20. Maturana, H.R. and Varela F.J (1992) The Tree of Knowledge: the Biological Roots of Human Understanding. , Boston, MA: Shambhala.
21. Mayer A, Kocher T D, Basisibwaki P and Wilson A C (1990) DNA divergence of the cichlid flock in Lake Victoria. Nature 347 550-553
22. Mayr E (1970) Population, Species and Evolution Cambridge, MA: Harvard University Press
23. Ode, P J and Hunter, M S (2002) Sex ratios of parasitic Hymenoptera with unusual life-histories. In Sex Ratios; concepts and research methods (Ed I C W Hardy). Cambridge: University Press. 218-34
24. Pratchett, T, Stewart, I And Cohen, J (1999) The Science of Discworld. London: Ebury Press
25. Pratchett, T, Stewart, I and Cohen, J (2002) The Science of Discworld 2: The Globe. London: Ebury Press
26. Rapport, D J (1991) Myths in the foundations of economics and ecology. Biol J Linn Soc 44 185 -202
27. Schrodinger E (1944) What is Life? The physical aspect of the living cell. Cambridge: University Press.

28. Stewart I (2002) The second law of gravitics and the fourth law of thermodynamics In From Complexity
     to Life: Explaining the Emergence of Life and Meaning: Proceedings of Templeton Symposium on Com-
     plexity, Information, and Design, Santa Fe 1999 (ed. N.H.Gregsen). Oxford: Oxford University Press. To
     appear.
29. Stewart, I and Cohen, J (1997) Figments of Reality; the origins of the curious mind Cambridge: University
     Press
30. Stewart, I and Cohen, J. (2000) Wheelers. New York:Warner-Aspect
31. Stewart I, Elmhirst T and Cohen J (2002) Symmetry-breaking as an origin of species, Conference on
     Bifurcations, Symmetry, Patterns, Porto 2000. To appear.
32. Varela, F J, Maturana H R and Uribe R (1974) Organization of living systems, its characterization and a
     model. Biosystems 5 187-196
33. Wachtershauser G (1992) Groundworks for an evolutionary biochemistry: the iron-sulphur world.
     Progress in Biophysics and Molecular Biology 58 85-201

ADAPTIVE EVOLUTION OF COMPLEX SYSTEMS UNDER UNCERTAIN ENVIRONMENTAL CONSTRAINTS: A VIABILITY APPROACH

JEAN-PIERRE AUBIN

14, rue Domat, F-75005 Paris

Abstract. The main purpose of viability theory is to explain the evolution of the state of a control system, governed by nondeterministic dynamics and subjected to viability constraints, to reveal the concealed feedbacks which allow the system to be regulated and provide selection mechanisms for implementing them. It assumes implicitly an "opportunistic" and "conservative" behavior of the system: a behavior which enables the system to keep viable solutions as long as its potential for exploration (or its lack of determinism) — described by the availability of several evolutions — makes possible its regulation.

Examples of viability constraints are provided by architectures of networks imposing constraints described by connectionist tensors operating on coalitions of actors linked by the network. The question raises how to modify a giving dynamical system governing the evolution of the signals, the connectionist tensors and the coalitions in such a way that the architecture remains viable.

1. Introduction

The purpose of this survey paper is to present a brief introduction of viability theory that studies adaptive type evolution of Complex Systems under Uncertain Environment and Viability Constraints that are found in many domains involving living beings, from biology to cognitive sciences, from ecology to sociology and economics. It involves polyscientific investigations spanning fields that have traditionally developed in isolation.

Instead of applying only known mathematical and algorithmic techniques, most of them motivated by physics and not necessarily adapted to such problems, viability theory designs and develops mathematical and algorithmic methods for studying the evolution of such systems, organizations and networks of systems,

1. constrained to adapt to a (possibly co-evolving) environment,
2. evolving under contingent, stochastic or tychastic uncertainty,
3. using for this purpose regulation controls, and in the case of networks, connectionist matrices or tensors,
4. the evolution of which is governed by regulation laws that are then "computed" according to given principles such as the inertia principle,
5. the evolution being either continuous, discrete, of an "hybrid" of the two when impulses are involved,
6. the evolution concerning both the variables and the environmental constraints (mutational viability),
7. the nonviable dynamics being corrected by introducing adequate controls when necessary,

J. Nation et al. (eds.), Formal Descriptions of Developing Systems, 165–184.

8. or by introducing the "viability kernel" of a constrained set under a nonlinear controlled system (either continuous or hybrid), that is the set of initial states from which starts at least one evolution reaching a target in finite time while obeying state (viability) constraints.

It is by now a consensus that the evolution of many variables describing systems, organizations, networks arising in biology and human and social sciences do not evolve in a deterministic way, and may be, not even in a stochastic way as it is usually understood, but with a Darwinian flavor, where intertemporal optimality selection mechanisms are replaced by several forms of "viability", a word encompassing polysemous concepts as stability, confinement, homeostasis, etc., expressing the idea that some variables must obey some constraints. Intertemporal optimization is replaced by myopic selection mechanisms that involve present knowledge, sometimes the knowledge of the history (or the path) of the evolution, instead of anticipations or knowledge of the future (whenever the evolution of these systems cannot be reproduced experimentally). Uncertainty does not necessarily obey statistical laws, but only unforcastable rare events (tyches, or perturbations, disturbances) that obey no statistical law, that must be avoided at all costs (precaution principle or robust control). These systems can be regulated by using regulation (or cybernetical) controls that have to be chosen as feedbacks for guaranteeing the viability of a system and/or the capturability of targets and objectives, possibly against tyches (perturbations played by Nature).

The purpose of viability theory is to attempt to answer directly the question that some economists or biologists ask: Complex organizations, systems and networks, yes, but for what purpose? One can propose the following answer: to adapt to the environment. This is the case in biology, since the Claude Bernard's "constance du milieu intérieur" and the "homeostasis" of Walter Cannon. This is naturally the case in ecology and environmental studies. This is also the case in economics when we have to adapt to scarcity constraints, balances between supply and demand, and many other ones.

The environment is described by constraints of various kinds (representing objectives, physical and economic constraints, "stability" constraints, etc.) that can never be violated. In the same time, the actions, the messages, the coalitions of actors and connectionist operators do evolve, and their evolution must be consistent with the constraints, with objectives reached at (successive) finite times (and/or must be selected through intertemporal criteria).

There is no reason why collective constraints are satisfied at each instant by evolutions under uncertainty governed by stochastic or thychastic control dynamical systems. This leads to the study of how to correct either the dynamics, and/or the constraints in order to reestablish this consistency. This may allow us to provide an explanation of the formation and the evolution of the architecture of the system and of their variables.

Presented in such an evolutionary perspective, this approach of (complex) evolution departs from the main stream of modeling studying static networks with graph theory and dynamical complex systems by ordinary or partial differential equations, a task difficult outside the physical sciences.

For dealing with these issues, one needs innovative concepts and formal tools, algorithms and even mathematical techniques motivated by complex systems evolving under uncertainty. For instance, and without entering into the details, we can mention systems sharing such common features arising in

economics, where the viability constraints are the scarcity constraints. We can replace the fundamental Walrasian model of resource allocations by decentralized dynamical model in which the role of the controls is played by the prices or other economic decentralizing messages (as well as coalitions of consumers, interest rates, and so forth). The regulation law can be interpreted as the behavior of Adam Smith's invisible hand choosing the prices as a function of the allocations,

dynamical connectionist networks and/or dynamical cooperative games, in which coalitions of players may play the role of controls: each coalition acts on the environment by changing it through dynamical systems. The viability constraints are given by the architecture of the network allowed to evolve,

genetics and population genetics, where the viability constraints are the ecological constraints, the state describes the phenotype and the controls are genotypes or fitness matrices.

sociological sciences, where a society can be interpreted as a set of individuals subject to viability constraints. They correspond to what is necessary to the survival of the social organization. Laws and other cultural codes are then devised to provide each individual with psychological and economical means of survival as well as guidelines for avoiding conflicts. These cultural codes play the role of regulation controls.

cognitive sciences, where, at least at one level of investigation, the variables describe the sensory-motor activities of the cognitive system, while the controls translate into what could be called a conceptual control (which is the synaptic matrix in neural networks.)

control theory and differential games, conveniently revisited, can provide many metaphors and tools for grasping the above problems. Many problems in control design, stability, reachability, intertemporal optimality, viability and capturability, observability and set-valued estimation can be formulated in terms of viability kernels. The Viability Kernel Algorithm computes this set.

The ways of using these concepts and tools dealing with viability issues are very different whether the domain of applications is socio-economic, biologic or technological.

— In the first case, the theoretical results about the above ways to think long term viability are useful for the (re)designing of regulatory Institutions: regulated markets when political convention must exist for global purpose; mediation or metamediation of all kinds, including law, social conflicts in production and in the cities; institutions for sustainable development. And progressively, when more data gathered by these institutions will be available, qualitative (and sometimes quantitative) prescriptions of viability theory will be useful.

— In the second case, the technological systems (robot of all kinds, from very little ones including nanorobots, to animats and to very big large ones (drones, underwater vehicles, etc) need embarked systems autonomous enough to regulate viability/capturability problems by adequate regulation (feedback) control laws. Modular and/or object-oriented general portable and distributed software flexible enough for integrating new problems (hybrid, distributed systems, dynamical games) are needed and could be approached in the viability framework.

It is time to cross the interdisciplinary gap and to confront and hopefully to merge the points of view rooted in different disciplines. Mathematics, thanks to its abstraction power by isolating only few key features of a class of problems, can help to bridge these barriers as long as it proposes new methods motivated by these new problems instead of applying the classical ones only motivated until now by physical sciences. Paradoxically, the very fact that the mathematical tools useful for social and biological sciences are and have to be quite sophisticated impairs their acceptance by many social scientists, economists and biologists, and the gap menaces to widen.

1.1. THE CONSTRAINED SET AND THE TARGET

The state variables of the systems under investigations range over the constrained subset K — the environment — of the state space $X := \mathbf{R}^n$. One can also introduce a subset $C \subset K$ — the target — that can be taken empty if no objective is assigned to the system. We denote by $K \backslash C$ the subset of elements of the constrained set K outside the target C.

An evolution $t \in [0, T[\mapsto x(t) \in X$ of the state of the system is said to be viable in K on the interval $[0, T[$ before it reaches the target C if

$$\forall\, t \in [0, T[, \quad x(t) \in K \backslash C$$

The evolution is captured (in finite time T) if $x(T) \in C$. When $C := \emptyset$ and $T := +\infty$, we simply say that the evolution is viable in K.

1.2. THE DYNAMICAL BEHAVIOR

Next, we provide the mathematical description of the "engine" governing the evolution of the state. We assume that there exists a control parameter, or, better, a regulatory parameter, called a regulon, that influences the evolution of the state of the system. This dynamical system takes the form of a control system with (multivalued) feedbacks :

$$\begin{cases} i) & x'(t) = f(x(t), u(t)) \ \ \text{(action)} \\ ii) & u(t) \in U(x(t)) \ \ \text{(contingent retroaction)} \end{cases} \tag{1}$$

taking into account the *a priori* availability of several regulons $u(t) \in U(x(t))$ chosen in a subset $U(x(t)) \subset Y$ of another finite dimensional vector-space Y subjected to state-dependent constraints.

Once the initial state is fixed, the first equation describes how the regulon acts on the velocities of the system whereas the second inclusion shows how the state (or an observation on the state) can retroact through (several) regulons in a vicariant way.

We observe that there are many evolutions starting form a given initial state x_0, one for each time-dependent regulon $t \mapsto u(t)$. The set-valued map $U : X \rightsquigarrow Y$ also describes the state-dependent constraints on the regulons. In this case, system (1) can no longer be regarded as a parametrized family of differential equations, as in the case when $U(x) \equiv U$ does not depend upon the state, but as a differential inclusion (see Aubin & Cellina [12] for example). Fortunately, differential inclusions enjoy most of the properties of differential equations.

An evolution of system (1) is a function $t \rightarrow x(t)$ satisfying this system for some (measurable) open-loop control $t \rightarrow u(t)$ (almost everywhere).

For networks defined by connectionist tensors and coalitions of actors, the regulons are not given *a priori* but built as "viability multipliers" in such a way that the architecture of the network is viable.

1.3. THE SIMPLEST EXAMPLE

We take $X = Y = \mathbf{R}$, $K := [a, b]$ where $0 < a < b$, $C = \emptyset$ and a system of the form

$$\begin{cases} i) & x'(t) = u(t)x(t) \\ ii) & u(t) \in [-\underline{u}, +\overline{u}] \end{cases} \tag{2}$$

where the regulon is simply chosen to be the growth rate of the system, only subjected to vary between lower and upper bounds. We shall pursue the study of this example along the way.

For other low-dimensional examples, see Chapter 2 of Aubin [3].

2. Characterization of Viability and/or Capturability

The problems we shall study are all related to the viability of the constrained set and/or the capturability of a target under the dynamical system modeling the dynamic behavior of the cognitive system.

2.1. DEFINITIONS

The viability of a subset K under a control system is a consistency property of the dynamics of the system confronted to the constraints it must obey during some length of time.

Namely, a subset $K \backslash C$ is locally viable under the control system described by (f, U) if for every initial state $x_0 \in K$, there exist a positive time $T_{x_0} > 0$ and at least one solution to the system starting at x_0 and which is viable in K on the interval $[0, T_{x_0}]$ before it reaches the target C at time $T \geq T_{x_0}$. When $C := \emptyset$, we shall say that K is viable if we can take $T_{x_0} := +\infty$ for all $x_0 \in K$.

To say that a singleton $\{c\}$ is viable amounts to saying that the state c is an equilibrium — sometime called a fixed point. The trajectory of a periodic solution is also viable.

Contrary to the century-old tradition going back to Lyapunov, we require the system to capture the target C in finite time, and not in an asymptotic way, as in mathematical models of physical systems. This is in particular mandatory for cognitive systems that must achieve tasks in finite time. However, there are close mathematical links between the various concepts of stability and viability. For instance, Lyapunov functions can be constructed using tools of viability theory. This needs much more space to be described: we refer to Chapter 8 of Aubin [1] and Chapter 8 of Aubin [3] for more details on this topic.

The first task is to characterize the subsets having this viability/capturability property. To be of value, this task must be done without solving the system for checking the existence of viable solutions for each initial state.

An immediate intuitive idea jumps to the mind: at each point on the boundary of the constrained set outside the target, where the viability of the system is at stake, there should exist a velocity which is in some sense tangent to the viability domain and serves to allow the solution to bounce back and remain inside it. This is, in essence, what the viability theorem below states. Before stating it, the mathematical implementation of the concept of tangency must be made.

We cannot be content with viability sets that are smooth manifolds (such as spheres, which have no interior), because inequality constraints would thereby be ruled out (as for balls, that possess distinct boundaries). So, we need to "implement" the concept of a direction v tangent to K at $x \in K$, which should mean that starting from x in the direction v, we do not go too far from K: The adequate definition due to G. Bouligand and F. Severi proposed in 1930 states that a direction v is tangent to K at $x \in K$ if it is a limit of a sequence of directions v_n such that $x + h_n v_n$ belongs to K for some sequence $h_n \to 0+$. The collection of such directions, which are in some sense "inward", constitutes a closed cone $T_K(x)$, called the tangent cone[1] to K at x. Naturally, except if K is a smooth manifold, we lose the fact that the set of tangent vectors is a vector-space, but this discomfort in not unbearable, since advances in set-valued analysis built a calculus of these cones allowing us to compute them. See Aubin & Frankowska [15] and Rockafellar & Wets [31] for instance.

2.2. THE ADAPTIVE MAP

We then associate with the dynamical system (described by (f, U)) and with the viability constraints (described by K) the (set-valued) adaptive or regulation map R_K. It maps any state $x \in K \backslash C$ to the subset $R_K(x)$ (possibly empty) consisting of regulons $u \in U(x)$ which are viable in the sense that $f(x, u)$ is tangent to K at x:

$$R_K(x) := \{u \in U(x) \mid f(x, u) \in T_K(x)\}$$

We can for instance *compute the* adaptive map in many instances.

Example: The regulation map is equal to

$$R_K(x) := \begin{cases} [0, +\overline{u}] & \text{if } x = a \\ [-\underline{u}, +\overline{u}] & \text{if } x \in]a, b[\\ [-\underline{u}, 0] & \text{if } x = b \end{cases}$$

It is set-valued, has nonempty values, but has too poor continuity property, a source of mathematical difficulties, as we shall see.

2.3. THE VIABILITY THEOREM

The Viability Theorem states that *the target C can be reached in finite time from each initial condition $x \in K \backslash C$ by* at least *one evolution of the control system viable in K* if and only if *for every $x \in K \backslash C$, there exists at least one viable control $u \in R_K(x)$.*

This Viability Theorem holds true when both C and K are closed and for a rather large class of systems, called Marchaud systems: Beyond imposing some weak technical

---

[1] Replacing the linear structure underlying the use of tangent spaces by the tangent cone is at the root of Set-Valued Analysis.

conditions, the only severe restriction is that, for each state x, the set of velocities $f(x, u)$ when u ranges over $U(x)$ is convex[2].

Curiously enough, when the constrained set is assumed to be viable, convex and compact, then one can prove that there exists a (viable) equilibrium. Without convexity, we deduce only the existence of minimal viable closed subsets.

The proofs of the above Viability Theorem and the Equilibrium Theorem are difficult, but their consequences are much easier to obtain and can be handled with moderate mathematical competence.

Example: The interval $[a, b]$ is viable under system (2).

2.4. THE ADAPTATION LAW

Once this is done, and whenever a constrained subset is viable for a cognitive system, the second task is to show how to govern the evolution of viable evolutions. We thus prove that an evolution of system (1) is viable if and only if it is governed by

$$\begin{cases} i) & x'(t) = f(x(t), u(t)) \\ ii) & u(t) \in R_K(x(t)) \ \text{(adaptation law)} \end{cases} \tag{3}$$

until the state reaches the target C.

We observe that the initial set-valued map U involved in (1)ii) is replaced by the adaptive map R_K in (3)ii). The inclusion $u(t) \in R_K(x(t))$ can be regarded as an adaptation law (rather than a learning law, since there is no storage of information at this stage of modeling).

2.5. PLANNING TASKS: QUALITATIVE DYNAMICS

Reaching a target is not enough for studying the behavior of cognitive systems, that have to plan tasks in a given order. This issue has been recently revisited in Aubin & Dordan [14] in the framework of qualitative physics (see Dordan [23] and Aubin [2] for more details on this topic). We describe the sequence of tasks or objectives by a family of subsets regarded as qualitative cells. Giving an order of visit of these cells, the problem is to find an evolution visiting these cells in the prescribed order.

2.6. THE METASYSTEM

In order to bound the chattering (rapid oscillations or discontinuities) of the regulons, we set *a priori* constraints on the velocities of the form

$$\forall t \geq 0, \quad u'(t) \in \Phi$$

---

[2] This happens for the class of control systems of the form

$$x'(t) = f(x(t)) + G(x(t)) u(t)$$

where $G(x)$ are linear operators from the control space to the state space, when the maps $f : X \mapsto X$ and $G : X \mapsto \mathcal{L}(Y, X)$ are continuous and when the control set U (or the images $U(x)$) are convex.

where $\Phi \subset Y$ is a bounded subset of the space of regulons (that may depend on both the state and the regulon). The metasystem associated with the initial viability problem is the system

$$\begin{cases} i) & x'(t) = f(x(t), u(t)) \\ ii) & u'(t) \in \Phi \end{cases} \qquad (4)$$

subjected to the metaconstraints

$$\forall t \geq 0, \quad x(t) \in K \ \& \ u(t) \in U(x(t)) \qquad (5)$$

Unfortunately, the above metaconstraints may no longer be viable under the metasystem.

Example: We require that the velocity of the regulon (equal here to the growth rate) must remain between bounds $-d$ and $+c$. Therefore, the metasystem is

$$\begin{cases} i) & x'(t) = u(t)x(t) \\ ii) & u'(t) \in \Phi := [-d, +c] \end{cases} \qquad (6)$$

subjected to metaconstraints

$$\forall t \geq 0, \quad (x(t), u(t)) \in [a, b] \times \left[-\underline{u}, +\overline{u} \right] \qquad (7)$$

One can prove that these metaconstraints are not viable under the metasystem.

2.7. RESTORING VIABILITY

The above example shows that there are no reasons why an arbitrary subset K should be viable under a control system. Therefore, the problem of reestablishing viability arises. One can imagine several methods for this purpose:

1. Keep the constraints and change initial dynamics by introducing regulons that are "viability multipliers"
2. Keep the same dynamics and looking for viable constrained subsets
3. Change either the dynamics or the set of constraints

   a) either by changing the regulons according to feedbacks or dynamic feedbacks that can be constructed (see for instance Aubin [1, 3]),
   b) or by letting the set of constraints evolve according to mutational equations, as in Aubin [6].

4. or change the initial conditions by introducing a reset map R mapping any state of K to a (possibly empty) set $R(x) \subset X$ of new "initialized states" (impulse control).

We shall describe succinctly these methods.

3. Viability Kernels and Capture basins

When a closed subset K is not viable under a control system, then two questions arise naturally: find solutions starting from K which remain viable in K as long as possible, and starting outside of K, find solutions which return to K as soon as possible to restore the viability. These natural questions justify the introduction of the following concepts:

1. The subset Viab(K) of initial states $x_0 \in K$ such that one solution $x(\cdot)$ to system (1)ii) starting at x_0 is viable in K for all $t \geq 0$ is called the viability kernel of K under the control system. A subset K is a repeller if its viability kernel is empty.

2. The subset Capt(K, C) of initial states $x_0 \in K$ such that the target $C \subset K$ is reached in finite time before possibly leaving K by one solution $x(\cdot)$ to system (1)ii) starting at x_0 is called the viable-capture basin of C in K. A subset $C \subset K$ such that Capt(K, C) $= C$ is said to be isolated in K.

One can prove that if the system is Marchaud and if K is closed, *the viability kernel* Viab(K) *of the subset K is the* largest *closed subset of K viable under the control system.* Hence, all interesting features such as equilibria, trajectories of periodic solutions, limit sets and attractors, if any, are all contained in the viability kernel.

Example: Viability kernel of metaconstraints Since the metaconstraints (7) are not viable metasystem (6), one can compute the viability kernel of $[a, b] \times [-\underline{u}, +\overline{u}]$. It is the subset of pairs (x, u) such that

$$u \in G_K(x) := \left[-\sqrt{2c \log\left(\frac{x}{a}\right)}, +\sqrt{2d \log\left(\frac{b}{x}\right)} \right] \square \tag{8}$$

One can also prove that the viability kernel is the **unique** closed subset $D \subset K$ viable and isolated in K such that $K \backslash D$ is a repeller. If K is a repeller and $C \subset K$ is closed, one can prove that the capture basin Capt(K, C) of $C \subset K$ is the **unique** closed subset D between C and D such that D is isolated in K and $D \backslash C$ is locally viable.

The viability kernels of a subset and the capture basins of a target can thus be characterized in diverse ways through tangential conditions thanks to the viability theorems. They play a crucial role in viability theory, since many interesting concepts are often viability kernels or capture basins.

Furthermore, algorithms designed in Saint-Pierre [32]) allow us to compute viability kernels and capture basins (see also Cardaliaguet, Quincampoix & Saint-Pierre [21] and Quincampoix & Saint-Pierre [30]). In general, there are no explicit formulas providing the viability kernel and capture basins.

Remark: The Crisis Function — If the solution is allowed to leave the constrained set K, one can use the crisis function introduced in Doyen & Saint-Pierre [24], which measures the minimal time spent by the evolutions $x(\cdot)$ outside K.

4. Selecting Viable Feedbacks

4.1. STATIC FEEDBACKS

A (static) feedback r is a map $x \in K \mapsto r(x) \in X$ which is used to pilot evolutions governed by the differential equation $x'(t) = f(x(t), r(x(t)))$. A feedback r is said to be viable if the solutions to the differential equation $x' = f(x, r(x))$ are viable in K. The most celebrated examples of linear feedbacks in linear control theory designed to control a system have no reason to be viable for an arbitrary constrained set K, and, according to the constrained set K, the viable feedbacks are not necessarily linear.

However, the Viability Theorem implies that a feedback r is viable if and only if r is a selection of the adaptive map R_K in the sense that

$$\forall x \in K \backslash C, \quad r(x) \in R_K(x) \tag{9}$$

Hence, the method for designing feedbacks for cognitive systems to evolve in a constrained subset amounts to find selections $r(x)$. One can design a "factory" for designing selections (see Chapter 6 of Aubin [1], for instance). Ideally, a feedback should be continuous to guarantee the existence of a solution to the differential equation $x' = f(x, r(x))$. But this is not always possible. This is the case of slow selection r° of R_K of minimal norm, governing the evolution of slow viable evolutions (despite its lack of continuity).

Example: The slow selection r° for the system (2) is equal to $r^\circ(a) = 0$ and $r^\circ(x) = -\underline{u}$ when $a < x \ \square\ b$. It is discontinuous at $x = a$.

4.2. DYNAMIC FEEDBACKS

One can also look for dynamic feedbacks $g_K : X : K \times Y \mapsto Y$ that governs the evolution of both the states and the regulons through the metasystem of differential equations

$$\begin{cases} i) & x'(t) = f(x(t), u(t)) \\ ii) & u'(t) = g_K(x(t), u(t)) \end{cases} \tag{10}$$

A dynamic feedback g_k is viable if the metaconstraints (5) are viable under the metasystem (10).

As for the (static) feedbacks, one can prove that all the viable dynamic feedbacks are selections of a dynamical adaptive map $G_K : K \times Y \rightsquigarrow Y$ obtained by "differentiating" the adaptation law (1)ii) thanks to the differential calculus of set-valued maps see Aubin & Frankowska [15]).

4.3. HEAVY EVOLUTIONS AND THE INERTIA PRINCIPLE

Among the viable dynamic feedbacks, one can choose the heavy viable dynamic feedback $g_K^\circ \in G_K$ with minimal norm that governs the evolution of heavy viable solutions, i.e., viable evolutions with minimal velocity. They are called "heavy" viable evolutions[3] in the sense of heavy trends in economics.

Heavy viable evolutions offer convincing metaphors of the evolution of biological, economic, social and cognitive systems that obey the inertia principle. It states in essence that "the regulons are kept constant as long as viability of the system is not at stake". Heavy viable evolutions can be viewed as providing mathematical metaphors for the concept of punctuated equilibrium introduced in paleontology by Eldredge and Gould in 1972. In our opinion, this is a mode of regulation of cognitive systems (see Chapter 8 of Aubin [2] for further justifications).

Indeed, as long as the state of the system lies in the interior of the constrained set (i.e., away of its boundary), any regulon will do. Therefore, the system can maintain the

---

[3] When the regulons are the velocities, heavy solutions are the ones with minimal acceleration, i.e., maximal inertia.

regulon inherited from the past. This happens if the system obeys the inertia principle. Since the state of the system evolves while the regulon remains constant, it may reach the viability boundary with an "outward" velocity. This event corresponds to a period of viability crisis: To survive, the system must find other regulons such that the new associated velocity forces the solution back inside the viability set until the time when a regulon can remain constant for some time.

Example: One can prove that the set-valued map G_K for the system (2) is given by the formula (8). Hence the heavy dynamic feedback g_K° is given by the formula

$$g_K^\circ(x, u) := \min\left(0, +\sqrt{2d \log\left(\frac{b}{x}\right)}\right)$$

if $u > 0$, $g_K^\circ(x, 0) = 0$ and, if $u < 0$,

$$g_K^\circ(x, u) := \max\left(0, -\sqrt{2c \log\left(\frac{x}{a}\right)}\right)$$

Starting from (x_0, u_0) where $u_0 \in \left]0, \sqrt{2d \log\left(\frac{b}{x_0}\right)}\right]$, the heavy solution $x(t) = x_0 e^{u_0 t}$ is associated with the constant growth rate $u(t) = u_0$ until the time $t_1 = \frac{\log\left(\frac{b}{x_0}\right)}{u_0} - \frac{u_0}{2d}$ when $x(t_1) = b e^{\frac{-u_0^2}{2d}}$ because $g_K^\circ(x(t_1)) = u_0$. Then *this is the last moment when we have to change the regulon* by taking $u(t) = u_0 - d(t - t_1)$ that decreases with minimal velocity until the time $t_2 = t_1 + \frac{u_0}{d}$ when $x(t) = b e^{\frac{-u(t)^2}{2d}}$ reaches the equilibrium $x(t_2) = b$ where $u(t_2) = 0$. One can prove that all solutions $(x(t), u(t))$ of the metasystem starting from $(x(t_1), u_0)$ at time t_1 either range the boundary of the viability kernel until t_2 (this is the case when $u(t) = u_0 - d(t - t_1)$) or leave $[a, b] \times [-\underline{u}, +\overline{u}]$ in finite time. This illustrates a general property of the boundary of the viability kernel, the barrier or semipermeability property discovered by Marc Quincampoix : no solution starting from the boundary of the viability kernel can enter its interior. One of them at least remains on the boundary, other evolutions may leave the constrained set in finite time.

4.4. WARNING SIGNAL

Imposing a speed limit on regulons conceals a warning signal to regulons, which must start to evolve as soon as the boundary of the viability kernel is reached.

But one can design other ways to implement warning signals for avoiding a impulsive change of regulon when the state is about to leave the constrained set K. A solution can be provided by replacing the constrained set K by a fuzzy set $\gamma_K : K \mapsto [0, 1]$, that is a membership function equal to 0 outside K, the membership values $\gamma_K(x)$ ranging between 0 and 1 inside K. The case when $\chi_K(x) = 1$ whenever $x \in K$ provides the original (usual) set K. A standard way for building fuzzy sets is to take for membership function

$$\delta_K(x) := \min\left(1, \varepsilon \inf_{y \notin K} \|x - y\|^2\right)$$

(see Aubin & Dordan [13] for instance). The fuzzy set γ_K being given, one can replace the original viability constraints by

$$\forall t \geq 0, \quad \gamma_K(x(t)) \geq \gamma_K(x_0)e^{-at}$$

where $\alpha > 0$. Assuming that the membership function γ_K is differentiable for simplicity, we introduce the fuzzy regulation map R_{γ_K} defined by

$$R_{\gamma_K}(x) := \{u \in U(x) \mid \langle \gamma_K'(x), f(x,u)\rangle + \alpha\gamma_K(x) \geq 0\}$$

We do not need to assume the membership function to be differentiable. In the case of the function δ_K, one can prove that whenever $\delta_K(x) < 1$,

$$R_{\delta_K}(x) := \{u \in U(x) \mid \langle x - \pi x, f(x,u)\rangle + 2\varepsilon\alpha\delta_K(x) \geq 0\}$$

where π is the projection onto the complement of K.

Then one can prove that fuzzy constrained set γ_K is viable if and only if for any $x \in K$, $R_{\gamma_K}(x)$ is not empty and that the fuzzy viable evolutions are regulated by the regulation law

$$\forall t \geq 0, \quad u(t) \in R_{\gamma_K}(x(t))$$

If the above necessary and sufficient condition is not satisfied, one can also prove the existence of a fuzzy viability kernel.

One can also assume that the set-valued map U is actually fuzzy, in the sense that it maps every x to a fuzzy set of controls (see Aubin & Dordan [13]).

4.5. IMPULSE SYSTEMS

There are many other dynamics that obey the inertia principle, among which heavy viable evolutions are the smoothest ones. At the other extreme, on can study also the (discontinuous) "impulsive" variations of the regulon. Instead of waiting the system to find a regulon that remains constant for some length of time, as in the case of heavy solutions, one can introduce another (static) system that resets a new constant regulon whenever the viability is at stakes, in such a way that the system evolves until the next time when the viability is again at stakes.

This regulation mode is a particular case of what are called impulse control in control theory (see for instance Aubin [7], Aubin, Lygeros, Quincampoix, Sastry & Seube [18]), hybrid systems in computer sciences (see for instance Shaft & Schumacher [33]) and Integrate and Fire models in neurobiology (see Bressloff & Coombes [20] and Shimokawa, Pakdaman & Sato [34, 35]), etc. Impulse systems are described by a control system governing the continuous evolution of the state between two impulsions, and a reset map resetting new initial conditions whenever the state enters the domain of the reset map.

An evolution governed by an impulse dynamical system, called a run in the control literature, is defined by a sequence of cadences (periods between two consecutive impulse times), of reinitialized states and of motives describing the "continuous" evolution along a given cadence, the value of a motive at the end of a cadence being reset as the next reinitialized state of the next cadence.

Given an impulse system, one can characterize the map providing both the next cadence and the next reinitialized state without computing the impulse system, as a set-
valued solution of a system of partial differential inclusions. It provides a "summary" of
the behavior of the impulse system from which one can then reconstitute the evolutions
of the continuous part of the run by solving the motives of the run that are the solutions
to the dynamical system starting at a given reinitialized state.

A cadenced run is defined by constant cadence, initial state and motive, where the
value at the end of the cadence is reset at the same reinitialized state. It plays the role of
"discontinuous" periodic solutions of a control system.

We prove in Aubin & Haddad [16] that if the sequence of reinitialized states of a run
converges to some state, then the run converges to a cadenced run starting from this state,
and that, under convexity assumptions, that a cadenced run does exist.

4.6. MUTATIONAL EQUATIONS GOVERNING THE EVOLUTION OF THE CONSTRAINED SETS

Alternatively, if the viability constraints can evolve, another way to resolve a viability
crisis is to relax the constraints so that the state of the system remains inside the new
viability set. For that purpose, kind of differential equation governing the evolution of
subsets, called mutational equations, have been designed. This requires an adequate def-
inition of the velocity $\overset{\circ}{K}(t)$ of a "tube" $t \rightsquigarrow K(t)$, called mutation, that makes sense
and allows us to prove results analogous to the ones obtained in the domain of differential
equations. This can be done, but cannot be described in few lines.

Hence the viability problem amounts to find evolutions of both the state $x(t)$ and the
subset $K(t)$ to the system

$$\begin{cases} i) \quad x'(t) = f(x(t), K(t)) \text{ (differential equation)} \\ ii) \quad \overset{\circ}{K}(t) \ni m(x(t), K(t)) \text{ (mutational equation)} \end{cases} \tag{11}$$

viable in the sense that for every t, $x(t) \in K(t)$. For more details, see Aubin [6].

5. Designing Regulons

When an arbitrary subset is not viable under a "intrinsic" system $x'(t) = f(x(t))$, the
question arises to modify the dynamics by introducing regulons and designing feedbacks
so that the constrained subset K becomes viable under the new system. Using the above
results and characterizations, one can design several mechanisms. We just describe three
of them, that are described in more details in Aubin [3]:

5.1. VIABILITY MULTIPLIERS

If the constrained set K is of the form

$$K := \{x \in X \text{ such that } h(x) \in M\}$$

where $h : X \mapsto Z := \mathbf{R}^m$ and $M \subset Z$, we regard elements $u \in Z$ as viability multipliers,
since they play a role analogues to Lagrange multipliers in optimization under constraints.

They are candidates to the role of regulons regulating such constraints. Indeed, we can prove that K is viable under the control system

$$x'_j(t) = f_j(x(t)) + \sum_{k=1}^{m} \frac{\partial h_k(x(t))}{\partial x_j} u_k(t)$$

in the same way than the minimization of a function $x \mapsto J(x)$ over a constrained set K is equivalent to the minimization without constraints of the function

$$x \mapsto J(x) + \sum_{k=1}^{m} \frac{\partial h_k(x)}{\partial x_j} u_k$$

for an adequate Lagrange multiplier $u \in Z$.

5.2. CONNECTION MATRICES

Instead of introducing viability multipliers, we can use a connection matrix $W \in \mathcal{L}(X, X)$ as in neural networks (see Aubin [2] for instance). We replace the intrinsic system $x' = \mathbf{I}f(x)$ (where $\mathbf{I}$ denotes the identity) by the system

$$x'(t) = W(t)f(x(t))$$

and choose the connection matrices $W(t)$ in such a way that the solutions of the above system are viable in K.

The evolution of the state no longer derives from intrinsic dynamical laws valid in the absence of constraints, but requires some "self-organization" — described by connection matrices — that evolves together with the state of the system in order to adapt to the viability constraints, the velocity of connection matrices describing the concept of "emergence". The evolution law of both the state and the connection matrix results from the confrontation of the intrinsic dynamics to the viability constraints.

One can prove that the regulation by viability multipliers u is a particular case of the regulation by connection matrices W: We associate with x and u the matrix W the matrix the entries of which are equal to

$$w_{i,j} := -\frac{f_i(x)}{\|f(x)\|^2} \sum_{k=1}^{m} \frac{\partial h_k(x(t))}{\partial x_j} u_k(t) \text{ if } i \neq j$$

and, of $j = i$,

$$w_{i,i} := 1 - \frac{f_i(x)}{\|f(x)\|^2} \sum_{k=1}^{m} \frac{\partial h_k(x(t))}{\partial x_i} u_k(t)$$

The converse is false in general. However, if we introduce the connectionist complexity index $\|\mathbf{I} - W\|$, one can prove that *the viable evolutions governed with connection matrices minimizing at each instant the connectionist complexity index are actually governed by the viability multipliers with minimal norms.*

The concept of heavy evolution when the regulon is a connection matrix amounts to minimize the norm of the velocity $W'(t)$ of the connection matrix $W(t)$ starting from the identity matrix, that can be used as measure of dynamical connectionist complexity. Such a velocity could encapsulate the concept of "emergence" in the systems theory literature. The connection matrix remains constant — without emergence — as long as the viability of the system is not at stakes, and evolves as slowly as possible otherwise.

5.3. HIERARCHICAL ORGANIZATION

One can also design dynamic feedbacks for obeying constraints of the form

$$\forall t \geq 0, \quad W^{m-1}(t) \cdots W^{j}(t) \ldots W^{0}(t)x(t) \in M \subset Y$$

is satisfied at each instant. Such constraints can be regarded as describing a sequence of m planning procedures.

Introducing at each "level of such a hierarchical organization" $x_i(t) := W^i(t)x_{i-1}(t)$, on can design dynamical systems modifying the evolution of the intermediate states $x_i(t)$ governed

$$x'_j(t) = g_j(x_j(t))$$

and the entries of the matrices $W^i(t)$ by

$$w^{j'}_{k,l}(t) = e^j_{k,l}(W^j(t))$$

Using *viability multipliers*, one can prove that dynamical systems of the form

$$\begin{cases} (1)^0 & x'_0(t) = g_0(x_0(t)) - W^{0*}(t)p_1(t)(j = 0) \\ (1)^j & x'_j(t) = g_j(x_j(t)) + p_j(t) - W^{j*}(t)p_{j+1}(t) \\ & (j = 1, \ldots, m-1) \\ (1)^m & x'_m(t) = g_m(x_m(t)) + p_m(t) \quad (j = m) \\ (2)^j_{(k,l)} & wj'_{k,l}(t) = e^j_{k,l}e(W^j(t)) - x_{jk}(t)p_{(j+1)l}(t) \\ & (j = 0, \ldots, m-1, \; k = 1, \ldots, n, \; l = 1, \ldots m) \end{cases}$$

govern viable solutions. Here, *the viability multipliers p_j are used as messages to both modify the dynamics of the jth level state $x_j(t)$ and to link to consecutive levels $j+1$ and j.*

Furthermore, the connection matrices evolve in a Hebbian way, since the correction of the velocity $W j'_{k,l}$ of the entry is the product of the kth component of the j-level intermediate state x_j and the lth component of the $(j+1)$-level viability multiplier p_{j+1}.

5.4. EVOLUTION OF THE ARCHITECTURE OF A NETWORK

This hierarchical organization is a particular case of a network. Indeed, the simplest general form of coordination is to require that a relation between actions of the form $g(A(x_1, \ldots, x_n)) \in M$ must be satisfied. Here $A : \prod_{i=1}^{n} X_i \mapsto Y$ is a connectionist operator relating the individual actions in a collective way. Here $M \subset Y$ is the subset of the resource space Y and g is a map, regarded as a propagation map.

We shall study this coordination problem in a dynamic environment, by allowing actions $x(t)$ and connectionist operators $A(t)$ to evolve according to dynamical systems we shall construct later. In this case, the coordination problem takes the form

$$\forall t \geq 0, \quad g(A(t)(x_1(t), \ldots, x_n(t))) \in M$$

However, in the fields of motivation under investigation, the number n of variables may be very large. Even though the connectionist operators $A(t)$ defining the "architecture" of the network are allowed to operate *a priori* on all variables $x_i(t)$, they actually operate at each instant t on a coalition $S(t) \subset N := \{1, \ldots, n\}$ of such variables, varying naturally with time according to the nature of the coordination problem.

Therefore, our coordination problem in a dynamic environment involves the evolution

1. of actions $x(t) := (x_1(t), \ldots, x_n(t)) \in \prod_{i=1}^{n} X_i$,
2. of connectionist operators $A_{S(t)}(t) : \prod_{i=1}^{n} X_i \mapsto Y$,
3. acting on coalitions $S(t) \subset N := \{1, \ldots, n\}$ of the n actors

and requires that

$$\forall t \geq 0, \quad g\left(\{A_S(t)(x(t))\}_{S \subset N}\right) \in M$$

where $g : \prod_{S \subset N} Y_S \mapsto Y$.

The question we raise is the following. Assume that we may know the intrinsic laws of evolution of the variables x_i (independently of the constraints), of the connectionist operator $A_S(t)$ and of the coalitions $S(t)$, there is no reason why collective constraints defining the above architecture are viable under these dynamics, i.e, satisfied at each instant.

One may be able, with a lot of ingeniosity and the intimate knowledge of a given problem, and for "simple constraints", to derive dynamics under which the constraints are viable.

However, we can investigate whether there is a kind of mathematical factory providing classes of dynamics "correcting" the initial (intrinsic) ones in such a way that the viability of the constraints is guaranteed. One way to achieve this aim is to use the concept of "viability multipliers" $q(t)$ ranging over the dual Y^* of the resource space Y that can be used as "controls" involved for modifying the initial dynamics.

This may allow us to provide an explanation of the formation and the evolution of the architecture of the network and of the active coalitions as well as the evolution of the actions themselves.

In order to tackle mathematically this problem, we shall

1. restrict the connectionist operators to be multiaffine, and thus, involve tensor products,
2. next, allow coalitions S to become fuzzy coalitions so that they can evolve continuously.

Fuzzy coalitions $\chi = (\chi_1, \ldots, \chi_n)$ are defined by memberships $\chi_i \in [0, 1]$ between 0 and 1, instead of being equal to either 0 or 1 as in the case of usual coalitions. The membership $\gamma_S(\chi) := \prod_{i \in S} \chi_i$ is by definition the product of the memberships of the members $i \in S$ of the coalitions. Using fuzzy coalitions allows us to defined their velocities and study their evolution.

The viability multipliers $q(t) \in Y^*$ can be regarded as regulons, i.e., regulation controls or parameters, or virtual prices in the language of economists. They are chosen adequately at each instant in order that the viability constraints describing the network can be satisfied at each instant, and the main theorem of this paper guarantees that it is possible. Another one tells us how to choose at each instant such regulons (the regulation law).

For each actor i, the velocities $x_i'(t)$ of the state and the velocities $\chi_i'(t)$ of its membership in the fuzzy coalition $\chi(t)$ are corrected by subtracting

1. the sum over all coalitions S to which he belongs of adequate functions weighted by the membership $\gamma_S(\chi(t))$:
2. the sum over all coalitions S to which he belongs of the costs of the constraints associated with connectionist tensor A_S of the coalition S weighted by the membership $\gamma_{S\backslash i}(\chi(t))$. This type of dynamics describes a *panurgean effect*. The (algebraic) increase of actor i's membership in the fuzzy coalition aggregates over all coalitions to which he belongs the cost of their constraints weighted by the products of memberships of the actors of the coalition other than him.

As for the correction of the velocities of the connectionist tensors A_S, their correction is a weighted "multi-Hebbian" rule: for each component of the connectionist tensor, the correction term is the product of the membership $\gamma S(\chi(t))$ of the coalition S, of the components $x_{i_k}(t)$ and of the component $q^j(t)$ of the regulon.

In other words, the viability multipliers appear in the regulation of the multiaffine connectionist operators under the form of tensor products, implementing the Hebbian rule for affine constraints (see Aubin [2])), and "multi-Hebbian" rules for the multiaffine ones, as in Aubin & Burnod [11].

Even though viability multipliers do not provide all the dynamics under which a constrained set is viable, they provide classes of them exhibiting interesting structures that deserve to be investigated and tested in concrete situations.

Remark: Learning Laws and Supply and Demand Law — It is curious that both the standard supply and demand law, known as the Walrasian tâtonnement process, in economics and the Hebbian learning law in cognitive sciences were the starting points of the Walras General Equilibrium Theory and Neural Networks. In both theories, this choice of putting such adaptation laws as a prerequisite led to the same cul de sacs. Starting instead from dynamic laws of agents, viability theory provides "dedicated adaptation laws", so to speak, as the conclusion of the theory instead as the primitive feature. In both cases, the point is to maintain the viability of the system, that allocation of scarce commodities satisfy the scarcity constraints in economics, that the viability of the neural network is maintained in the congitive sciences. For neural networks, this approach provides learning rules that possess the features meeting the Hebbian criterion. For the general networks studied here, these features are still satisfied in spirit. □

We refer to Aubin [10] for more details on this topic.

6. Viability and Optimality

Interestingly enough, viability theory implies the dynamical programming approach for optimal control.

Denote by $S(x)$ the set of pairs $(x(\cdot), u(\cdot))$ solutions to the control problem (1) starting from x at time 0. We consider the minimization problem

$$
\begin{cases}
V(T, x) = \inf_{(x(\cdot), u(\cdot)) \in S(x)} \inf_{t \in [0, T]} \\
\left(c(T - t, x(t)) + \int_0^t l(x(\tau), u(\tau)) d\tau \right)
\end{cases}
$$

where c and l are cost functions. We can prove that the graph of the value function $(T, x) \mapsto V(T, x)$ of this optimal control problem is the capture basin of the graph of the cost function c under an auxiliary system involving (f, U) and the cost function l. The regulation map of this auxiliary system provides the optimal solutions and the tangential conditions furnish Hamilton-Jacobi-Bellman equations of which the value function is the solution (see for instance Frankowska [25, 26, 27, 28] and more recently, Aubin [9]). This is a very general method covering numerous other dynamic optimization problems.

However, contrary to optimal control theory, viability theory does not require any single decision-maker (or actor, or player) to "guide" the system by optimizing an intertemporal optimality criterion[4].

Furthermore, the choice (even conditional) of the controls is not made once and for all at some initial time, but they can be changed at each instant so as to take into account possible modifications of the environment of the system, allowing therefore for adaptation to viability constraints.

Finally, by not appealing to intertemporal criteria, viability theory does not require any knowledge of the future[5] (even of a stochastic nature.) This is of particular importance when experimentation[6] is not possible or when the phenomenon under study is not periodic. For example, in biological evolution as well as in economics and in the other systems we shall investigate, the dynamics of the system disappear and cannot be recreated.

Hence, forecasting or prediction of the future are not the issues which we shall address in viability theory.

However, the conclusions of the theorems allow us to reduce the choice of possible evolutions, or to single out impossible future events, or to provide explanation of some behaviors which do not fit any reasonable optimality criterion.

Therefore, instead of using intertemporal optimization[7] that involves the future, viability theory provides selection procedures of viable evolutions obeying, at each instant, state constraints which depend upon the present or the past. (This does not exclude antic-

---

[4] the choice of which is open to question even in static models, even when multicriteria or several decision makers are involved in the model.

[5] Most systems we investigate do involve myopic behavior; while they cannot take into account the future, they are certainly constrained by the past.

[6] Experimentation, by assuming that the evolution of the state of the system starting from a given initial state for a same period of time will be the same whatever the initial time, allows one to translate the time interval back and forth, and, thus, to "know" the future evolution of the system.

[7] which can be traced back to Sumerian mythology which is at the origin of Genesis: one Decision-Maker, deciding what is good and bad and choosing the best (fortunately, on an intertemporal basis, thus wisely postponing to eternity the verification of optimality), knowing the future, and having taken the optimal decisions, well, during one week...

ipations, which are extrapolations of past evolutions, constraining in the last analysis the evolution of the system to be a function of its history.)

References

1. Aubin J.-P. (1991) *Viability Theory* (second edition in preparation) Birkhäuser, Boston, Basel, Berlin.
2. Aubin J.-P. (1996) *Neural Networks and Qualitative Physics: A Viability Approach*, Cambridge University Press.
3. Aubin J.-P. (1997) *Dynamic Economic Theory: A Viability Approach*, Springer-Verlag (second edition in preparation).
4. Aubin J.-P. (1998) Minimal Complexity and Maximal Decentralization, in Beckmann H.J., Johansson B., Snickars F. & Thord D. (eds.), *Knowledge and Information in a Dynamic Economy*, Springer, 83-104.
5. Aubin J.-P. (1998) *Optima and Equilibria*, Springer-Verlag (second edition).
6. Aubin J.-P. (1999) *Mutational and morphological analysis: tools for shape regulation and morphogenesis*, Birkhäuser.
7. Aubin J.-P. (1999) *Impulse Differential Inclusions and Hybrid Systems: A Viability Approach*, Lecture Notes, University of California at Berkeley.
8. Aubin J.-P. (2000) *Applied functional analysis*, Wiley-Interscience
9. Aubin J.-P. (2001) *A Concise Introduction to Viability Theory, Optimal Control and Robotics*, cours DEA MVA, Ecole Normale Supérieure de Cachan.
10. Aubin J.-P. (2001) Regulation of the Evolution of the Architecture of a Network by Connectionist Tensors Operating on Coalitions of Actors, preprint.
11. Aubin J.-P. & Burnod Y. (1998) *Hebbian Learning in Neural Networks with Gates*, Cahiers du Centre de Recherche Viabilité, Jeux, Contrôle # 981.
12. Aubin J.-P. & Cellina A. (1984) *Differential inclusions*, Grundlehren der math. Wiss. # 264, Springer-Verlag.
13. Aubin J.-P. & Dordan O. (1996) Fuzzy Systems, Viability Theory and Toll Sets, in Hung Nguyen (ed.), *Handbook of Fuzzy Systems, Modeling and Control*, Kluwer, 461–488.
14. Aubin J.-P. & Dordan O. (2001) Dynamical Qualitative Analysis of Evolutionary Systems, *Proceedings of the ECC 2001 Conference*.
15. Aubin J.-P. & Frankowska H. (1990) *Set-Valued Analysis*, Birkhäuser, (second edition in preparation).
16. Aubin J.-P. & Haddad G. (1991) Cadenced runs of impulse and hybrid control systems, *International Journal Robust and Nonlinear Control*.
17. Aubin J.-P. & Haddad G. (2001) Detectability under Impulse Differential Inclusions, *Proceedings of the ECC 2001 Conference*.
18. Aubin J.-P., Lygeros J., Quincampoix. M., Sastry S. & Seube N. (2001) Impulse Differential Inclusions: A Viability Approach to Hybrid Systems, *IEEE Transactions on Automatic Control*.
19. Aubin J.-P., Pujal D. & Saint-Pierre P. (2001) Dynamic Management of Portfolios with Transaction Costs under Contingent Uncertainty.
20. Bressloff P.C. & Coombes S. (1998) A dynamical theory of spike train transitions in networks of Integrate and Fire Oscillators, *SIAM J. Applied Analysis*, 98.
21. Cardaliaguet P., Quincampoix M. & Saint-Pierre P. (1999) Set-valued numerical methods for optimal control and differential games, in *Stochastic and differential games. Theory and numerical methods*, Annals of the International Society of Dynamical Games, Birkhäuser, 177–247.
22. Destexhe A. (1997) Conductance-based Integrate and Fire models, *Neural Computation*, 9, 503–514.
23. Dordan O. (1995) *Analyse qualitative*, Masson.
24. Doyen L. & Saint-Pierre P. (1997) Scale of viability and minimal time of crisis, *Set-Valued Analysis*, 5, 227–246.
25. Frankowska H. (1989) Optimal trajectories associated to a solution of tangent Hamilton-Jacobi equations, *Applied Mathematics and Optimization*, 19, 291–311.
26. Frankowska H. (1989) Hamilton-Jacobi equation: viscosity solutions and generalized gradients, *J. of Math. Analysis and Appl.* 141, 21–26.

27. Frankowska H. (1991) Lower semicontinuous solutions to Hamilton-Jacobi-Bellman equations, *Proceedings of 30th CDC Conference, IEEE, Brighton, December 11–13*.

28. Frankowska H. (1993) Lower semicontinuous solutions of Hamilton-Jacobi-Bellman equation, *SIAM J. on Control and Optimization*.

29. Meyer P.-A. (1996) Artificial life and the animat approach to Artificial Intelligence, in Boden (ed.), *Artificial Intelligence*, Academic Press, 325–354.

30. Quincampoix M. & Saint-Pierre P. (1998) An algorithm for viability kernels in Hölderian case: Approximation by discrete viability kernels, *J. Math. Syst. Est. and Control*, 8, 17–29.

31. Rockafellar R.T. & Wets R. (1997) *Variational Analysis*, Springer-Verlag.

32. Saint-Pierre P. (1994) Approximation of the viability kernel, *Applied Mathematics & Optimisation*, 29, 187–209.

33. Shaft (van der) A. & Schumacher H. (1999) *An introduction to hybrid dynamical systems*, Lecture Notes in Control, 251, Springer-Verlag.

34. Shimokawa T., Pakdaman K. & Sato S. (1999) Time-scale matching in the response of a leaky integrate-and-fire neuron model to periodic stimulation with additive noise, *Physical Review* E, 59, 3427–3443.

35. Shimokawa T., Pakdaman K. & Sato S. (1999) Coherence resonance in a noisy leaky integrate-and-fire model.

36. Shimokawa T., Pakdaman K., Takahata T., Tanabe S. & Sato S. (to appear) A first-passage-time analysis of the periodically forced noisy leaky integrate-and-fire model, *Biological Cybernetics*.

ARCHETYPAL DYNAMICS: AN APPROACH TO THE STUDY OF EMERGENCE

WILLIAM H. SULIS
Collective Intelligence Laboratory
McMaster University
Hamilton, Ontario, Canada

Abstract. This paper introduces the main concepts and constructs of archetypal dynamics, a formal approach to the study of emergence based upon the analysis of coherent, meaning-laden information flows within complex systems. The fundamental triad of archetypal dynamics consists of a semantic frame, which defines entities their meaningful interactions, realizations of this frame as systems, and interpretations of this frame as users and agents. A formal system of representation by tapestries and constructors is described. Emergence is the recognition that a given system serves as the realization of multiple independent semantic frames, a situation which can studied through archetypal dynamics.

1. Preamble

Developing systems offer significant challenges as well as opportunities to theorists and experimentalists alike. Whatever the discipline, one is faced with the fact that the basic constituents of any developing system will change over time and space in both their structural and functional relationships. This poses deep problems for any attempt to describe such development using formal theories, models, or simulations.

Formal methods have generally been applied to invariant simple systems, whose components, and structural and functional relationships are stationary in time. In principle, for such systems one can construct a phase space representing the totality of states of these individual components and describe the structural and functional relationships among them by means of equations that can be solved exactly, approximated, or simulated.

The phase space approach does not work as well when dealing with complex, developing systems. The lack of structural and functional stationarity implies that not only do the boundary conditions and dynamical equations evolve over time, but that the phase space itself evolves as well. The modeler is faced with describing both the dynamics within a particular phase space, as well as the dynamics between phase spaces, and the dynamics within a phase space will depend upon the dynamics between these spaces. To date there have been very few attempts to create a dynamics of phase spaces, notably Isham [19] and Aubin [2].

Simulation provides one way around the problem of nonstationarity. Dynamic programming methods allow for the creation or destruction of components, and of structural and functional relationships. Simulation is limited, however, by resource restrictions, which limit the number of components that can be handled concurrently, and by limitations in our knowledge of component states and their dynamical dependencies and

J. Nation et al. (eds.), Formal Descriptions of Developing Systems, 185–228.

interactions. Moreover, there lurks the ever present problem of non-simulatability described by Rasmussen [22], which limits the general applicability of simulation. The conditions under which non-simulatability is likely to arise are exactly those that appear in the context of a developing system.

The study of developing systems entails deeper issues than the mere complexity of their descriptions. Three of the most critical are contextuality, contingency, and emergence. Most formal models assume that the locus of causation resides within the system under study, and therefore that the role of context is primarily to determine the choice of initial conditions. Cohen and Stewart [11] provide many examples, however, that demonstrate that context plays a crucial role in development. For example, gastrulation in frogs does not occur unless the developing embryo is situated within a flowing current that literally curls the frog embryo so as to create the gastrointestinal tract. The capacity for gastrulation might be genetically determined, but gastrulation will not occur unless the environmental context is exactly right. Implicit in the action of the frog DNA is the assumption that the appropriate environmental conditions will exist at exactly the proper time in the developmental process to permit gastrulation, otherwise the developmental process will run astray. Another example occurs among the cichlid fish of Lake Victoria, where the sex of a developing embryo is determined by the distribution of sexes in the population of cichlids in its immediate environment. It is not predetermined by its genetic structure. In both of these cases, an understanding of the processes underlying development will be incomplete unless contextual factors are also taken into account. Effective models of development must be able to incorporate both local as well as global features.

Contingency also plays a prominent role in development. Many models treat development as if it were a deterministic process. In reality, biological, psychological and socio-economic development occur within a stochastic environment. The subsequent development of a system may depend critically upon what has occurred previously, and these critical events may occur more or less at random. This is certainly true of evolution and of psychological and socio-economic development where late features are built upon a foundation of previously acquired features. The stochasticity inherent in developmental processes can result in critical features appearing too early, or too late, thus altering the subsequent developmental trajectory. These contingent events generally manifest themselves locally, though they may have widespread effects subsequently. Development begins with reproduction, but as Cohen [10] points out, reproduction is not about gene selection and expression, it is about selective wastage. Organisms do not mate to produce the perfect offspring. Rather, organisms produce enormous numbers of latent or nascent offspring, almost all of which die, leaving only a handful of survivors. Chance and context play central roles in reproduction, not determinism or selection. Contingency also plays a preeminent role in driving adaptation and the creation of diversity, allowing development to become a creative process and not merely a reproductive process.

Emergence completes our triumvirate. The intuitive meaning of emergence is that of "a coming into view, or existence". Taken in this sense, emergence is a feature common to all developing systems. In the technical literature, emergence has become associated with the appearance of global order from local disorder, or as the appearance of global properties unforseen based upon an understanding of only local interactions. These formal concepts of emergence generally involve the relationship of behaviors or properties across levels, and some element of uncertainty, novelty, or surprise. Order and disorder, however, are relative concepts, whose interpretation depends very much upon which aspects of a system's behavior are most salient to the problem at hand. For example, in the study of transient induced global response stabilization [27] it has been found that for some systems, certain "random" appearing stimuli are better able to induce stable patterns of response than are "patterned or ordered" stimuli. Generally one expects to see order induce order and disorder induce disorder. Here, counter to intuition, disorder induces order and order induces disorder. Order, per se, is not a defining characteristic of emegence.

Some authors relate emergence to the appearance of surprising or unexpected behaviors which cannot be predicted on the basis of an understanding of the system. While it seems problematic to base the conception of emergence upon the idea of ignorance, there is value in recognizing that some complete descriptive systems fail to possess sufficient descriptive and explanatory power to capture all possible aspects of some systems. This seemingly paradoxical idea will be elaborated in a later section.

There is a subtle alternative view of emergence, which appears in the literature on emergent computation. Forrest [15] defines emergent computation as follows: 1) there exists a collection of agents, each following explicit instructions, 2) interactions among the agents form implicit global patterns macroscopically (epiphenomena), 3) there exists a natural interpretation of the epiphenomena as computation. This definition is of value because it shifts the focus away from the system per se and onto its context, and particularly upon the role played by information and semantics. Unfortunately it suffers from an implicit reductionist bias, referring to these global phenomena as epiphenomena, thus ascribing a greater significance (or indeed reality) to one level over another. This is unjustified. Considering neural representations of information [28], one observes neural activity giving rise to meaningful macroscopic patterns of activity, while simultaneously, global patterns of activity induced through sensory registration impose changes at the neural level through the induction of receptor modification (for example, long term potentiation). These global patterns are significant because they are efficacious, possessing the capacity to induce physical changes at the neuronal level. Simultaneously, they possess the capacity to inform and to influence action at the psychological level. Their efficacy gives them a causal status and justifies their being considered as true phenomena in their own right.

The purpose of this paper is to propose a mathematical framework within which to situate these problems of globality, locality, contingency and contextuality for the purpose of constructing and analyzing formal models of developing systems. In particular, it is meant as a tool to help elucidate the phenomenon of emergence, which, compelling though it may be as a formal construct, still lacks a formal theory. It is centered around notions of efficacious information and semantics, and the organization of these through the concept of the semantic frame.

This framework is presented in the same spirit as category theory, a branch of mathematics developed during the 1930's that attempted to develop a common language for describing mathematical objects and constructions. Initially dismissed as being merely descriptive, category theory has since developed into a field of study in its own right, facilitating rapid advances in many areas of mathematics. Its power stems from its universality, permitting results in one area to be quickly applied to other areas. The existence of various forms of mathematical constructions can be quickly proven once certain categorical features of the models under study have been identified. These constructions can subsequently be utilized even if they cannot be explicitly constructed, providing a powerful tool for furthering research. The framework to be described below is presented with the same goal in mind - to provide a common language and a formal framework in which universal models of emergence can be studied and their insights subsequently applied to specific situations.

2. The Philosophical Framework

2.1. EMERGENCE

The examples of emergence that are commonly cited in the literature tend to be drawn from mathematics and physics. These examples are notable for their striking clarity but suffer from their tendency to bias thinking about emergence in favor of physical constructs. Indeed, virutally all examples of emergence that appear in the literature illustrate one form of emergence only, that of *vertical emergence*. This type of emergence arises across distinct ontological levels, usually separated by a change of scale, generally of space though this may involve time as well. The emergent phenomena to be explained are thought of as arising from averaging or mixing processes, or from the breaking of dynamical symmetries, as in apparent low level disorder giving rise to higher level order. Excessive attention is paid to the formal or mathematical impact of a change of scale on the system dynamics. Little or no attention is paid to the semantic effects of such scale changes, yet as we shall see, these are fundamental to understanding the nature of emergence.

The game of Life provides an example of emergence within a mathematical setting. Life, invented by the mathematician John H. Conway [12], is played out on a two dimensional grid. Each square of the grid may be colored black or white. A neighborhood of a square consists of the eight squares immediately adjacent to it. During each play of the game, the colors of all of the squares are simultaneously changed according to the following rules. If the square to be changed is black, and if the total number of white squares in its neighborhood is exactly three, then it becomes white, otherwise it remains black. If the square to be changed is white, and if the total number of white squares in its neighborhood is either two or three, then it remains white, otherwise it becomes black. An initial configuration of black and white squares is selected and the game consists of successively applying the rules to the grid, watching the unfolding of the resulting patterns. Understood in this manner, Life consists of a collection of squares whose individual states are updated according to a prescribed deterministic set of rules, and our only option for interacting with the game lies in our choice of the initial configuration.

In the Seventies, Conway and his colleagues discovered something altogether remarkable. They found an initial condition whose subsequent evolution under the rules of the game never repeated itself. This initial condition consisted of a specific local configuration of colors which formed a pattern termed a "glider". As the rules were successively applied to this initial condition, they observed that the neighboring squares underwent a series of color changes until a new local pattern was finally created whose form was identical to that of the original glider, but which was displaced by one square from its original position on the grid. They realized that this pattern could itself be interpreted as an entity in its own right, a glider. Moreover, they could interpret its behavior under the automaton rules as its becoming displaced or moving one square adjacent to its original position.

Later, they discovered another initial configuration termed a "glider gun" whose subsequent evolution could be interpreted as the glider gun producing a never ending stream of gliders which propagated along a straight line away from the gun. Another configuration termed an "eater" was found which appeared to destroy any gliders that interacted locally with it. New configurations were found which could absorb a stream of gliders moving along one line and emit a new stream of gliders moving along a different line. By suitably arranging guns and eaters and the directions and timing of these streams of gliders, they showed that it was possible to simulate the flow of bits within a digital computer. Certain stable patterns could be interpreted as logical gates, the gliders as bits, and the evolution of the overall pattern as the carrying out of a computation. The game of life thus exhibits emergent behavior that meets Forrest's definition of emergent computation.

It is essential to understand that even though the dynamics of the patterns that gives rise to the computer simulation is an example of emergence behavior, this behavior does not simply "arise out of" the behavior of the underlying automaton cells. Indeed, a simple examination of the behavior of Life under random initial conditions reveals that the appearance of this particular form of emergent behavior is rare indeed. In order to obtain a meaningful computation from the game of Life one must interact with it in a specific manner, ascribing precise meanings to particular configurations of colors and patterns of interaction. The conditions of the computational problem must be realized in terms of specific initial conditions of the game that can be meaningfully interpreted as providing the initial data and configuration of the computational device. It is necessary to specify certain final conditions which indicate the termination of the computation. It is necessary that meaningful patterns of colors, interacting coherently according to meaningful rules, do so in a manner which is consistent with the meanings ascribed to them, and that the final pattern obtained can be meaningfully interpreted as the outcome of the computation. At any step, a loss of coherence in the dynamics of the interacting pattern elements and of consistency with the interpretation of such interactions as information transformations will result in a failure of the simulation.

We see therefore that the game of Life admits at least two distinct ways of understanding, interpreting, and interacting. In the first, the squares, their color states and their subsequent temporal evolution form the objects of interest. In the second, the squares form merely a backdrop, while the objects of interest are color patterns, forming the gliders and logic gates that evolve according to their own rules in a consistent manner, provided that we now interact with the game in a manner which is consistent with this second interpretive scheme. If we do not interact with the game in a manner which is consistent with its interpretation as a computational system, then the meaningful coherence among all of the various entities may be lost, and in general it will not be possible to interpret the final configuration of the game as the outcome of a calculation. We cannot, for example, adjust the states of squares at random, as we could within the first interpretation. The persistence of the emergent phenomenon requires the cooperation of the environment. It is not an intrinsic feature of the game of Life but a co-creation of the game and of its environment.

The emergence of mental phenomena from neural interactions provides a more psychologically accessible example. The brain consists of approximately 30 billion neurons and 100 billion glial cells. The neurons, connected to one another through various forms of synapses, are arranged in a complex network. The glial cells provide the structural support that maintains the physical integrity of the network. The function of the brain is often described by analogy to the circuitry of a computer. Neurons are viewed as conduits for signals, the synapses as swithcing elements. The reality is vastly more complicated. Each neuron is a complex dynamical entity in itself, whose dynamics depends not only on the states of their various receptors but on circulating neurohumoral factors as well. Unlike a computer, the nervous system is not hardwired but rather is capable of functional rewiring [18] depending upon the environmental context. There is no direct representation of perceptual reality by neurons. Rather, the patterns of neural activity in response to environmental stimuli depend upon immediate context and prior history [28]. Russian psychologists have long recognized the involement of the body as a whole in mental phenomena through complex feedback relationships [1]. The environment per se has been recognized to play a role in mental phenomena, especially memory [25].

Though there is little doubt that the existence of mind requires the existence of an intact neuropil, the latter is not sufficent for the expression of mind. Like a cell in the game of Life, a single neuron does not carry out specific tasks relevant to the higher order expression of mind. The dynamics of mind cannot be reduced to the dynamics of neurons. The dynamics of an individual neuron can be fully understood in terms of the dynamics of its membrane, the embedded receptors, and the environment of circulating neurohumoral factors. Such dynamics appear to be capable of mathematical description. A specification of all of these for the network as a whole would enable us to describe the observed dynamics, and in principle the entire network could be described mathematically. Nevertheless we would not have captured a description of the mental phenomena being expressed.

Mental phenomena possess ontological properties that are wholly distinct from those of the neurons supporting them. Neurons can be localized in space and time. Neurons have quantifiably measurable properties, such as membrane potentials, electrolyte concentrations, glucose concentrations, thermal production, and so on. Mental phenomena possess none of these. Mental phenomena arise from the interactions between entire networks of neurons and their environments [27, 16]. They involve global patterns of neural activity, not the activity of individual neurons. Like the case of Life, the observation of mental phenomena requires a change in perception from that appropriate to neurons. It requires the existence of a frame of reference in which mental phenomena are recognized, defined, and given meaning, in which their interactions can be recognized and given meaning, and in which our interactions with such phenomena are defined and given meaning.

Examples of vertical emergence appear everywhere in life. The existence of a single celled organism as an expression of the dynamics of its constituent biomolecules is readily appreciated. Likewise, the existence of a multicellular organism as the expression of the collective dynamics of its individual cells is familiar to us all. Societies and cultures exist through the collective actions of the individuals within them.

It is essential to note that in all of these cases, the emergent phenomenon is not merely "contructed out of" or "reducible to" the lower level phenomena. The existence of the emergent phenomena depends upon the existence of factors that lie external to the system per se. The emergent pheneomenon consists of distinct entities, having distinct properties and characteristics, having distinct forms of dynamics, and distinct patterns of interactions among themselves. Our understanding of and the meanings that we attach to these two classes of entities are quite distinct. One class cannot conceptually be understood as merely being entailed in some fashion by the other. The relationships that we form with the entities of each class are fundamentally distinct from one another. The ability to recognize and to interact with the emergent phenomena requires, in all cases, the existence of a new conceptual framework, within which the emergent phenomena are given definition and meaning. The phenomenon and its emergent phenomenon constitute ontologically distinct classes of entities, distinguished by more than mere ontological level, or scale.

This aspect of emergence becomes more apparent once we begin to consider forms of emergence other than vertical emergence. A second category of emergence which typifies emergence in the social sciences is that which I have termed *horizontal emergence*. In horizontal emergence, one observes emergent phenomena arising out of and residing at the same ontological level as the underlying phenomena.

The emergence of culture provides an example of horizontal emergence familiar to us all. Although it is possible in theory for a new culture to emerge de novo from the random self organization of a disparate group of previously unrelated individuals, it is much more likely that a novel culture will emerge over time from a preexisting culture. Individuals within a preexisting culture form a subculture which gradually separates out from the original culture. This new culture may so develop as to become completely foreign to members of the original culture. This is especially true if this development takes place over a sufficiently long period of time so that no individuals possessing living memories of the transitional process remain alive. The referents necessary to understand the point of view of an individual from a different culture may be therefore be lacking from personal experience, resulting in a state of incomprehension when confronted with such a person. It is possible for individuals born into these two distinct cultures to develop such differences in attitudes, values, rituals, and goals that encounters between them may be accompanied by feelings of unfamiliarity, suspicion, derision, xenophobia and prejudice. Such feelings often underlie the overt excuses that give rise to cultural clashes, and ultimately to war.

The emergence of cultures results in the appearance of new meanings to be assigned to individuals, roles, and events, and new ways of interacting with other individuals and with the environment. Culture provides us with shared ways of thinking and of feeling, of ascribing meaning to that which happens around us, and prescribes for us acceptable ways for interacting with our fellows. These distinct points of view serve to distinguish one culture from another.

Within a particular culture, social roles provide another example of horizontal emergence. Social roles come associated with particular attitudes, and expectations, both in terms of how the person manifesting a role behaves towards us, as well as how we are to behave towards them. Social roles evolve as environmental, cultural and economic conditions evolve. Although social roles sometimes exhibit a hierarchical organization within a culture, this is not essential to their nature, and social roles emerge at more or less the same ontological level within a culture, which is typical of horizontal emergence. Individuals within a culture frequently manifest multiple social roles, sometimes in distinct times and places, but sometimes simultaneously, and confusions concerning such roles frequently occur, resulting in misunderstandings and interpersonal difficulties.

Other examples of horizontal emergence include the diversification of languages, in which the necessity of specific frames of meaning is self evident, and speciation in evolution.

Horizontal emergence appears in physics as well, though it is not generally recognized as such. Perhaps the most fundamental example of horizontal emergence that appears in physics is that in which a hadron, such as a proton, arises out of interactions among a collective of quarks and gluons [23]. Here the change of scale is insignificant. What is most essential is a change of ontological status. Quarks and gluons, by their essential character, are unobservable particles forever trapped in bound states that constitute a hadron. Nevertheless, they possess physical characteristics that distinguish types from one another, and engage in distinct patterns of interaction among themselves, these interactions between quarks being mediated by gluons. As presently conceptualized quarks, through the intermediary gluons, engage in a complex dynamic dance, continually forming and reforming, arranging and rearranging. Out of this continual dynamical process there emerges a stable hadron such as a proton, with its own unique pattern of interactions with the other fundamental particles. These hadrons emerge out the quark-gluon interactions.

Quarks, being forever bound, are unobservable. Our understanding of them derives from inference and theory. No phenomenon occurs which can be interpreted as being due to a single quark, nor do we have any means at our disposal for interacting with a single quark. The hadrons on the other hand, though they are formed out of interactions among these quarks, are observable, and indeed do produce phenomena which can be interpreted as due to single hadrons, and we do possess the means to interact with these as such. The conceptual system through which we apprehend the concept of quark is subtly different from that through which we apprehend the concept of hadron, though they are linked.

The phenomenon of wave-particle duality [17] provides another example of horizontal emergence in physics. Classically, waves and particles form ontologically distinct modes of being, with distinct properties, dynamics, and formal descriptions. The fundamental constituents of matter may exhibit behavior that is consistent with either of these two modes, depending upon the particular context that the entity finds itself within. Conceptually, each such mode requires a distinct frame of meaning to apprehend it, though the formalism of quantum mechanics tends to obscure this.

Emergent phenomena may appear at the same or at different ontological levels from that of the supporting phenomena. A change of scale is not a critical feature of emergence. The efficacy of emergent phenomena shows that they are not merely epiphenomena but phenomena in their own right. Emergence cannot simply be dismissed as a misdirection of our critical attention. Surprise and ignorance are not critical features of emergence, since the phenomena persist long after our recognition of them. Nevertheless, emergent phenomena do transcend the description of the source phenomena. Emergent phenomena force us to step outside of the original system, and the frame work of definition and meaning surrounding it. Emergent phenomena require new definitions, the recognition of novel modes of behavior, of interaction, of being. The recognition of emergent phenomena requires the adoption of a novel conceptual system, a system with new entities, new dynamics.

2.2. SEMANTIC FRAMES

The conceptual system that I have been alluding to in the previous section cannot be a purely formal system, like a theory in mathematical logic. A theory works well enough when one is already handed a set and a collection of interactions and asked to describe them. But what is one to do when confronted with something truly novel? The phenomena being considered must be parsed into units, and how is this to take place without ascribing to them some minimal meaning, even if it no more than "this is interesting"? Relations must be established between these minimal units. These relations in turn come equipped with meaning. For example, to assert that a succession of basic units comprises the successive states of some entity requires a conception of said entity, and that requires attaching at least the meaning "this is a state of that entity" to each unit. Moreover, in order to describe interactions between entities, one must be able to assert that a basic unit represents "this" entity, and not "that" entity. Again, this requires attaching meaning to each basic unit.

In mathematics and physics, meaning is either eschewed completely, or else hidden in the background and subsequently ignored. Formal considerations are pre-eminent, while semantics is relegated to philosophers and linguists. Physics alone, however, does not provide the semantics needed to interact with living systems. It does not inform us that cats should not be confused with footballs, that food should not be confused with fuel, that a refrigerator magnet is not is equivalent to a CD ROM of the Beatles. The game of Life shows clearly that the physics, that is, the rules of the game, is insufficient to enable us to play the game of computation. Put another way, the syntax of the game does not suffice to enable us to play the game of compuation; we need the semantics as well.

The conceptual system that I am conceiving is termed a *semantic frame*. It is termed a frame because it is a conceptual organizing principle, one which parses phenomena into distinct entities, modes of being, modes of behaving, modes of acting and interacting, and ascribes meaning to all of these in a coherent and consistent manner. Moreover, this semantic frame shapes and guides our interactions with these phenomena so as to maintain this thread of meaning. I am particularly interested in complete semantic frames, those which offer complete descriptions of the phenomena at hand, regardless of whether or not they offer complete explanations of the phenomena.

The semantic frame provides answers, in whole or in part, to the six basic epistemic questions:

— Who are the relevant entities?
— What are their actions, reactions and interactions?
— Where does this all play out?
— When do these events occur?
— How do these events manifest themselves and function?
— Why do these events transpire?

The semantic frame is *not* a cognitive construct, though we may try to understand it by relating it to other similar ideas in cognitive psychology. The semantic frame is a much more general construct, and applies to non sentient entities just as much as to sentient ones. For emergent phenomena to exist, they must have some description within a semantic frame which gives meaning to the various entities and interactions. Morever, it is also necessary that the environment within which such phenomena appear be able to interact with these phenomena in a manner which is consistent with the semantic frame, thus preserving their integrity. This is true is the case of the game of Life. We must interact with Life in ways that preserve the emergent dynamics, otherwise it will be lost. This is true of frog gastrulation as well. The wrong flow, no frog. It is true of life generally. We generally do not do very well by cats if we confuse them with footballs. Their continued existence as living creatures requires that we understand such differences. That in turn, requires that our actions be meaningful and consistent with an understanding of what it means to be living, and what is necessary to sustain that living state.

The semantic frame is a primitive construct. It is a conceptual, ideational construct. As such, I do not believe that it is describable in formal terms. In particular I do not believe that it possesses a formal expression in mathematical terms. Unlike theories in logic, which are expressions of form and structure, the semantic frame governs action and interaction; it is much more a descriptor of process. Moreover it does not describe one process in particular, but a meaningful coherence of processes more generally. Like the concept of set, which is also a primitive construct, I expect that the construct can be conceptually motivated through experience, even if it cannot be fully rendered in formal terms, and that such motivation can be used to help guide our efforts to formalize aspects of the concept and to work with them in our research. It is possible to work with the concept of set even though it is not possible to express precisely what a set is. Formal attempts to do so have resulted in many different theories of set. The concept of set is simply greater than can be encompassed by any formal system. I believe that the same will prove true of the concept of the semantic frame, though I also believe that that will not hinder our ability to utilize it as a organizing principle. Indeed, the example of the concept of set is an example of a semantic frame.

The concept of the semantic frame extends to the study of emergence the notion of the archetype in psychology, hence the origin of the name *Archetypal Dynamics*. The notion of the archetype was conceived by the Swiss psychologist Carl Jung at the turn of the Twentieth Century. Jung faced a similar problem is trying to describe the notion that he had in mind. He wrote:

> *The archetypal representations (images and ideas) mediated to us by the unconscious should not be confused with the archetype as such. They are very varied structures which all point back to one essentially "irrepresentable" basic form. The latter is characterized by certain formal elements and by certain fundamental meanings, although these can be grasped only approximately. The archetype as such is a psychoid factor that belongs, as it were, to the invisible, ultraviolet end of the psychic spectrum. It does not appear, in itself, to be capable of reaching consciousness* ([13] pg. 83).

Later, Jung writes:

> *We must, however, constantly bear in mind that what we mean by "archetypes" is itself irrepresentable but has effects which make visualizations of it possible, i.e. the archetypal images and ideas* ([13] pg. 84).

This aspect of semantic frames will be discussed in a later section.

Semantic frames abound in life. Outside of formal systems such as those appearing in science, philosophy or religion, we tend to take them for granted. Indeed, our brains are so skilled at accessing and utilizing a variety of distinct semantic frames simultaneously, we hardly recognize that we are doing so. Worse, we tend to confound these frames with one another, making it difficult if not impossible to see their distinctiveness and separateness. This ease of use is critical for our survival, since we may need to rapidly shift from one semantic frame to another as the salient features of a situation demand. However this facility leaves us with the impression that distinct semantic frames do not actually exist, but are merely ad hoc categorizations drawn upon as needed and discarded when not. Nothing could be further from the truth.

Although in practice we use multiple frames simultaneously, for purposes of understanding, analysis and clarity, it is necessary to exhibit these distinct frames in stark detail. Indeed, it was only through the device of imagining a universe without noisy elements that a sense of the deep mathematical symmetries underlying natural interactions was realized. Even though systems may not adhere to these symmetries in detail when occurring naturally, they nevertheless are influenced and constrained by their presence, resulting in behavior which may be noisy in the small but predictable in the large. This understanding was made possible by isolating certain aspects of reality and emphasizing their study over others. From this process has derived all of the great theories of mathematics and physics. The life and social sciences have proven to be more of a challenge because emergence abounds in these subjects, meaning the existence of multiple distinct semantic frames, and so it has proven difficult, if not impossible, to develop a theory based on a study of a single semantic frame. The coordinated and simultaneous study of multiple semantic frames is required to gain an understanding of these vastly more complex systems.

In what follows an emphasis will be placed upon the idea of a *complete* semantic frame, one which provides a complete description of particular class of phenomena, their actions and interactions, and our interactions with them, and ascribes a coherent and consistent set of meanings to these. A complete semantic frame is considered to be a self contained construct, not requiring reference to other constructs for meaning. These complete semantic frames may be conceived of as the building blocks for more elaborate frames upon which more complex entities, structures and processes are built. This requires an act of conceptual deconstruction on the part of the reader, who must pay close attention to when multiple conflicting frames are accessed simultaneously and then confounded in the mind as if they were a single frame, thus hiding their distinct identities. The reader shall come to see that the world reflects not a single overarching semantic frame, as hoped for by those adherents of reductionism, but instead reflects a multiplicity of irreducible, distinct semantic frames. The universe, metaphorically speaking, is less like a diamond, crystalline and pure, and more like an opal, amorphous, richly textured and hued, and subject to contextual effects.

2.3. INFORMATION

If, as I have suggested, it is not possible to describe a semantic frame directly, that it is not possible to say what a semantic frame is, and to formalize it in mathematical or linguistic terms, then what can be said? At first glance, the semantic frame would appear to be a useless or empty construct. But before passing judgment too early, let us shift our focus away from what a semantic *IS*, and consider what a semantic frame *DOES*. A semantic frame organizes our experience of reality into meaningful entities, actions, and interactions, and our interactions with these. What exactly is this experience of reality? For living systems, reality is experienced in two modes. First of all, there is the direct physical mode, involving the immediate exchange of physical energy and momentum through interactions with other organisms and physical entities. The response of the organism to such interactions is primarily reactive, without decision making or choice. Secondly, there is the semantic mode, in which each such interaction conveys meaning to the organism, which enables it to select its next actions. The response of the organism in this case is active, involving decision making and choice. Such meaning is generated internal to the organism, but it depends upon the expression of meaningful signals and signs by the other organisms with which it interacts. These signs contribute to the generation of meaning by the organism. The process by which this occurs is termed *information*.

Before exploring this approach to the concept of information, it is important to note that sentience is not required for the expression of a semantic frame; physical systems also respond to meaning as well. Consider again the wave-particle duality issue for fundamental entities. It is possible to construct a two path apparatus which enables a photon to exhibit either wave-like or particle-like behavior depending on the choice of a switch which selects out either one path or two path information [17]. This switch is placed at a sufficient distance from the entry of individual photons into the apparatus so that they have long passed the point of choosing which path or paths to have entered. Moreover the switch is activated at a time after which a photon is thought to have entered the apparatus. Regardless of which position the switch is placed in, the end result always conforms to the choice of the switch - wave-like for two path, particle-like for one path. Physicists accept that the behavior is subtly determined by the availability of information about the path choice. This information is somehow encoded in the physical structure of the apparatus, though it is not the specific form which is important, since any such two choice apparati will yield the same result. The two path experiment holds true so long as the meaning associated with the idea of a choice of path information is preserved, no matter how this is implemented in any given apparatus.

The usual approach to infromation is based upon the work of Shannon, who was concerned with the fidelity of coded information transfer along communication channels. The amount of information that could be accurately encoded and passed along a channel was a central issue, and this led to the idea of associating the amount of information with the reduction in uncertainty resulting from the organization of the signs. This in turn revealed close ties to the concept of entropy and as such, it has garnered wide acceepatance in the physical sciences. However, this work did not concern itself with the content or meaning of information per se, only the form. Living systems respond to meaning, not merely form. Indeed as the example above shows, so do physical systems.

There is a weak sense in which the reduction of uncertainty idea holds in the two path experiment. It can be argued that selecting for one path reduces the uncertainty associated with selecting for two paths and so contains more information. Yet both choices yield meaningful and useful knowledge about the system. The choices can be made at will, with equivalent energy costs to the system, so the quantity of the information has no physical consequence. The one path choice might be said to yield more information since it localizes the photon more, but this is not exactly the case. The fact that the choice can be made at any time after the photon enters the apparatus shows that in reality it is not possible to ascertain whether or not a photon actually traversed a particular path. The photon is not forced to follow a particular path, merely to exhibit a particular behavior at the detector. In either case, this results in a probability distribution of registration sites. Only the shape of the probability distribution changes, from unimodal to multimodal. Apparently, the only consequence of the quantity of the information is a change in shape in a probability distribution. But this quantity produces its effect only by virtue of its meaning. A similar two choice experiment involving photon polarization (either of two directions versus one direction) would yield either two narrow peaks or one narrow peak at the detector, quite different distributions from above. The two slit experiment applied to water for example would produce a different result. It is not the quantity of information that is critical here, rather, it is the meaning asosciated with it for the particular system.

The meaning attached to information is critical for determining the future behavior or actions of both living and non living systems. Moreover, the act of informing is, in a very real sense, an active rather than a passive act. It is dynamic rather than static, and thus process rather than form. To deal with this I needed a conception of information which moved away from a dependence merely upon form. Webster's dictionary defines information in two ways: as "data", i.e. as form, or as "the act of informing or the state of being informed", i.e. as process. In the latter definition, information becomes an active, rather than a passive construct. The verb "to inform" is defined as "to give a special quality or character, to imbue". To imbue is in turn defined as "to permeate or pervade, to instill", while "to instill" is defined as "to impart by gradual instruction or effort, to implant". This chain of associations suggests that information is to be understood as an efficacious agency that is capable of affecting the characteristics, and ultimately the behavior, of some other. Although information is here being given an active status, that does not mean that information is being given a causal status. Information may influence the course of events but it does not necessarily cause them to occur.

Information is generally considered to be a secondary aspect of reality. It is always assumed that there exists some thing out there for which the information is "about". If we are to have knowledge of some thing, then it is necessary that such a thing be able to inform us about such knowledge. If some thing in reality were unable to inform us we would never have knowledge of it. Even if we learn indirectly of the existence of some thing it is necessary for that thing to have informed some other thing prior to our being informed. An intermediary is required. In any case, to learn about some thing, some other thing must be informed. Thus we are led to the first principle of archetypal dynamics:

1) **The Existence Principle:***I am, therefore I inform.* In the section on emergence I made much of the fact that emergent phenomena, regardless of their physicality, were efficacious, in that they were able to influence the behavior and actions of other entities. The efficacy of such phenomena suggested to me that these were far from being epiphenomenal but deserved to be considered as true phenomena in their own right. In many, if not all cases of emergent phenomena, the phenomena in question possess the capacity to inform. This is true of Life, the game. It is true of life more generally. The capacity to inform automatically imparts to a phenomenon an efficacious aspect, and if efficacy is an attribute of existing entities, then the capacity to inform imparts existence to those entities which possess it. This argument leads to the second fundamental principle of archetypal dynamics:

2) **The Equivalence Principle:***I inform, therefore I am.* Taken together, these two principles assert that there is essentially no distinction between reality, and meaningful

effective information about that reality. Although such a view might seem extreme, it has been dominant in physics for much of the past century. Bohr and Wheeler have both asserted that no elementary phenomenon is a phenomenon until it is a registered phenomenon. In other words, a phenomenon cannot be held to exist unless it is has first informed about its existence. This suggests that an elementary phenomenon is a form of information. Moreover, it follows that there is no fundamental difference in kind between the varied phenomena of reality, only differences in form and organization. In a sense, this means that epistemology begets ontology. If this appears to be true of elementary physical phenomena, the presumed substrate of all physical reality, then it makes perfect sense to accept it for the non-physical aspects of reality as well.

Archetypal dynamics thus considers information as being the more fundamental construct, and concerns itself with the organization and flow of information that constitutes a system. Archetypal dynamics is concerned with information that is content laden, meaning laden, and effective, and the role that such information plays in the ontology, epistemology and dynamics of complex systems. In the ontology of archetypal dynamics, the most basic unit of reality is termed an *Informon*. An informon simply refers to any aspect of reality that possesses information, or better yet, any aspect of reality that possesses the capacity to inform. Archetypal dynamics concerns the myriad ways in which informons become linked together to form the phenomena that consitute reality. An informon is an aspect of reality that exists prior to any semantic frame. An informon must be attached to, coupled to, or in resonance with, a semantic frame in order to be given form, meaning, and behaviour.

This might seem strange at first glance. But consider an analogy. Suppose that one had in their possession two reels of film from two movies. The ideas of the movies to which these two reels belong constitute the semantic frames for these reels. They give meaning, coherence and consistency to the reels. Now suppose that each reel is unrolled and cut up into individual cells, and that the cells are then scattered on the floor. One is now faced with seemingly unrelated bits of experience. Each cell represents a single informon. Left as such one has only informational chaos. However, the existence of the ideas of the movies allows one to give meaning to each cell and to the relationships between cells, to organize the cells into coherent, consistent and meaningful patterns, thereby separating both cells and patterns according to the degree to which they correspond to aspects of each movie. Even if some cells were missing from each reel the ideas of the movies would allow them to be organized into a close facsimilie of a portion of the original film. Obviously this process works best if one has actual copies of the movies in hand, but one could proceed with copies of the scripts, or even just verbal descriptions of each movie. Thus one can proceed given either a complete realization of the semantic frame, i.e a copy of the movie, or merely an interpretation of the movie, i.e. the script or verbal description, or indeed a memory of its viewing. The realizations and interpretations constitute differing aspects of the semantic frame, which will be discussed in more detail in the next section.

Information and informons manifest in two ontologically and epistemologically distinct forms, *public* and *private*. In epistemology the distinction is frequently made between objective and subjective. These terms are best applied to conscious entities, whereas the public - private distinction holds across all ontological domains. Those familiar with computer simulations are well used to dealing with information that is specific to the internal dynamics of the system being modeled, and information that can be exchanged with other systems, whether modeled or real, such as the user of the simulation. Private information is information that is perceived by and unique to a single entity, whereas public information is information that is accessible by and shared among a collection of entities. The computer analogy might make one think that the distinction is merely one of perspective, exo versus endo, and that the difference between them is merely one of practice rather than principle. Current experience at the quantum level shows that this is not the case and that there are fundamental differences between public and private information. The existence of private information can be seen in the existence of quarks, which appear to be forever unobservable even though their influence is observable. It is seen in the influence of the quantum potential [8], providing the information needed to determine the choice of wave function in entanglement and delayed choice experiments [17]. It is seen in the influence of the so-called exchange of virtual particles in particle interactions [23]. In each of these cases, information is passed between members of the system under study, but that information cannot be made explicit, cannot be manifested independent of the system. Such information is efficacious, but is not publicly available. Indeed, the transfer of information at the quantum level seems to violate fundamental attributes of normally perceived reality, such as the special theory of relativity. *Private information cannot be directly observed, measured or signed. Nevertheless, private information can be hypothesized, represented, and simulated.*

Public information, on the other hand, *does* satisfy the usual constraints imposed upon it, and can be made explicit through observation and measurement. It is fair to say that public information comprises that which we commonly refer to as objective reality. Those readers knowledgeable in quantum mechanics might argue that what is being proposed is a "hidden variables" theory, and that such theories have been proven not to exist by virtue of the work of Von Neumann and Bell [5]. More precisely though, that line of research has only shown that local hidden variables theories cannot describe the quantum mechanical realm. Nonlocal hidden variables theories are permissible, and indeed, most likely essential, for understanding quantum phenomena. Moreover, the notion of information cannot be equated with that of variable or of state. A variable already involves a decomposition of an aspect of reality into more basic units or dimensions, and thus is a derived entity. The existence of information is logically prior to the existence of variables, and thus is a more fundamental notion. Information need not require variables for its manifestation. Indeed information must exhibit a particular organization in order that it admit a representation by means of a variable, and thus the absence of hidden variables does not force the absence of hidden information.

The concept of private information is important for understanding the organization of information flows in complex systems. If one likes, private information can be thought of as information that meaningfully influences the dynamics of a system but which is not realizable at the level of the system per se. It is information whose presence requires us to step outside of the framework within which the system is conceptualized and described. Examples would include the neurotransmitter hypothesis for depression, the gene theory of development and so. This relative distinction between public and private information, while less broad than the ontological distinction described above is nevertheless adequate for those who find themselves unwilling or unable to make the cognitive leap required of the full conception.

2.4. REALIZATION, INTERPRETATION AND REPRESENTATION

Semantic frames have been described as a primitive construct utilized to describe the organization of informons into meaningfully coherent patterns. They have been described as inherently non-formal and non-mathematical. How then are we to utilize the construct in any meaningful or useful manner? For the concept of a semantic frame to be useful it is necessary to be able to delimit it somehow. To accomplish this, it is first necessary to understand the ways in which real systems might relate to their semantic frames. Three different modes of relating come readily to mind. These are realization, interpretation, and representation.

First of all, in order for information to be of value it must be information about something. Similarly, a semantic frame, as an organizing principle for information, must too be about something. That something is termed a *realization*. A realization need not be real in the material sense, but it is necessary that a realization be efficacious, that is, it is able to make a meaningful difference. A realization of a semantic frame is termed a *system*.

A semantic frame may have many different realizations, just as a theory in mathematical logic may have many different models. Likewise, a system may realize several different semantic frames. A semantic frame whose realization is some system is termed an *external frame* for the system. The external frame tells us how to cut the system into meaningful units, and then provides a meaningful description of how those units relate to one another. Archetypal dynamics is particularly interested in complete external frames. These are frames that, in principle, provide a complete description of all possible relationships between the identified units.

It is important to note, however, that a system need not fully realize a complete semantic frame. Moreover, it is possible for a given system to realize a complete semantic frame throughout some aspects but not throughout all aspects. Although most isolated systems may fully realize a complete semantic frame, most systems in reality are not isolated, and systems that are in interaction with other systems may encounter situations that are not fully accounted for by one semantic frame and thus may experience shifts in behavior that render the original frame incomplete. Indeed, the appearance of learning or adaptation or development will create such a circumstance.

The connection between a system and its external frame is weaker than those considered in more traditional approaches. It is local in nature, and multiple semantic frames may be required to cover all possibilities for the system. The degree to which a system can be described by, or coupled to a particular semantic frame is termed its degree of *resonance or coupling*. A strong coupling or strong resonance implies that a greater proportion of its history can be linked to the semantic frame. Thus the stronger the resonance, the closer the frame comes to providing a global description of the system. Similarly, the weaker the resonance, the more the frame becomes an idiosyncratic or coincidental description.

These relationships can be better observed through a reconsideration of the game of Life. The basic definition of Life, involving cells, states, grids, and rules, together with an understanding of game and how the game is to be played, provide a complete external semantic frame for the game. This frame provides a complete description of the game, the possible moves, and how we are to relate to it. Any possible configuration of the game can be found working within this external frame. Life provides a complete realization of the external frame of the game of Life.

The computational frame also constitutes a complete semantic frame, providing a complete definition and description of all of the necessary components: bit streams, logic gates, their configurations and interactions so as to be able to carry out any computation, at least as the term is currently understood. Life, when initialized using patterns of gliders, eaters, glider guns, and so on, provides a complete realization of the computational frame, which now comprises an external semantic frame for Life. This relationship between Life and semantic frame does not hold for all possible configurations of Life, however, but only for those select configurations that preserve the meaningful connection. Indeed, the presence of other configurations on the Life grid would weaken the connection between Life and the computational frame. Moreover, our interacting with the game according to the prescriptions of the computational frame is also necessary to preserve the correspondence. Failure to do so would also weaken the coupling to the computational frame. A perfect correspondence provides a maximal resonance; a complete abscence, such as with a random initial configuration, would provide a minimal resonance.

Life is able to realize two distinct, complete external semantic frames. Indeed, one might surmise the existence of many more complete semantic frames with which Life could couple. This leads us to the following definition:

Definition: A *complex system* is a system that realizes multiple, independent, complete semantic frames.

Realization, however, is only one half of the dyad required to render a semantic frame efficacious. The other half of this dyad is *interpretation*. Realization actualizes the descriptive aspects of a semantic frame, providing a something for the semantic frame to organize. Although meaningful, realization does not provide meaning per se. A realization provides a stimulus to the creation of meaning, but meaning itself arises within an agency, which co-creates it and which utilizes it to inform and to direct subsequent action. Meaning arises when a semantic frame is interpreted. Interpretations occur in two subtly distinct forms. Given a semantic frame and a specific realization, an interpretation is termed a *user* if it lies external to the realization, and an *agent* if it lies internal to the realization. The semantic frame is then termed an *internal frame* for the users/agents.

Meaning provides choice, and meaning is needed to understand behavior whenever choice is possible. In a strictly deterministic system, meaning holds little relevance since there is no choice possible. All future actions are predetermined by any given present or past action. It is only in situations in which each action is followed by a choice of possible reactions that meaning becomes a relevant factor. Of course, this is true in most real situations.

The realization-interpretation duality can be illustrated through an example from computer science. We wish to study a naturally occurring system through the process of simulation. Natural objects are encoded data structures and their interactions are encoded as program elements, whose actions on the data structures reflect or mimic those actions that would occur naturally among the original objects. We in turn interact with the simulation, creating various scenarios for study. The data structures, and any representation of such through the use of icons or animated images, together with the observed interactions, constitute a realization of the natural system. The program elements which govern the interactions must process data elements, interpreting them in terms of objects and interactions, thereby enabling them to make action decisions. These program elements, residing as it were inside the realization, are agents. We in turn must interpet what we perceive arising from the simulation, the realization of the natural objects, and make choices ourselves for how to proceed to interact with it. We, being outside the realization, are users. We thus have a semantic frame (the natural system), its realization (the simulation), and two modes of interpretation, program elements (agents), and us (users).

As users or agents, we interpret a semantic frame to provide meaning for what we observe and to help determine how we will act. We interpret a semantic frame, not to understand ourselves, but to understand that which lies outside of ourselves. Whether we are users or agents depends upon whether or not we ourselves are part of the system with which we are interacting.

We thus have the following basic modes of interaction with semantic frames.

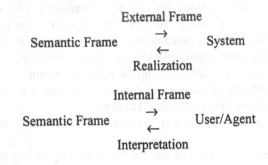

The triple of semantic frame, realization and interpretation is termed the Fundamental Triad.

The third component of this basic structure is the representation. I have stated my belief previously that semantic frames are irrepresentable. So how can I speak now of representations? Although a semantic frame is irrepresentable in its totality and in its essence, there are nevertheless aspects of semantic frames that are sufficiently delimited in scope and sufficiently regular in function so as to permit limited representations. Moreover, although a semantic frame may be essentailly irrepresentable, it is still necessary if we are to make any headway towards understanding the concept to work with representations, always bearing in mind their limitations.

Representation is a subtle concept, but an essential one. Realizations and interpretations are ontological constructs, asserting the existence of systems, users, and agents, their actions, reactions and interactions, and their ontological relationship to semantic frames. Representations, on the other hand, are epistemological constructs. They provide knowledge and information about semantic frames, realizations and interpretations, but their ontological aspects are determined otherwise. In general, representations are realizations of a representational semantic frame, which imparts only a skeletal form and dynamics, the specifics requiring a user in order to be fully expressed. The user, simultaneously interpreting both the semantic frame and the representational semantic frame, imparts form and dynamics to the representation in such a way that interactions with the representation impart to the user experiences which mimic, stimulate or echo, those derived from the semantic frame. Thus the representation stands in for some aspect of the fundamental triad. A representation of the fundamental triad is neither semantic frame, realization of the frame nor interpretation of the frame in itself. It may, however, stand in for or take the place of one of these, in effect re-presenting to a user aspects of the original situation. A representation thus serves as a reservoir, capable of eliciting particular meanings in a reproducible manner. It is akin to a memory of some aspect of a fundamental triad, but not that aspect itself. A representation is a symbolic picture, a caricature, a cartoon, to help us to visualize the real extant connections. A representation is a meta-construct.

The representation requires a separate representational semantic frame which it realizes and whose user interprets, and which conveys the sense of representing something. A representational semantic frame is, in a sense, a kind of universal frame, since any other frame may be linked to it, just as any other realization or interpretation can be associated with some representation of the representational frame. The physicality of a representation is irrelevant to its representation of a given semantic frame. It is the linkage between semantic frame/realization/interpretation and the representation that is important.

A representation is a system whose intrinsic dynamics can be ignored, and which possesses stable, static attributes. An example would be a sheet of paper and ink markings on the paper. Once in place, the ink markings are both stable and static. Moreover, the dynamics of the paper and ink are irrelevant to the understanding of any system represented, and are important only in so far as it is necessary to preserve the stability and stasis of the markings. A user, through an interpretation, establishes linkages between certain attributes (markings) and the varied components of a realization and/or interpretation of some semantic frame.

Representations are absolutely essential to the pursuit of any formal study. It is necessary to be careful when using them, however, and essential to always bear in mind that they do not reflect a direct linkage with the subject. They do not necessarily provide realizations or interpretations of the semantic frame being represented, and thus the information that they provide may not always be referrable back to that frame. In the last century, Godel showed, within the setting of mathematical logic, that is was possible to have true statements about a system that could not be proven within the system [6]. Systems of formal logic, while they possessed the necessary representative power, did not possess sufficient probative power to enable them alone to be used to explore a given system. Moreover, since representative systems possess a universal character, it is possible to formulate information within such a system which does not apply to the system being represented. For example, while within logic, true statements about a system can be formulated, false statements may also be formulated. It may not always be possible to determine whether or not a given statement about a system is true or false [6]. Moreover, it is also possible to formulate statements which may be consistent within the representational system, i.e. consistent with the representational semantic frame, but wholly inconsistent with the system or semantic frame being represented. This can be observed in the so-called logical paradoxes, in which perfectly proper statements about a system can be formulated that bear no relationship at all to the system per se (for example, "The Cretan said, "All Cretans are liars""). They are vacuous at best. The representational process, powerful as it may be, nevertheless suffers from many deep weaknesses, which must always be borne in mind.

The fixed attributes of a representational system constitute signs. Certain specific patterns constitute symbols. A sign reflects an attribute as part of a realization, whereas a symbol reflects an attribute as part of an interpretation. Whether an attribute is to be considered as a sign or a symbol depends upon its functional relationship to the semantic frame. Indeed, the distinction between a realization and an interpretation is a functional one, just as is that of representation. It is important, however, to keep these distinctions in mind, since they result in differing efficacies, and are not merely arguments about terminology. A semiotician will likely take me to task for the simple-minded approach that I have outlined here, but my goal is to construct a practical formal system for the description and analysis of emergence, not to provide an all encompassing ontology, epistemology, semiotics or psychology. The distinction between symbol and sign has led to considerable debate over the centuries and the approach taken here is rooted in a belief in the need for conceptual clarity. Humans have an almost pernicious tendency to confound semantic frames and their structural and functional aspects, resulting in semantic, epistemological, and ultimately causal and ontological confusion. Archetypal dynamics is an attempt to consciously distinguish between the various aspects of semantic frames through which we come to apprehend and comprehend the reality around us. Its emphasis on clarity and on rigor is essential, even at the expense of comprehensiveness.

2.5. MECHANISMS: SALIENCE AND DYNAMICAL STRUCTURAL STABILITY

2.5.1. *Salience and TIGoRS*

So far I have constructed a rather cumbersome philosophical apparatus, which offers a great deal of description but little else. There are many descriptive frameworks out there. Why this one? How is the semantic frame to be linked to the dynamics of some system? A first step in this direction is provided by the concept of *salience*indexTIGoRS.

In common parlance, something is salient if it is conspicious or prominent. That means that it catches our attention. Having caught our attention, we may seek to extract information and meaning from the experience, and then use what we have learned in contemplating present and future action. Whether something is salient or not depends upon the current situation. For example, the white blood cell count might be salient in treatment planning for a patient with leukemia, whereas the color of their hair would not be. On the other hand, their hair color would be more salient to their hairdresser, while their white cell count might be merely notable.

To a user/agent acting within an interpretive framework provided by a semantic frame, salient informons are those that possess, or elicit, meaning-laden interpretations referenced to the internal semantic frame. The user/agent attends to those salient informons that exert an influence over the choice of subsequent actions. Note that they do not determine such actions, since multiple choices are possible. Not all informons are salient, and so the saliency induced by the internal semantic frame results in a meaning-laden parsing of experience by the user/agent. The concept of salience has been applied in the field of archetypal psychology by Robertson [24].

Sentience is not necessary for the expression of salience. Salience can be observed in the phenomenon of TIGoRS, or Transient Induced Global Response Stabilization. Since salience is such a fundamental notion, essential to understanding how a system/user/agent can couple to a semantic frame, I shall describe TIGoRS in some detail. To understand TIGoRS I must first describe an approach to complex dynamical systems based on the transient language formulation of a dynamical system [27]. The standard formulation of a dynamical system does not readily yield information about its computational capability under natural ecological constraints. The transient language/dynamical automaton formulation of a dynamical system was introduced to rectify this problem. In the standard approach to a dynamical system, the set of all instantaneous states in which the system can exist forms a state space S. For any time t, and any initial state x, the dynamic yields a new state x_t which is the state in which the system is found at time t having begun in state x at time 0. We can express this concisely by defining a mapping $f_t : S \rightarrow S$ such that $f_t(x) = x_t$. Each map which takes S to itself is termed a selfmap of S. The set of selfmaps of S is denoted S^S.

The full dynamic is thus given as a map $f : T \rightarrow S^S$ where T is some temporal set. A trajectory from an initial point x is just the set $\{f_t(x) \mid t \in T\}$ which corresponds to a continuous path in the state space. The standard formulation readily yields information about the asymptotic behaviour of the system, in particular whether or not it possesses attractors. However it does not readily yield information about the short duration transients of the system and it is precisely these transients which correspond to our usual concept of behaviour. A different approach is required in order that information about these transients can be derived.

In the transient language formulation, the set P of all short duration transients forms the state space and the trajectory of a dynamical system is represented as a sequence of finite duration transients. The dynamic is constructed from the set of selfmaps of P as in the standard formulation. These short duration transients can be thought of as letters in an alphabet so that a trajectory becomes a sequence of letters, much like a word. In this way changes in the system such as those due to external inputs give rise to transformations on these "words". These transformations in turn are generated by a "grammar" and in this way a computational representation of the dynamical system is obtained, a dynamical automaton.

A complete formulation is given elsewhere [26]. A simplified version suffices here. A transient language (S, T, d) consists of a semigroup S, a positive abelian semigroup T and a mapping (duration) from $S \overset{d}{\rightarrow} T$ such that $d(ss') = d(s) + d(s')$. As a prototypical example, consider the set of all maps of the form $[0, a) \overset{f}{\rightarrow} \mathcal{R}$ where $a \in \mathcal{R}$, the real line. Given $[0, b) \overset{g}{\rightarrow} \mathcal{R}$, define fg as $[0, a + b) \overset{fg}{\rightarrow} \mathcal{R}$ where $fg(x) = f(x)$ if $x \in [0, a)$ and $fg(x) = g(x)$ if $x \in [a, a + b)$. Each such map represents a transient of a dynamical system, hence the name. A dynamical automaton $(S, \mathcal{E}, T, \Delta)$ consists of two transient languages $S = (S, T, d)$ and $\mathcal{E} = (E, T, d')$ and a dynamic Δ such that given $\psi \in S$ and $\eta \in E$, we have $\Delta(\psi, \eta) = \psi'\psi_\eta$ where $d(\psi) = d(\psi')$ and $d(\psi_\eta) = d'(\eta)$ and $\Delta(\psi, \eta\eta') = \Delta(\Delta(\psi, \eta), \eta')$. Every dynamical system gives rise to a corresponding dynamical automaton.

TIGoRS occurs when a transient stimulus applied to a dynamical system produces a set of responses which cluster closely in pattern space. Assume that a suitable metric ρ has been imposed on the transient language. In order to avoid an erroneous attribution of clustering to TIGoRS when actually due to mere statistical coincidence, we say that a stimulus η induces TIGoRS if, given any two initial histories ψ, ψ', it follows that $\rho(\psi_\eta, \psi'_\eta) < 1/2\rho(|\text{rand}(\psi_\eta)|, |\text{rand}(\psi'_\eta)|)$ where $\text{rand}(\psi_\eta)$ and $\text{rand}(\psi'_\eta)$ are randomly generated patterns of same norm as ψ_η, ψ'_η.

TIGoRS has been detected in a variety of complex systems models including tempered neural networks, coupled map lattices with input and driven cellular automata [26].

A series of simulations have been carried out in order to assess the robustness of the TIGoRS phenomenon using the cocktail party automaton [27]. This is an adaptive cellular automaton which can be controlled for the degree of adaptive response, inhomogeneity and asynchrony. Each cell is provided with both a state and a rule. Updating can be done either synchronously, or via a fixed asynchronous scheme, or via a stochastic asynchronous scheme. The state of the cell is first updated, then any input is applied to the cell according to the particular input mode. The rule of the cell is then updated. The rule can remain fixed or be updated according to the following adaptive scheme. Whenever the state of a cell is updated, a comparison is made between the response of the cell and that of all other cells possessing the same local neighborhood state configuration. The difference in the number of cells disagreeing and agreeing is calculated and the cell modifies its transition table entry to the opposite value if this difference exceeds a predetermined, individualized, fixed threshold. The cycle is then repeated.

The metric chosen for these studies is the Hamming distance. Two finite duration transients are compared cell for cell at each time step and the number of sites at which the two patterns disagree is summed. The total number of discordant sites is the Hamming distance between the two patterns.

Each input to the host automaton was a complex spatiotemporal pattern derived from the output of a second, input automaton having an identical lattice structure as the host. A fixed correspondence was established between the input and host cells. This provided a mapping between the output pattern of the input automaton and the cells of the host. The input pattern thus consisted of an array of state values, indexed by cell and by time. At time n for the host automaton, the row of the pattern corresponding to time n was sampled cell by cell and applied, according to the input mode, to the corresponding cell of the host automaton. The input automaton was chosen at random using varying combinations: homogeneous/inhomogeneous, linear/complex/chaotic rules, synchronous/asynchronous (random 20%), fixed/adaptive.

Inputs were presented in the recognition input mode, in which a single complete output from the input automaton was stored as the pattern. This was applied to the host as follows: the pattern was sampled as described above. The cells of the pattern which were to be applied to the host were again sampled randomly, this time at a 10% rate. This time, given a chosen pattern cell, the corresponding host cell had its state changed to match that of the pattern cell, regardless of value. Each presentation of the stimulus again varied between trials, constituting a random sampling of the original input. The automaton was studied for its ability to match its response to the pattern. Thus its capacity for pattern completion was studied. This provides one means of pattern recognition, hence the choice of the descriptor: recognition mode.

In the table are displayed the results of four trials, each consisting of 100 presentations of a stimulus to a single inhomogeneous, asynchronous, adaptive cocktail party automaton using recognition mode. The stimulus varied between the trials. Displayed are the mean (variance) of the Hamming distances between responses and a fixed response template (S), between responses and input pattern (P), between each individual pair of responses (DP) and between each individual pair of randomly generated surrogate patterns having an identical distribution of pattern norms to that of the original sample (RP).

TIGoRS is selective. Not all stimuli are capable of inducing TIGoRS. Moreover, TIGoRS is stable over time. A simulation was carried out in which the automaton was presented with a stimulus followed by a random stimulus of equal duration followed by a second presentation of the stimulus, all via recognition mode. This was repeated for 100 different initial conditions. The mean (variance) Hamming distance between individual states and between states and input pattern respectively were 8 (2.5), 6 (2.5) for the initial stimulus, 48 (1.2), 48 (1.1) for the subsequent random stimulus and 7 (3.2), 5 (3.1) for the repeat presentation of the initial stimulus. The same template was used for all three cases. The strong matching to the template shows that all of the responses to the fixed input stimulus lie within the same neighborhood about the template. Thus transient stabilization occurred on both presentations of the stimulus and the response set was the same in both cases. Thus the response set remains stable over time and bound to the stimulus and therefore can serve as a stable representation of the input stimulus.

It is interesting and extremely important to note that the final rule configurations of the automaton following each presentation of the stimulus are different, even if the same initial configuration of rules is used. A fixed correspondence between the rule governing a given cell and the stimulus does not exist, in spite of the fact that a high degree of pattern matching is taking place in each case. In other words, the correlation between the stimulus and the system response exists only at the level of the entire system. The action of the system at the cell level is stochastic. Invariance of computation is an attribute of the system level, not the cell level. One can have high level computation without low level computation.

The importance of TIGoRS is that it provides a primitive mechanism for salience, without recourse to an external oberserver. Instead, the system itself parses reality, as observed via its input stream, into salient units and salient patterns. These patterns are recognized by virtue of their ability to stabilize the response of the system around a central pattern. That is, the output of the system becomes organized as a result of the stimulus. The system, by virtue of its dynamical behavior, creates the conditions for the existence of a primitive semantic frame. TIGoRS demonstrates how even a simple complex system model does not interact with the world in a random mmanner, but rather imposes it own structure on the world, as well as its own set of actions. TIGoRS provides a primitive mechanism by means of which a system can couple through its dynamical behavior to a

TABLE I.   Results of TIGoRS in a cocktail party automaton subject to four different inputs

Ham	TIGoRS			
Dist	Strong	Weak	Pseudo	Absent
S	4 (1.4)	31 (3.7)	18 (2.3)	48 (0.9)
P	3 (1.2)	28 (2.0)	21 (4.2)	44 (4.7)
DP	4 (2.1)	31 (5.5)	16 (3.9)	47 (5.1)
RP	39 (6.0)	49 (5.0)	17 (4.4)	49 (4.9)

semantic frame. In the case of the cocktail party automaton, these systems show a high salience for input patterns that have been generated by other cocktail party automata, and thus any interactng collection of such automata can fruitfully be understood as acting according to the tenets of a shared semantic frame. Indeed, I suggest that semantic frames come into existence whenever a collective of agents parse reality into a shared set of salient units and begin to act in coherent, consistent ways in response to the appearance of these units. TIGoRS provides just one possible mechanism whereby even a simplex complex system model can begin to expresses computational behavior, and therefore a capacity to act as a potential processor of information.

2.5.2. *Dynamical Structural Stability*

TIGoRS applies to pre-configured complex systems, that is, systems with agents that are already bound together in some informationally relevant fashion. How does such a collective arise in the first place? One possible mechanism is through *dynamical structural stability*. This provides a mechanism by means of which a loosely coupled collective can bind together to form a stable coherent unit. The formation of Bénard convection cells in a heated fluid is an example well known to most readers. If a fluid is placed into a narrow gap between two glass plates and slowly heated from below, after a period of time there will appear a network of hexagonal cells. These cells consist of fluid that is flowing from the bottom plate where it has been heated, to the top plate where it cools, only to flow back down to the bottom again. Thus a rolling motion is induced in the fluid, but curiously it is organized into convection cells rather than appearing as merely random. In fact, the motion of any individual fluid element appears quite random; it is only at the macroscopic level that it appears to be organized. This is dynamical structural stability. Small perturbations to the cells disrupt slightly their placement but not their overall form. This has been termed self-organizing but I consider that a misnomer since it ignores the context without which the phenomenon does not appear.

The Belousov-Zhabotinsky reaction in which two chemical species interact so as to produce spirals of color in a reaction vessel provides another example of dynamical structural stabilty. As Cohen has demonstrated [9], this reaction is quite robust to perturbation - a slight shaking will destroy the patterns, which re-emerge after a few moments. These patterns again demonstrate structure that is dynamic, that is, the constituents are in a constant state of flux, yet which is stable under perturbation.

These two examples involve interactions among a very large collection of interacting agents. An example involving a small population of agents has been provided by Trofimova et.al. [30, 31, 32] in her work on random graphical dynamical systems. Several different models have demonstrated that even a small collection of agents (<2000), interacting in a variety of ways to form connections among themselves, can form dynamically stable structures. Indeed, it has been found that a parameter termed *sociability*, being the number of connections that a given agent can maintain at any given time, determines the distribution of sizes of interconnected clusters. These distributions are stable across initial conditions, depending for the most part upon the maximum sociability permitted to the agents. These clusters thus exhibit a stable structural feature, size, even though their constituents, that is the interconnected agents, change over time.

Dynamical structural stability appears to underlie the formation of many types of complex entities, and provides another mechanism whereby stable structures available for information exchange can arise.

2.6. CAUSALITY: STOCHASTICITY, NOISE, SYNCHRONICITY, INFLUENCE

Mathematics utilizes three approaches to describe the temporal (causal) relationships among events. First of all, there is *determinism* or *functionalism*, in which a given event is followed by precisely one event. Second, there is *indeterminism*, or *randomicity* or *stochasticity*, in which a given event may be followed by any of many events, and under repeated trials the frequency of occurrence of these possible events is given as a probability distribution function. These two approaches are most familiar to natural scientists. Determinism is the approach of classical Newtonian mechanics, and precludes any aspect of decision or choice in the actions of systems. Once the initial conditions have been chosen, the subsequent behavior of a system is pre-determined, fixed for all time. Indeterminism is the approach of quantum mechanics. Quantum mechanics provides for multiple possibility but not choice. States are not chosen, which implies some form of decision, rather they simply appear. However, their frequency of appearance is constrained by a probability distribution function whose temporal evolution is strictly determined via the deterministic Schrodinger equation.

Non-determinism is the approach of computer science and logic, where choice is indeed possible. Multiple possibilites follow from any given event, but the appearance of a particular event requires an act of choice, since no other mechanism is specified in advance. Non-determinism is the approach necessary for dealing with many games, such as Chess. Any given move can be followed by many possible moves. Which move is made though depends upon an act of will. It is not really sensible to specify a probability distribution for these moves. Each configuration on the board imposes constraints upon any possible distribution for any given piece. These distributions may also depend upon the experience of the player and their ability to see future possibilities, and so may not be stationary in time.

There are many situations involving apparent chance but where probabilities cannot readily be defined. For example, the space of possibilities might simply be too large to define a measure, such as may be the case in calculating interaction probabilities using Feynman diagrams, or in calculating distributions in quantum gravity theory involving possible alternate space-time configurations. A situation may fluctuate so rapidly in time that it is not possible to establish stationary probabilities. An appropriate context may not be provided in which the space of possibilities can actually be defined, allowing probabilites to be calculated. This may be true for systems in interaction with an external, uncontrolled enviornment. It may occur in systems undergoing developmental processes. Indeed whenever development, adaptation or learning occur, the space of possibilities will change as the system evolves, and may never be known in advance, making the calculation of probabilities impossible.

Non-determinism is the approach of choice for archetypal dynamics. The fundamental triad conception is irrelevant for understanding isolated deterministic and stochastic systems since the absence of choice obviates the need for meaning. Archetypal dynamics is less relevant to such systems. Archetypal dynamics comes to the fore, however, in the study of non-autonomous systems, and especially complex or emergent systems. Given a complex system interacting with a complex environment, possessing a varying mix of salient and non-salient features, it is all too likely that in such systems, the environment with which the system interacts provides a source of variation and indeterminism. In many cases this variability will be stochastic, but in others it will relate to the intrinsic variability of the environment, which will possess both salient and non-salient features in a stochastic but complexly correlated mix. Since all too often the distributions of possible events will be unknown (and in some case even uncalculable), the dynamics acquires a non-deterministic character, making non-determinism the appropriate approach for archetypal dynamics.

Causality and its elucidation is not a central concern of archetypal dynamics. The role of events is to influence, not to determine, behavior. Causality in archetypal dynamics can best be understood by an example. Consider again the game of Chess. The rules of the game are fixed and immutable. There are no probabilities attached to the use of these rules, and moves are non-deterministic. Each move involves multiple possibilities, but the weighting of moves depends upon the individual players, their prior experience, and the current configuration. In general, all available moves are possible, and the use of probabilities does provide any insight into how moves should be selected so as to win the game. Indeed, the concept of winning implies a goal directedness that in turn depends upon the existence of a semantic frame giving meaning to the game playing experience. Moves are not made randomly, but are selected by the individual players, each holding the goal of winning in mind. Every move creates a new context for the next move, influencing that move but not determining it. Each move alters the set of possible future moves leading to a win, but the choice of any given move is not determined. No player can ever know what the next player will do, and so the actual final sequence of moves can never be predicted or determined in advance. At almost every step of the game, the next move will always be non-deterministic. The relationship between moves is acausal, at least in the usual sense of the term, yet meaningful. Nevertheless, each move will be influenced by the totality of preceeding moves. Archetypal dynamics concerns itself with systems exhibiting a non-determinstic dynamics in at least some aspect. Sometimes this will be a feature of the environment, sometimes the system in question, sometimes both. For example, in the computational game of Life, the environment is non-determinstic but Life itself is determinstic.

As an aside, one should note that non-determinism is the most general of such temporal associations. Determinism is simply non-determinism in which each choice involves but a single possibility. Indeterminism is simply non-determinism in a setting in which it is possible to define a probability measure. Thus archetypal dynamics implicitly includes all possibilites within its rubric.

As archetypal dynamics concerns itself with meaning laden information and its non-deterministic dynamics, it provides a major role for synchronicity, that is, influences which are meaning-laden but acausal. Synchronicity is a form of dynamics that has hitherto been neglected in the mainstream literature, but which is essential to the study of contextual and contingent effects needed in a study of emergence and development, as well as that of singular events, which science has for the most part eschewed .

Randomness enters into archetypal dynamics in the form of *noise*. Webster's dictionary gives several definitions of noise but two stand out: 1) random, persistent disturbance of a signal, 2) meaningless data generated along with desired data. In archetypal dynamics, noise refers to all events, actions, reaction, or interactions that lack meaning within the context of the governing semantic frame. Noise may be facilitatory or destructive. It may be causal, acausal, or non-causal. Its only essential defining characteristic is that it lacks meaning. As such, it provides an important generative, adaptive or creative influence. Noise is frequently thought of as a destructive force, but it is merely an agent of change. Its destructive character stems from its ability to alter a meaningful event and render it into a meaningless event, but it is equally possible, as in stochastic resonance, for the opposite to occur, the rendering of an apparently meaningless event into a meaningful one.

The existence of noisy events for a semantic frame is critical if that frame is to permit development and creativity. After all, if something truly new is to emerge out of something old, then not all possibilities could have been contained within the old. Some possibilities must have gone unrecognized, in other words, without prescribed meaning. Noise provides the room necessary within a complete semantic frame for novelty, adaptation, learning, evolution. Otherwise, any system realizing that frame would be forever trapped repeating the same patterns over and over again. That is the problem with determinism. The same state always leads to the same consequent. The only way this can change is through external agencies.

Complex systems do undergo modifcation. Learning can occur. Previously meaningless events acquire meaning, either becoming incorporated within the pre-existing semantic frame or forcing the creation of a new frame. Likewise, noise introduced into a prior interpretation or realization can stimulate the creation of novel interpretations or realizations, potentially modifying the original semantic frame, altering the resonance, or creating a new frame. Noise is thus a critical and fundamental ingredient in any semantic frame. Noise provides the freedom and flexibility which enables semantic frames, through the mechanism of semantic resonance, to influence the behavior of complex systems.

Before leaving this brief foray into metaphysics, I would like to conjecture the following.

Law of Semantic Inertia: *A system remains indefinitely coupled to a semantic frame unless it interacts with a system coupled to a distinct semantic frame.*

An interaction with a system acting under a different semantic frame gives rise to what could be described as a semantic force. Such interactions, through the mechanisms of non-determinism, and salience, could result in couplings to new semantic frames. Indeed, I would like to conjecture that semantic frames may in fact arise out of synchronistic interactions among systems that, through their individual saliences and adaptations, mutually reinforce one another so as to create a shared semantic frame.

3. The Mathematical Formalism: Loci, Nexi, Tapestries and Links

3.1. COHERENCE AND SEMANTIC FRAMES

In this second part, I will outline the details of the mathematical formalism with which I shall approach the subject of archetypal dynamics. There has long been a effort in mathematical logic, linguistics, and computer science, to formalize the idea of semantics. The most successful of these attempts would be Montague semantics [29] and Situation Theory [3]. Both of these have their origins in logic and linguistics and treat semantics as a formal system, using representations based in language.

Archetypal dynamics views meaning as arising from interpretation, that is, out of the relationship between semantic frame and user/agent. As such, it has a distinctly dynamical flavour, meaning being closely linked to action, reaction, interaction, choice. Having asserted that semantic frames are fundamentally irrepresentable, I prefer to avoid the strictly formal language based appraoches to meaning. But how then am I to formulate a formal model for their study? The approach that I have taken is to focus, not upon the semantic frame per se, but upon one of its consequences, the establishment of meaningful links between events or informons. The term used to describe the meaningful linkages between informons is *coherence*. According to Webster's dictionary, coherence refers to sticking together, or to maintaining logical consistency. Both senses are relevant here. Coherence refers to the manner in which informons relate to one another. In particular, coherence refers to those relationships between informons that convey or preserve specific meaning or meanings. That is to say, coherent informons are bound together by a common thread of meaning, and therefore are bound together through the intermediary of a semantic frame. Even if we cannot represent the semantic frame that induces the coherence, we can represent the coherence itself, being a relationship between the informons.

Mathematics is ideally suited as a language for the expression of relationships. Indeed, mathematical logic is wholly concerned with the study of relationships that can be described in formal terms. There are many different logics available for the study of relations. Two branches of logic relevant here are modal logic [7] and circular logic [4]. Both of these logics will play a role as meta-logics in the subsequent analysis of the coherence formalism. It is important to emphasize that they are not equivalent to the coherence formalism, which has different goals, operations, and applications, and that they have different underlying semantic frames.

Successive informational events can be classified into two basic categories - coherent and incoherent. In a coherent collection of information events, individual events appear to relate to one another in meaningful ways. For example, if we are watching a mother feed her child, we will observe a sequence of actions - taking a spoon, placing it within a jar, stirring it, removing it with food, placing it in the child's mouth, removing it and then the pattern repeating. All of the elements of this sequence relate to one another meaningfully as part of the process of feeding, and so form a coherent pattern. If the mother should suddenly throw the spoon down and run out of the room, we would not interpret that as being part of the feeding pattern. It would not cohere with the previous sequencing and we would consider it incoherent according to the feeding intepretation. A new interpretation is thus required. The presence of too much incoherence between events results in a pattern which cannot be interpreted and thus can only be considered as either random, senseless, or simply noise, meaning simply that is precludes being informative. Departures from coherence are generally thought to signify the prescence of some additional influence or force. This may be external or it may be internal through the activation of some additional control function not active previously.

In a coherent succession of information events, the relationships that exist between events can be assigned meanings that hold in a consistent manner across the collection and can themselves be construed as information events. In an incoherent collection of information events, one finds that it is not possible to assign globally consistent meanings to these relationships, and so there will exist pairs of information events to which no meaningful relationships can be ascribed, at least according to the provisions of the specific semantic frame within which all of this interpretation takes place. Relationships between informons can thus be classified into two basic categories - meaningful, and senseless or noisy. A noisy patterning implies that we are unable to discern or to impose a meaningful pattern or organization to the collection of informons. Noisy collection of informons are thus characterized by their incoherence, by their inability to bind together into some logically consistent and meaningful pattern. Meaningful collections of informons, on the other hand, are noted for their coherence, by their ability to sustain a meaningful pattern of interrelatedness. The relationships between coherently linked informons are themselves capable of admitting meaningful interpretations and these interpretations hold in a logically consistent manner throughout the collection.

In general, if one observes a system over a prolonged period of time, one will note that coherence holds only in a local manner. Coherences extend only for finite times and across finite spaces. Nevertheless, one often finds that the history of a system will decompose into collections that are locally coherent but globally incoherent, and yet these individual collections will be found to relate in meaningful ways, indicating the presence of coherence operating at another level.

Based upon this understanding of coherence we have the following definitions.

Entity: *A generator of informons is termed an entity. Relative to the notion of generation and generated, an entity generates a globally coherent collection of informons termed a history of the entity.*

State: *Suppose that we are given an entity and a corresponding history. Relative to any meaningful notion, a state is a locally coherent collection of informons. Note that the notion of state is always relative to some frame of meaning.*

The approach to semantic frames taken here is through a study of the coherence relations imposed by a semantic frame upon a collection of informons. The specification of these informons derives from the semantic frame, but that is an aspect that is assumed to have already been implemented and is not addressed here. One might argue that that is perhaps the most difficult, troubling,and vexing question related to the entire conception of the semantic frame,and one would be correct in that judgment. That is why I have chosen instead to beg that question and begin with something simpler. Hopefully, experience gained from a study of cohernce relationships will eventually lead to insights into the deeper question of how a semantic frame parses reality and then imposes that parsing onto reality. That aside, the study of coherence and its representation is still quite powerful and of interest in its own right.

3.2. LOCI

A semantic frame determines a framework of meaning, establishing a parsing of reality into various classes of entities, actions, reactions and interactions. The medium of exchange of meaning is information, and information in turn is conveyed through informons. An informon is any aspect of reality that is capable of conveying and eliciting meaning. I have noted above that an entity, as an ontological construct, is a generator of informons. Recall the metaphor of an informon as being a cel from a film spool, the film being the entity generating the cels. If one prefers, one may think of an informon as akin to a sense perception or a measurement. An informon, as a conveyance of information, may be simple or complex, depending upon the nature of the information conveyed. A *locus* is an irreducible informon. As such, it is a single, irreducible instance of information. This does not make any implication as to the possible properties that might be attached to a locus. A locus may be spatially or temporally irreducible, as in the case of a space-time event. A locus may be spatially or temporally extended, as in the case of a state of an organism, or a mental state. A locus may have no material properties at all, though its representation or realization might. The properties of a locus depend upon the specific semantic frame that designates it as such. The indivisibility or primitivity of a locus is imparted by the particular semantic frame, and not by any intrinsic feature. The concept of locus is relative to, local to, and referenced by, a particular semantic frame. More complex informons are thus constituted out of primitive loci.

It is also important to keep in mind that a semantic frame does *not* determine all possible loci in advance. There is no set of all loci. In reality, loci appear continually, many of them arising de novo. The semantic frame enables one to recognize these instances as loci in the first place, and to place them within an appropriate context of meaning. Thus I shall refer to collections of loci, rather than sets of loci, to emphasize this open-ended aspect.

Just as information, and therefore informons, may be public or private, so may be the loci that convey such information.

3.3. NEXI

Loci, while of interest individually, achieve significance primarily through their relationships with other loci. These relationships may be coherent or incoherent, that is, meaningful or meaningless. Since the study of coherence is the main focus of this framework, the meaningful relationships are of central interest. The presence of a meaningful relationship between two loci implies that such a relationship is, in itself, informative, and thus an informon. An irreducible instance of such an informon is termed a *nexus*. Just as a locus is a basic unit of information, a nexus is a basic unit of relatedness. Meaningful relationships between loci are generally ordered. Many examples come readily to mind: "is father to", "is child of", "is left of", "belongs to", "is a member of", "is preferable to", :"logically entails", "creates", "belongs with as part of a larger informon", and so. Of course some nexi may be unordered, such as " is an instance of some entity". In this case, a nexus conveying such meaning will link both loci in both orders.

A nexus, being an informon, possesses an ontological status in its own right. Note, however, that many nexi will convey a similar general meaning above and beyond the meaning imparted to the specific instance represented. Thus nexi will often be grouped together into larger classes of shared meaning.

Nexi, being informons, may be public or private. This gives rise to several possible combinations of type between loci and nexi. Not all combinations will be ontologically possible, or epistemologically meaningful or consistent. The following combinations are suggested as being the only ontologically realizable combinations.

Public Nexus : Public Locus	→	Public Locus
Private Nexus : Private Locus	→	Private Locus
Private Nexus : Private Locus	→	Public Locus
Private Nexus : Public Locus	→	Private Locus

Given two loci, p,q, we may denote a nexus n linking them as either

$$n(p, q) \text{ or } p \overset{n}{\rightarrow} r$$

In what follows, the private/public distinction will be suppressed in the interest of clarity. Nevertheless, the reader should understand that all elements in the constructions to be described in the next sections come in either public or private forms and that these distinct forms must satisfy the coherence relations noted above.

3.4. PRIMITIVE LAYERS

Nexi that share a similar meaning can be gathered together with the loci that they associate into a structure that expresses not only the shared meaning but also the coherences that this meaning induces among the loci. This basic structure is termed a *primitive layer*. Mathematically, a primitive layer consists of a pair (L, N) where L is a collection of loci and N is a collection of nexi linking loci in L. Since each nexus links exactly two loci in an ordered manner, the pair (L, N) is simply a directed graph having vertex set L and edge set N. In order to emphasize the existence of a shared meaning among these nexi, we shall label each nexus with a common label, a, thus obtaining a labelled directed graph (L, N_a).

Any graph can also be thought of as simply a pair (L, R) where L is a collection and R a relation on L, that is, $R \subset L \times L$, in which we identify each element $(p, q) \in R$ with its corresponding nexus $n(p, q)$. The resulting structure (L, R) is an example of a *frame* in modal logic. This correspondence will prove useful in later analysis, but the difference in terminology is to emphasize the difference in interpretations and usage in these two approaches. In modal logic, the frame constitutes a complete ontology, which is used to validate formal modal theories. In archetypal dynamics, the primitive layer and its generalizations are merely generalized states of a complex dynamical system, which we interact with in a specific manner, not to prove general truths, but to generate possible histories. Modal logic and archetypal dynamics, though they utilize a similar mathematical structure, nevertheless involve different interpretations of this structure, define different actions, reactions, and interactions within this structure, and different interactions with it. This reflects the different semantic frames underlying these two approaches.

3.5. TAPESTRY LAYERS

In general, a collection of loci will have coherence relationships established by a multiplicity of meanings. Each such meaning will have associated with it a primitive layer. Putting these all together yields a *tapestry layer*. A tapestry layer, therefore, is a labelled, directed multigraph of the form $(L, N_a)_{a \in \Gamma}$ where as before the vertices L consist of a collection of loci, the edges consist of collections of nexi, each labelled by the index set Γ referencing the associated meaning. For notational simplicity we denote $\{N_a \,|\, a \in \Gamma\}$ by N_Γ.

3.6. TAPESTRIES

Each nexus is an irreducible informon with respect to dyadic relationships. As such, a nexus constitutes a locus relative to the conception of a dyadic relationship. Nexi are themselves organized in meaningful ways, and the dyadic relationships between nexi, now viewed as loci, will themselves constitute a next level of nexi. For example, a nexus relating a pair of nexi is termed an *adnexus*. Note that since nexi arise from specific collections which represent a shared meaning, the adnexi will relate nexi within one meaning collection to those within another meaning collection. This is necessary in order to maintain consistency among these various meanings. Upon repetition this process generates a countably infinite hierarchy of relationships. Each collection of loci generates a collection of nexi, which, when considered as loci, generates another level of nexi, and so on. Formally though we have treated loci as vertices within certain graphical structures, and nexi as edges. If we now treat nexi as loci, we risk semantic confusions, so in order to keep everything straight, we might treat a primitive layer as a triple (L_1, L_2, f) where L_1, L_2 are collections of loci and the *edge function* $f : L_2 \mapsto L_1 \times L_1$, providing the edge relationships. Here, N is simply the image collection $f(L_2)$.

When the relationships are clearly understood we can abuse notation and treat L_2 and N as being one and the same.

In the case of a tapestry layer, we must now associate a particular edge function to each labelled collection of nexi. Thus we may denote a tapestry layer as $(L_1, L_2^\Gamma, f_{21}^\Gamma)$ where for each index $\alpha \in \Gamma$ we have that $f_{21}^\alpha : L_2^\alpha \mapsto L_1 \times L_1$, so that f_{21}^α is an edge function for L_2^α.

The successive construction of layers of nexi yields a countably infinite set of tapestry layers. This collection of tapestry layers is termed a *tapestry*. There are two general ways in which we can represent a tapestry, their use depending upon the circumstance. To illustrate these we shall use the explicit notation *locus* for collections whose members are viewed as loci and *nexus* for collections whose members are viewed as nexi.

In circumstances in which it is important to preserve the locus-nexus distinction, one may use the *ladder* representation:

$$
\begin{array}{c}
\vdots \\
\downarrow \\
(locus_2 \quad , \quad nexus_2^{\Gamma_2}) \\
\downarrow \\
(locus_1 \quad , \quad nexus_1^{\Gamma_1}) \\
\downarrow \\
(locus_0 \quad , \quad nexus_0^{\Gamma_0})
\end{array}
\tag{1}
$$

where each downwards arrow represents an isomorphism $locus_n \mapsto nexus_{n-1}$. The ladder representation makes explicit that each pair is a tapestry layer, and thus a labelled directed multigraph.

There are circumstances, however, in which the ladder representation is rather cumbersome and it will be useful later to consider different collections of loci whose relationships to one another are defined externally. That leads us to consider the *Web* representation:

$$(locus_0, locus_1^{\Gamma_1}, f_{10}^{\Gamma_1}, \ldots, locus_{n-1}^{\Gamma_{n-1}}, f_{n-1n-2}^{\Gamma_{n-1}}, locus_n^{\Gamma_n}, f_{nn-1}^{\Gamma_n}, \ldots) \qquad (2)$$

where for each $\alpha \in \Gamma_n$, $f_{nn-1}^{a} : locus_n^{a} \mapsto locus_{n-1}^{\Gamma_{n-1}} \times locus_{n-1}^{\Gamma_{n-1}}$.

The two representations are seen to be equivalent (note that $nexus_n^{a} = f_{nn-1}^{a}(locus_n^{a})$).

In any realization of a semantic frame it is expected that the individual loci that express the meaning-laden relationships provided by the semantic frame will possess individual properties. These properties can be expressed as unary relations on the collection of loci. For example, there are meanings such as "is a male", "is a child", "is red". Unary relations can be thought of as degenerate nexi, which can be represented as reflexive edges such as $a \to a$. Properties provide a kind of internal coherence relation on loci, or at least on whatever substrate underlies these loci. Thus it is not wholly unreasonable to consider properties as nexi. Thus properties, or at least the meaning relations associated with properties, can be subsumed within the tapestry structure, and therefore the tapestry provides a basic representation of any realization of a semantic frame. Within the tapestry structure we may thus represent both the properties of individual loci as well as the coherence relationships that exist among this collection of loci, and thus tapestries provide a basic representation of a realization of a semantic frame.

3.7. LINKAGES AND LINKS

In traditional mathematics, connections between structures are provided through the use of functional mappings, which express similarities in form and in internal functional relationships. Mappings are powerful tools, but they are too restricted in structure to capture the full potential of the semantic frame. Mappings relate individual elements of one set or class to another. Set (class) valued mappings generalize this notion by permitting individual elements to be mapped to subsets (subclasses) of elements rather than individual elements. Though more powerful, these extended mappings still do not capture the essential character of semantic frames, which is expressed in its imposition of meaning.

In order to be able to relate semantic frames to one another it is necessary to generalize the mapping relationship further, and to introduce the notion of a *linkage* between frames. In order to motivate this idea, consider the relationship that exists between classical and quantum mechanics. Classical mechanics expresses in formal terms the meanings provided for by the classical semantic frame, which imparts a deterministic and objective character to the various entities that make up the classical world. The quantum semantic frame, on the other hand, imparts a deterministic and subjective character to the various entities that make up the quantum world. Classical mechanics provides a powerful representation tool for describing the classical frame. As of yet, no such mechanics is available for describing the quantum frame, though it is possible that the attempts to create a theory of quantum gravity may someday provide such a tool. Quantum mechanics as currently implemented provides a tool that links events in the quantum realm to events in the classical realm (measurements). Quantum mechanics does not describe quantum events per se. Rather quantum mechanics provides a linkage between quantum and classical events, and thus between the quantum semantic frame and the classical semantic frame. Quantum mechanics is not a simple mapping between the events represented in one frame and those of another. Instead it is a theory, a collection of processes and procedures whereby one can interrelate the worlds described by these two frames. That is the conception of a linkage that I wish to convey. A linkage is more procedure or process rather than mere description of the type conveyed by a simple mapping.

3.8. $\aleph$-ADIC TAPESTRIES

The tapestry provides a basic representation of the coherence relations linking a collection of loci. Coherence relations, by definition are dyadic or binary relations. Archetypal dynamics focuses upon coherence relations because they appear to be the easiest to formulate mathematically and have a rich representational structure. The analysis of coherence requires access to a greater range of mathematical structures and relations than provided for by binary relations alone. These additional relations need not represent aspects of the system, nor need they be imposed by its external frame. Instead they constitute formal relations lying wholly within the mathematical domain. They are adjuncts which aid in analysis and therefore are worth mention here even if they do not constitute the main focus of this paper.

In the web representation a tapestry takes the form

$$(locus_0, locus_1^{\Gamma_1}, f_{10}^{\Gamma_1}, \ldots, locus_{n-1}^{\Gamma_{n-1}}, f_{n-1n-2}^{\Gamma_{n-1}}, locus_n^{\Gamma_n}, f_{nn-1}^{\Gamma_n}, \ldots) \qquad (3)$$

Note that there is no formal reason why the edge functions must be limited to binary functions. Indeed, for any cardinal $\aleph$, consider a function ${}^{\aleph}f_{nn-1}^a$ which maps ${}^{\aleph}locus_n^a$ to $({}^{\aleph}locus_{n-1}^{\Gamma_{n-1}})^{\aleph}$. Such a function may be thought of as associating to each generalized nexus, a generalized edge consisting of an element of a $\aleph$-ary relation defined on a collection of generalized loci. If we extend the construction as before, we obtain a generalized tapestry termed an $\aleph$-adic tapestry. In the web representation we have:

$$({}^{\aleph}locus_0, {}^{\aleph}locus_1^{\Gamma_1}, {}^{\aleph}f_{10}^{\Gamma_1}, \ldots, {}^{\aleph}locus_{n-1}^{\Gamma_{n-1}}, {}^{\aleph}f_{n-1n-2}^{\Gamma_{n-1}}, {}^{\aleph}locus_n^{\Gamma_n}, {}^{\aleph}f_{nn-1}^{\Gamma_n}, \ldots) \qquad (4)$$

3.9. TAPESTRY SCHEMA

It should be fairly intuitive to expect that the $\aleph$-adic tapestries should form a structured hierarchy of relations over a collection of loci, and furthermore one can expect that these relations will be linked in various ways. The $\aleph$-adic tapestries themselves will thus admit a variety of linkages, generally through traditional set-valued mappings between their various component structures, since these tapestries convey primarily formal mathematical relationships. A *tapestry schema* consists of a collection of $\aleph$-adic tapestries and the linkages between them. Here Here I use the term *linkage* to refer to conenctions between tapestries and semantic frames, and *link* to refer to connections between individual components of tapestries. Links are thus specific features of linkages. Denoting a linkage by $\Downarrow$, a tapestry schema takes the form:

$$(^{\aleph}locus_0, \, ^{\aleph}locus_1^{\Gamma 1}, \, ^{\aleph}f_{10}^{\Gamma 1}, \ldots, \, ^{\aleph}locus_{n-1}^{\Gamma n-1}, \, ^{\aleph}f_{n-1\,n-2}^{\Gamma n-1}, \, ^{\aleph}locus_n^{\Gamma n}, \, ^{\aleph}f_{nn-1}^{\Gamma n}, \ldots)$$

$$\Downarrow$$
$$\cdots$$
$$\Downarrow \qquad\qquad\qquad\qquad\qquad\qquad\qquad\qquad\qquad\qquad (5)$$

$$(^{1}locus_0, \, ^{1}locus_1^{\Gamma 1}, \, ^{1}f_{10}^{\Gamma 1}, \ldots, \, ^{1}locus_{n-1}^{\Gamma n-1}, \, ^{1}f_{n-1\,n-2}^{\Gamma n-1}, \, ^{1}locus_n^{\Gamma n}, \, ^{1}f_{nn-1}^{\Gamma n}, \ldots)$$

$$\Downarrow$$

$$(^{0}locus_0, \, ^{0}locus_1^{\Gamma 1}, \, ^{0}f_{10}^{\Gamma 1}, \ldots, \, ^{0}locus_{n-1}^{\Gamma n-1}, \, ^{0}f_{n-1\,n-2}^{\Gamma n-1}, \, ^{0}locus_n^{\Gamma n}, \, ^{0}f_{nn-1}^{\Gamma n}, \ldots)$$

3.10. GENERALIZED TAPESTRIES

$\aleph$-adic tapestries and tapestry schemata resemble *general frames* in modal logic and are similar to models in classical logic, and thus can be expected to possess considerable descriptive power as far as formal mathematical relationships are concerned. Nevertheless, these constructions suffers from several drawbacks when dealing with meaning more generally, particularly as I approach the problem of the representation of interpretation. They possesses only a relatively simple recursive structure based upon repeated applications of relations of similar type. Attempting to use this structure to represent the diverse mixed type relations than can be expected more generally results in a combinatorial nightmare. Self reference, the bane of semantics, is difficult to express in this structure. The subtleties involved in distinguishing between the roles of locus and nexus are not very apparent, their different usages within the formulation lack coherence and the multiplicity of functional relationships leads to a lack of transparency. Moreover, the relations involved in $\aleph$-adic tapestries are all set based relations. This proves problematic once the concepts of weaves and looms are introduced to deal with the generation of tapestries. The introduction of a dynamics of tapestries takes us outside the realm of normal forms of logic. There is a lack of easy generalizability. Finally the structure simply lacks elegance. In spite of these problems, tapestries per se play a fundamental role in the representation of realizations, which is why so much emphasis has been placed upon them.

A great deal may be done with the formulation as it is, but in order to rectify these problems and create a more elegant structure it is necessary to generalize the notion of tapestry slightly through a subtle, but significant, shift in point of view. Previously I emphasized the distinction between locus and nexus by focussing upon the dual nature of nexi as being both relations, that is, edges in a graph, as well as vertices (loci). This led to the ladder and web representations depending upon which aspect was most significant to emphasize. Note, however, that the assertion that nexi can sometimes be viewed as loci can be turned around to imply that loci can sometimes be viewed as nexi, that is, as denoting relations. That is, loci can be participants in relations, or can denote relations. Simple loci can only enter into relations, whereas loci that can be considered as nexi can both enter into as well as denote relations. A significant simplification occurs if we shift our focus away from locus as vertex and nexus as edge, to one of locus and nexus as vertices having different kinds of relations (edges). Loci support one class of edge, reflecting a passive participation in a relation, whereas nexi can support two classes of edge, one passive in which they participate in a relation, and one active, in which they impose or denote a relation between other loci. This distinction between loci and nexi is similar to the distinction between information as data and information as active informon. This shift in emphasis turns out to be surprisingly powerful.

Following this intuition, I introduce two different kinds of relations, one conveying an aspect of imposition, and one conveying an aspect of participation. Consider the following structure. Begin with a single collection L of loci. Next, introduce two different kinds of edges relating these loci. The first collection S consists of *struts* and the second R consists of *relators*. Both consist of labelled, directed edges between loci. The differences between them lies in the manner in which we relate to them. Struts convey the notion of imposing a relationship, which conveys a particular meaning represented by the label, upon a collection of loci. The relators convey the relationship that has been imposed. Thinking of a particular locus as a locus or a nexus then depends upon whether or not we focus upon its associated struts or relators.

For example, in a normal tapestry, suppose that we have a pair of loci a, b and a nexus $n : a \xrightarrow{a} b$. In the corresponding generalized tapestry we will have loci a, b, n, struts $n \xrightarrow{a} a$ and $n \xrightarrow{a} b$ and relator $a \xrightarrow{a} b$. Although the shift to the generalized tapestry involves the introduction of more edges, the end result is a single, directed, labelled multigraph, a much more tractable and conceptually coherent mathematical structure.

We define a *general tapestry* to be a 4-tuple (L, S, R, Ω) where

1. L is a collection of loci
2. S is a collection of struts
3. R is a collection of relators
4. Ω is a collection of meaning labels
5. (L, S) is an Ω-labelled, directed multigraph
6. (L, R) is an Ω-labelled, directed multigraph
7. Given any $a, b \in L$, and label α, there exists at most one α-labelled strut $s \in S$ linking a and b. That is, at most there is either $s : a \xrightarrow{a} b$ or $s : b \xrightarrow{a} a$.
8. Given any $a, b \in L$ and label α, there exist at most two α-labelled relators, $r, r' \in R$ with $r : a \xrightarrow{a} b$ and/or $r' : b \xrightarrow{a} a$.
9. For any $a, b \in L$ and any α-labelled relator $r \in R$ with $r : a \xrightarrow{a} b$, there exists exactly one $n \in L$ and α-labelled struts $s, s' \in S$ such that $s : n \xrightarrow{a} a$ and $s' : n \xrightarrow{a} b$.
10. There exists an equivalence relation $=$ on R defining an *equality* on R.

These conditions ensure coherence in the structure, ensuring that loci that act as nexi do so for only single meaningful relations imposed on other loci, and that any meaningful relation imposed on loci must have an associated locus serving as a nexus. The equivalence relation allows one to ascribe multiple meanings to single relators, which proves useful in constructing complex meaning based structures.

Definition: Let $n \in L$ and $\alpha \in \Omega$. The α-scaffold induced by n consists of the subgraph $sc(n, \alpha)$ of (L, S) whose edge set $e[sc(n, \alpha)]$ consists of all struts of the form $n \xrightarrow{a} a, a \in L$ and whose vertex set $v[sc(n, \alpha)]$ consists of all vertices of these struts. The α-scope of n is the subgraph $s(n, \alpha)$ of (L, R) whose vertex set $v[s(n, \alpha)] = v[sc(n, \alpha)]$

and whose edge set $e[s(n, \alpha)]$ consists of all relators of the form $a \xrightarrow{a} b$ where $a, b \in v[s(n, \alpha)]$.

The α-scope and α-scaffold provide the basic expression of meaning of a particular locus n. More generally though, one can study this expression within the larger context of a particular general tapestry. This gives us the following definitions.

Definition: Given a general tapestry $T = (L, S, R, \Omega)$ and a locus $n \in L$, the T-scaffolding of n, $sc(n, T)$ is the minimal subgraph of (L, S) closed under α-scaffolding as α extends over all of Ω. The T-scope of n, $s(n, T)$ is the minimal subgrap of (L, R) closed under α-scoping as α extends over all of Ω.

Note that a tapestry is simply a restricted form of general tapestry, in which the α-scaffold of any locus contains at most two struts.

3.11. WEAVES

A tapestry represents our knowledge of a particular system, or of reality if one prefers. Although it is assumed that the relevant semantic frame underlying this parsing is complete, and therefore capable of parsing *all* experience, the tapestry need not, and indeed in most cases, will not represent all of reality. Except in the ideal world of mathematics, it is most unlikely that we will ever experience any system, let alone all of reality, in its entirety. At best, we experience a small piece and build upon that experience over time. In other words, as time passes, we add to our knowledge of the system or of reality, and in so doing we are able to extend the knowledge that is represented by the tapestry. The process by which a given tapestry is extended is termed a *weave*.

Archetypal dynamics considers a given tapestry to be a "state" of reality, or at least a state of some complex system and its environment. The task of archetypal dynamics is to explore the varied ways in which a given tapestry may be extended, maintaining coherence and consistency with what has gone before, to explore the impact that shifts in semantic frame(s) may have on the weaving of the tapestry, and through these tapestries to explore the conenctions that exist between semantic frames, and in particular, to study the phenomenon of emergence, in which mutiple semantic frames co-exist for a given tapestry. The weaving of new tapestries out of old is thus a fundamental feature of archetypal dynamics and separates it from those approaches to knowledge based upon logic. Archetypal dynamics treats knowledge and knowledge acquisition as a dynamical system.

It is expected that there will always exist gaps in our knowledge of any aspect of reality. Neither the past, nor the future can be known in their entirety. At best we may add to our prexisting base of knowledge and sometimes even preexisting knowledge may be lost. The loss of parts of a tapestry will not concern us here. The future cannot be known in advance and like private informons, must be guessed at or simulated. The future cannot be predicted in detail, but the various possibilities can be enumerated, and in select situations, it may even be possible to assign probabilities to them. In general though, these future possibilites come with no weighting attached, no determination of plausibility or likelihood. The best that one can do is to enumerate them and perhaps, via the underlying semantic frame, establish some parsing of these possibilites based upon shared meaning or relevance, and perhaps to use such a parsing to generate a distribution. For example, when dealing with spontaneously generated animal behavior, the behaviors received are usually classified into broad categories - eating, nest building, feeding, preening, without specifying the details of the actual behavior. These categories parse the collection of behaviors into meaningful groupings, and at the same time bulk up these groupings so as to be able to provide them with a probability measure, which might not exist were the individual behaviors to be considered separately.

The approach to the weaving of tapestries endorsed here is similar to that of playing combinatorial games [20, 21]. A combinatorial game is essentially a game in which there is a set of configurations of the game, a set of rules specifying which moves are possible, a valuation on the set of configurations which specifies which configurations constitute wins, which losses, and which are merely transitional. There is no determinism or chance involved in such games. They are entirely non-deterministic.

The weaving of a tapestry is analogous to the playing of a combinatorial game. It consists of extensions of the tapestry (moves) governed by the contraints of maintaining coherence and consistency with a given semantic frame (rules), unless the frame changes. In general, valuations are not present explicitly but arise from the interpretations of the users/agents interacting without/within the system which establish individual goals which in turn form the basis for decision making. A goal can be thought of as a valuation which establishes which moves achieve the goal (win), fail to achieve the goal (loss) or are merely considered to be interim. The exact details in any particular case will depend upon the situation, but more will be said later.

3.12. REPRESENTATIONS OF REALIZATIONS

The fundamental triad consists of a semantic frame, a realization in terms of a system, and interpretations in terms of users and agents. Systems are realizations of external frames, which in turn provide, at the very least, meaning-laden descriptions of the systems that realize them. A system will generate informons, particular instances of which we have termed loci. These loci, by virtue of their association to a system, will possess coherences, some local, some global. These coherences in turn arise through the influence of the external frame, which imparts meaning to them. These loci, being expressed by a system, will possess properties, which as has been shown above, can be thought of as internal coherences. These loci, together with their coherences, internal and relational, can be represented as a tapestry. Thus a tapestry can be viewed as a representation of a realization of an external semantic frame. The particular representation, whether ladder, web, or general tapestry, may be chosen based upon convenience.

A tapestry represents our knowledge of a particular system. A tapestry schema, on the other hand, provides a tool for the analysis of that knowledge. A tapestry schema provides a meta-level description of the system, and may include knowledge that is not directly referrable to the external semantic frame for the system and which is not available to agents within the system. Thus while a tapestry reflects what takes place more or less internal to the system, the tapestry schema takes a point of view that is external to the system. A tapestry schema may reflect aspects of an external frame, but more often will be contaminated by meta-level considerations and so it is considerd to be an analytical tool rather than a representation per se.

3.13. CONSTRUCTORS AND REPRESENTATIONS OF INTERPRETATION

The problem of interpretation is more difficult. In many ways, a realization treats information in the static sense as data. A realization is what a semantic frame is "about", and so provides information that can be accessed, interpreted and manipulated. Interpretation, on the other hand, treats information as process. An interpretation expresses the essential aspect of the semantic frame - meaning. From the standpoint of archetypal dynamics, meaning is inextricably linked to action, reaction and interaction, influencing or determining decision and choice. Interpretation is thought of as the process by which information is given meaning, and that meaning appears in the subsequent actions and choices of the users/agents. Interpretation is active, whereas realization is passive.

The essential key to the representation of interpretation will lie, not so much in the particular formalism chosen, but rather in the manner in which one relates to that formalism. In particular, while one relates to the formalism of realization in a passive manner, as description, one relates to the formalism of interpretation in an active manner, as prescription.

A user or an agent *is* an interpretation by viture of their decision making capacity. A user or agent will be the realization of some other semantic frame, but that aspect - its purely physical, structural, or formal aspect, is separate from its role as an interpretation. Indeed, a user or an agent may be the realization of one semantic frame while being an interpretation of a wholly distinct semantic frame.

As an interpretation, a user or agent must recognize informons and interpret their significance and choose their subsequent actions accordingly. Thus an interpretation must not consist of informons per se, but rather it must be an organizer of informons. It must reside at a level immediately above that of the informons that it interprets. Moreover, interpretation, and the processes of decision and action that it supports, possesses both specific and general aspects. How might that be implemented within the tapestry framework?

Whatever the nature of the representation of interpretation, it must reflect the ability of a user/agent to meaningfully interpret a collection of loci, organized in a particular manner, and to select out a subsequent course of action, that is, to specify the next loci to appear, along with suitable connections between these and previous loci. In other words, this representation must involve patterns of loci in some manner. These patterns must themselves be meaningful, and must relate to other patterns in meaningful ways. Moreover, patterns of patterns can be expected to be of significance. The reader should begin to sense the necessity of some kind of recursivity underlying any such representational framework.

Previously I suggested that in order to construct an archetypal dynamics it is necessary to depart from traditional set based approaches. The approach to interpretation taken here involves the use of *constructors*. Meaning *imposes* relations upon collections of loci, and imposition by its very nature is an active process. Consider a traditional relation. We form a set of elements, and then a set of n-tuples of such elements. This new set is termed an n-ary relation on the set of elements. In traditional set based approaches, the set of elements is provided a priori, and relations are based upon this set. In archetypal dynamics, however, we are dealing with systems that develop over time, with tapestries that extend over time. These is no a priori base set with which we can work. Moreover, meaning imposes structure on these elements, and so the selection of elements to be included in any relation becomes an active process, not an a priori statement. Relations are to be constructed, not merely described. Moreover, relations do not themselves exist in isolation but rather in varied relationships to one another. There is a meta-level of structure that shapes each relation by virtue of its connections with other relations. So, relations have structure and connections.

A constructor is a process that imposes meaningful relationships upon collections of elements. How could a constructor be represented? Consider again the constructor associating n elements together in a subset. Suppose that we take a graph of the following form. Take a collection of n letters, $a_1, \ldots, a_n$, and a letter s, and we form a graph consisting of pairs (s, a_i). Now suppose that we associate each letter a_i with some real entity. The graph then binds these n entities together into something possessing minimal structure, since no element relates to any other element because there are no connections between them, only to the s letter which serves merely to bind them and does not take on any entity itself. s binds the a_i together into a set of n enttities. This graph $(\{a_1, \ldots, a_n, s\}, \{(s, a_i) | i = 1 \text{to n}\}$ is a representation of this constructor.

Definition: A *formal tapestry* is a general tapestry having the following form. Let L, S, R, Ω denote collections of letters, and let f, g, h, i denote maps $S \xrightarrow{f} L \times L, R \xrightarrow{g} L \times L, S \xrightarrow{h} \Omega, R \xrightarrow{i} L \times \Omega$. f, g provide a directed multigraphical structure on L, while h, i provide the labelling. All of the axioms of the general tapestry are presumed to be satisfied here. The tuple $(L, S, R, \Omega, f, g, h, i)$ is termed a formal tapestry.

The formal tapestry simply replaces the vertices in some general tapestry with place-holding letters, which now serve the role of variables. The formal tapestry now plays a role similar to that of a well founded propositional formula in logic, or of a free structure in mathematics. In logic there exists a notion of satisfiability of a formula with respect to a model or a class of models. There exist two such satisfiability notions in archetypal dynamics. Note that one may distinguish two classes of letters, variables and constants, which vary only in their satisfiability conditions relative to a class of models, constants having fixed values across all models in the class, while variables are free to take on arbitrary valus.

Definition: Given a general tapestry $T = (L', S', R', \Omega')$ and a formal tapestry $F = (L, S, R, \Omega, f, g, h, i)$, we say that F is *descriptively or postdictively satisfied* in T if one can find maps $L \to L', S \to S', R \to R', \Omega \to \Omega'$, which induce graph isomorphisms between F and its image in T.

The second notion of satisfiability is more subtle, but essential in connection with weaves.

Definition: Given a general tapestry $T = (L', S', R', \Omega')$ and a formal tapestry $F = (L, S, R, \Omega, f, g, h, i)$, we say that F is *prescriptively or predictively satisfied* in T if one can find subsets $L'' \subset L, S'' \subset S, R'' \subset R, \Omega'' \subset \Omega$ and maps $L'' \to L', S'' \to S', R'' \to R', \Omega'' \to \Omega'$, which induce graph isomorphisms between the subtapestry F'' of F restricted to these subsets, and its image in T. Moreover, there exists a weave $\mathcal{W}(T)$ of T such that F is descriptively satisfied in T.

A formal tapestry F is predictively satisfied in a general tapestry T provided that a portion of it is descriptively satisfied in T and there is an extension of T in which it is wholly satisfied. In other words, the formal tapestry F describes some portion of the present status of T and predicts some aspect of its future status. Predictive satisfaction provides the decision making function of interpretation.

Definition: A *constructor* is a formal tapestry C in which there exists exactly one locus n and one label α such that the α-scope of n in C, $s(n, \alpha) = C$. The locus n can be thought of as the root or handle of the constructor.

In many respects, the idea of a constructor is similar to notions in recursion theory and the Lamda calculus [14], except that a constructor is represented by a complex patterned structure, unlike the linear structures typical of logic.

Given a collection of constructors, it is possible to generate new constructors utilizing recursive operations. For example, given two constructors $A = (L, S, R, \Omega, f, g, h, i)$, $B = (L', S', R', \Omega', f', g', h', i')$, we may define a new constructor as follows. Choose a terminal letter $a \in L$. A terminal letter is simply a letter which is not the initial vertex of any edge in either S or R. Define $L'' = L/\{a\} \cup L'$. $R'' = r \cup R'$. $S'' = S \cup S' \cup \{n \rightarrow b |$ there exists $s \in S, n \xrightarrow{s} a$ and $b \in L'\}$.

As another example, given two constructors C, C' with roots n, n' we can form a new constructor D by simply adding a new locus d and struts $d \rightarrow n$ and $d \rightarrow n'$, and taking the other components to be unions of the appropriate sets. This gives a formal union of constructors.

The above constitutes a form of generalized recursion. Many other such constructions are possible.

Definition: The *depth* of a constructor is the cardinality of the maximal chain in the graph of struts.

Definition: A *constructor tapestry* is a formal tapestry in which each locus is a constructor and each relator represents a structural relationship between constructors. For example, a relator may define a recursive relation between constrcutors. We may unfold a constructor tapestry analogously to a web representation into a tuple $(S_1, S_2, S_3, \ldots)$ where each S_a is a collection of constructors of depth α recursively linked to other constructors. That is, each element of S_n is recursively constructed from elements of lower depth.

The constructor tapestry provides a collection of interrelated constructors. When used in a predictive fashion, such constructors provide a form of decision making in relation to some tapestry. This is precisely the capacity needed in an interpretation. For this reason, I suggest that a construction tapestry provides the natural representational structure for interpretation.

As an aside, one should note that it may be possible to simplify a constructor tapestry further, losing the high level constructor relationships and dropping down to the level of the internal loci and relations. Such a constructor tapestry is organized at the same formal level as the tapestries that it may serve to construct, allowing a more subtle mode of interrelationship to arise between realization and interpretation. Note further that any for any formal tapestry, any locus n and any index α, the α-scope of n, $s(n, \alpha)$ is, formally, a constructor, and so any formal tapestry may be considered as a constructor tapestry and decomposed as above. This, in turn, means that given any generalized tapestry, there exists a constructor tapestry that is satisfied by it.

This observation suggests that the formalism described above to represent interpretations is both reasonable and elegant, partially resolving some earlier concerns.

3.14. LOOMS

We now have at hand all of the components necessary to our enterprise. In any specific situation, it is unlikely that all of these components will be necessary but it is still good to have them at hand if they are needed. Suppose, therefore, that one wishes to explore a particular semantic frame or the interrelationships between two semantic frames which exist in an emergent relationship to one another. First of all, it is necessary to specify a realization of the semantic frame, or of the emergent pair of semantic frames. The unfolding of this realization will be represented by means of a suitable a tapestry. This tapestry expresses the influence of the external frame(s) of the realization.

Next, it is necessary to attach, to each user and to each agent, a constructor tapestry which represents the influence of its internal frame. In other words, to specify its interpretation of the frame, which is necessary in order for it to determine its own actions, reactions, and interactions. In general, it is necessary to attach such a formal tapestry to each locus in the tapestry, since a priori one does not know the underlying nature of each locus. The use of general tapestries for the representation of the realization unifies this overall structure. The parsing of the tapestry into entities simplifies this somewhat since a single constructor tapestry might be applied to each history, although the impact of learning may be to induce a dynamics on these constructor tapestries. That complexity will not be discussed here.

The resulting structure consisting of a general tapestry to which are attached constructor tapestries at each locus is termed a *loom*. The loom thus conveys all information necessary to carry out the process of weaving, and thus is the basis from which all extensions of the tapestry are generated. The interplay between the various constructors, expressing the multiple semantic frames of users, agents as well as systems, can be expected to give rise to the variability and diversity, the adaptability and creativity exhibited as the loom unfolds. The loom acts like a game environment: the constructor tapestries provide a collection of rules that each user or agent may apply, and the general tapestry of the system provides the game board upon which these various moves are played out. Each situation determines how these rules are to be played, how conflicts are to be resolved, how novelty is to be dealt with, what the goals of the various players are.

It is important to bear in mind that as a loom unfolds, especially under the influence of a user, the varied components may shift the degree of resonance with or coupling to the initial semantic frames. As this unfolds, these couplings may weaken and new couplings may arise. Thus the semantic frames that influence the loom and the weaving process will cover the loom in much the same manner as coordinate charts cover the surface of a manifold. That is, the effects of a particular semantic frame may only be local to the loom. This will certainly be the case in which emergence is being considered.

3.15. ENTITIES AND THREADS

The formalism described above is cumbersome because of its generality. In order to have flexibilty and descriptive power it is necessary to accept generality over specificity, at least at first. One way of reducing the generality and simplifying the framework is to presume the existence of entities, generators of informons as previously defined. The existence of entities enables one to parse the collection of informons into dynamically coherent histories, reducing the degrees of freedom in the framework. An entity, being a generator of informons, imposes a primary level meaning upon every such informon generated. This meaning amounts to expressing the fact that it was generated. Moroever, the weaving process imposes an ordering upon such informons, enabling one to express the fact that some informons arose subsequent to others during the weaving process. The generative meaning imposes an equivalence relation upon all of the informons associated with a particular entity. The weaving process imposes a second level ordering relation. Combining the two results in a parsing of the collection of informons in the tapestry into linearly ordered sets. These sets comprise the histories of the entities, and are termed threads. The tapestries which represent the effects of the internal frames attached to the individual loci comprising the thread generated by a given entity maybe collapsed into a single tapestry attached to the entity (or to the thread as a whole), thus simplifying the structure of the loom.

3.16. INTERACTIONS AND KNOTS

Interactions between entities can be described in myriad ways. Using the description of entity histories as threads, interactions between entities will induce patterns of interaction among the loci that comprise these threads. Such interactions can take one of two basic forms. First of all, they may be private. In this case, one might consider the threads to interact "en passant", meaning that the threads might touch one another through the appearance of nexi between associated loci, or wrap around one another in ways that alter the direction of the future extensions of the threads, but in a manner which leaves no permanent trace. In other words, the actual places on the threads which touch one another might be variable, they might slip over one another and thus there is no permanent interaction. Secondly, interactions might be public. In this case they would be expected to leave a permanent trace. Two possible kinds of interaction come immediately to mind. First, two threads might share a locus in common. Such an interaction results in a localized fusion of threads and a potential loss of identity of the underlying entities. Such interactions would be typical, for example, of interacting bosons. Secondly, they might share information without ever sharing a common informon, in which case, nexi would be established between the two threads. Such nexi would induce a permanent alteration in the trajectories of the threads as their histories unfold, and we can think of representing such interactions as knottings of the individual threads. Unlike the en passant interactions previously mentioned, a knot would leave a permanent trace, and so could easily represent a public informon representing the interaction. In such interactions, the trajectories are altered but individual identities are preserved. Such interactions are typical of interacting fermions.

The use of the thread representations allows the use of a broader range of mathematical tools, and links to more traditional modes of thinking about systems, without necessarily losing the full generality of the tapestry framework. It is, however, still speculative as this juncture.

References

1. Anochin, P. (1975) *Biology and Neurophysiology of the Conditional Reflex and its Role in Adaptive Behavior*, Oxford University Press.
2. Aubin, J-P. (1999) *Mutational and Morphological Analysis: Tools for Shape Evolution and Morphogenesis*, Birkhaüser.
3. Barwise, J. & Perry, J. (1983) *Situations and Attitudes*, MIT Press.
4. Barwise, J. & Moss, L. (1996) *Vicious Circles*, CSLI Publications.
5. Bell, J. (1987) *Speakable and Unspeakable in Quantum Mechanics*, Cambridge University Press.
6. Bell, J. & Machover, M. (1977) *A Course in Mathematical Logic*, North-Holland.
7. Blackburn, P., de Rijke, M. & Venema, P. (2001) *Modal Logic*, Cambridge University Press.
8. Bohm, D. & Hiley, B. (1993) *The Undivided Universe*, Routledge.
9. Cohen, J. (2000) in *Nonlinear Dynamics in the Life and Social Sciences* eds. W. Sulis & I. Trofimova, IOS Press.
10. Cohen, J. (2003) in *Formal Models of Developing Systems*, I.Trofimova, J. Rand, J. Nation & W. Sulis eds. Kluwer.
11. Cohen, J. & Stewart, I. (1994) *The Collapse of Chaos* Viking Press.
12. Berlekamp, E., Conway, J.H. & Guy, R. (1982) *Winning Ways for your Mathematical Plays*, Academic Press.
13. DeLaszlo, V. (1959) *The Basic Writings of C.G.Jung*, Modern Library.
14. Fenstad, J. (1980) *General Recursion Theory: An Axiomatic Approach*, Springer-Verlag.
15. Forrest, S. (1991) *Emergent Computation*, 2, MIT Press.
16. Freeman, W. (1999) *How Brains Make Up Their Minds*, Phoenix.
17. Greenstein, G. & Zajonc, A. (1997) *The Quantum Challenge*, Jones and Barlett.
18. Harris-Warrick, R., Marder, E., Selverston, A. & Moulins, M. eds. (1992) *Dynamic Biological Networks*, MIT Press.
19. Isham, C.J. (1989) *Classical and Quantum Gravity* 6(11), Nov, 1509-1534.
20. Nowakowski, R. (1996) *Games of No Chance*, Cambridge University Press.
21. Nowakowski, R. (2002) *More Games of No Chance*, Cambridge University Press.
22. Rasmussen, S. (1995) in *Advances in Artifical Life, Lectures Notes in Artificial Intelligence 929* eds. F. Moran et. al., Springer-Verlag, New York.
23. Rith, K. & Schaëfer, A. (1999) *The Mystery of Nucleon Spin*, Scientific American 281(1), 58-64.
24. Robertson, R. (2000) *Mining the Soul*, Nicolas Hays.
25. Rubin, D., ed. (1986) *Autobiographical Memory*, Cambridge University Press.
26. Sulis, W. (1994) in *1993 Lectures in Complex Systems*, L. Nadel & D. Stein eds., Addison-Wesley.
27. Sulis, W. (1995) in *Advances in Artifical Life, Lectures Notes in Artificial Intelligence 929* eds. F. Moran et. al., Springer-Verlag, New York.
28. Sulis, W. (2000) in *Nonlinear Dynamics in the Life and Social Sciences* eds. W. Sulis & I. Trofimova, 98-132, IOS Press.
29. Thomason, R. (1970) *Formal Philosophy: Selected Papers of Richard Montague*, Yale University Press.
30. Trofimova, I. (2000) in *Nonlinear Dynamics in the Life and Social Sciences* eds. W. Sulis & I. Trofimova, 217-231, IOS Press.
31. Trofimova, I., Potapov, A. & Sulis, W. (1998) *Collective Effects on Individual Behavior: In Search of Universality*, Chaos Theory and Applications, 35-45.
32. Trofimova, I. (2003) in *Formal Models of Developing Systems*, I. Trofimova, J. Rand, J. Nation, & W. Sulis eds. Kluwer.

Chapter 4.

Modeling

SOCIABILITY, DIVERSITY AND COMPATIBILITY IN DEVELOPING SYSTEMS: EVS APPROACH

IRINA N. TROFIMOVA
Collective Intelligence Laboratory,
McMaster University, Canada

1. How to cook systems

1.1. EVS APPROACH

There is something about being a women and thinking about cooking, even in spite of feminist protests. Suppose we would like "to cook a natural system", or at least to model it as realistically as possible. What tools do we need, and what should we do? Well, we need several elements, so let us take P elements and put them on the table. Are they a system? No, they don't do anything, and even if each of them would jump up and down on the table, they would still not be a system. We need them to relate to each other through their behavior. That is where the major differences exist between various models of multi-agent systems: their rules of relations, or interactions between agents.

There are thousands of examples of this in the model8ing literature. Here we use our favorite - Ensembles with Variable Structure (EVS) [14, 17, 18], which are based on the following principles:

1. Non-locality of connections between agents.
2. Agents *randomly* check other agents with respect to compatibility.
3. *"Mutual agreement"* principle: connections between agents appear only when both agents "agree" to establish them, and if one agent wants to terminate a connection then it breaks.
4. Population has a *diversity* of elements, defined via some parameters or vectors.
5. The number of connections to be checked/established is limited by the parameter of *sociability*.

In some EVS models we specified the diversity of these agents through the dynamics of a resource (such as limits on spending or receiving a resource, etc). In those models each agent received and spent some resource at each time step, allowing the simulation of resource flow through the agent and through the system [15, 18].

As you see from the Table 1, our EVS models differ from other popular models by the non-locality of potential connections between agents (Principle 1), combined with systematic disconnections. Earlier models, such as those of random graph theory, percolation theory, cellular automata theory or self-organized criticality drew inspiration not only from physics, but also from biology, which is a descriptive science of rather explicit phenomena. Biology, like classical mechanics, describes many local phenomena: interactions between objects that have immediate proximity and

J. Nation et al. (eds.), Formal Descriptions of Developing Systems, 231–248.

mechanisms based on such interactions. These local descriptions have received the greatest attention of scientists modeling natural systems. At the same time there are a number of examples of non-local, indirect interactions in natural systems, which do not easily fit into those models. Locality, though, is a special case of non-locality.

Models	Connections	Disconnection rule
Random graph theory [9]	Local	No disconnection
Percolation models [7]	Local	No disconnection
Cellular automata [4]	Local and fixed	By a state of neighbors
Random boolean networks (the best review is [2])	Potentially non-local	No disconnection, but there are weights of a connection
Self-organized criticality [3]	Local	Random disconnection
Ensembles with variable structures [14-18]	Potentially non-local, limited by sociability	By compatibility or holding time

Table 1. Comparison of multi-agent models in terms of interaction rules.

Thus, we have a collection of interacting agents, where an interaction between two agents means a synchronization or inter-dependence in the behavioral dynamics of the two agents. When two agents demonstrate such an interdependence they are said to establish a connection. While trying to establish a connection an agent randomly checks a number S_j of other elements of the population, even while holding some pre-existing connections. We must now decide when to say that, "a system has emerged" as opposed to, "no, it is still just a set of agents peacefully carrying out exactly those simple steps that we asked them to". Technically speaking, as soon as a pair of agents have "found each other", they form a little system. On the other hand, we want something dramatic in our cooking, something big and significant rather then mere pairs of connected agents. We want to see something such as clustering behavior among agents, which was not prescribed in the rules of interaction that we gave the agents. So now a few words about clusters.

1.2. WHAT IS COMMON BETWEEN BENARD CELLS AND THE UNIVERSITY OF HAWAII?

Our rational upbringing suggests to us that we should hold on to something valuable as soon as we get it. It feeds our illusion that in reality there are connections that once established, forever: connections to our parents, friends, colleagues, the city that we grew up in, the organization that we are associated with. This illusion informs, for example, random graph models, or modern models of the Internet [5], in which connections are fixed, so we can count them. In reality of course, nature does not sustain connections without using them. Consider first of two a very simple questions: How to describe our workshop? What constitutes it? Would we have it if we were to place the same 23 people in the same room for six days without a requirement to exchange their knowledge? Probably not, as they should work "on science". Would we have it if the location of the sessions were in some natural setting like the beach, rather than in the assigned conference room? Probably, yes. Does it depend upon our physical interactions, and on the exact choice of these very specialists? Probably, not. Would we have it if we were to do it on the Internet like some Internet Congresses? Probably yes. It seems that what is important here is having *interactions, scientific exchange*, while the physical arrangement of the workshop does not matter.

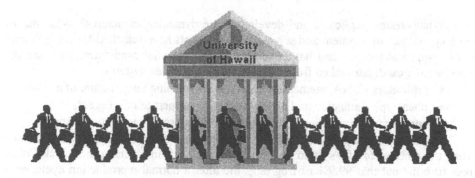

Figure 1. The dynamical structure of University of Hawaii.

I hope you remember Benard's experiment with the heating of oil in a fry pan, which at some point starts to produce nice geometrical and symmetric structures. What is common between these structures and the University of Hawaii, apart from the increasing heat outside? I would say that the common thing is *the dynamics of structure,* a sort of virtual reality of connections between agents. These structures are dynamic ones in terms of the list of agents that constitute them.

You have parents, friends, and colleagues in your life as long as you keep meeting them and updating those interests that hold these connections together. You can still hold these connections long after their death, talking to them in your mind, or you can lose these connections after you stop sharing your interests with them, even if they are alive. When you move out of your city you should know that you cannot come back several years from now: it will be a different city, with different people and different priorities of places around. Nevertheless, this city will still exist, as well as some network of family and friends, just the "players" will be different. Relationships, that is, interactions between agents, constitutes the structure of a natural system, be it the University of Hawaii, a government, a living cell or a Benard cell structure (Fig. 1).

In each given period of time these structures "update" the list of the agents which constitute it. Universities and governments regularly change their staff and associates, students and consultants. Cells update the chemicals that constitute their organelles on a constant and regular basis. In multicellular organisms, the environment of a single cell is more structured than the environment for a one-cell organism, but it is still an environment, providing building material for the cell, and changing "the list of actual elements" in the cell. We lose friends and family members from time to time, and we gain new ones, whether we like them or not.

Hopefully that was a convincing explanation why EVS models have such interaction principles as number 2 and 3: random checking for a possible connection and disconnection in case one of the connected agents loses interest in the connection. These principles make EVS models very dynamic, and make the clusters, i.e. the structures that appear within these simulations, very volatile. Other approaches that were mentioned above simulate equilibrium conditions, or, as in self-organized criticality models, certain special disequilibrium conditions.

1.3. PRESENTING DIVERSITY, COMPATIBILITY AND SOCIABILITY

What creates, replicates and develops these dynamical structures? What makes some agents part of a system and some – not? Scientists have searched for these factors within empirical settings and have identified many potential candidates. Here are the most popular candidates taken from the literature on complex systems:
- replicators - DNA, memes, and social rules keeping the structure of a system;
- phenotype – situation and history involving properties of an agent;
- environment – situation and history involving properties of environment;

Does this list exhaust the possibilities, or are there other global factors that could affect the emergence of a system and the clustering behavior of its agents? Jack Cohen likes to point out that 99.9% of frog eggs die after a normal reproduction cycle, so it seems that the first two factors are significant in 99.9% of cases. The success rate for any group of physical, chemical, social or economical agents to become a developed multi-element system is not any higher. Some environment takes over, whatever groups of elements remain, so we should focus on environmental factors. The environment expresses a much greater non-local impact than the other two factors. What global environmental factors are universal for the emergence of a system? Specifically, we are interested in the following questions: could such global factors of an environment as the diversity of a population, its sociability limits and limits of compatibility of elements have an impact on a system's development, and if so, what is this impact? First, we should say a few words about these factors.

Agents that might constitute a system could potentially, at some time, contact any other agent of similar nature, but physically there is always a limit on how many connections they could hold, or use, or check. This limit on immediate access to other members of the population, i.e. the number of connections to be checked/established per step is the parameter of *sociability* of an element in EVS modeling. For comparison, the concept of *connectivity*, which is popular in Internet modeling, refers to the number of actually extant links.

The other useful concept for studying interactions among agents is the *compatibility* concept, which was introduced in simulations of the interactions of diverse agents. These agents express their behavioral dynamics on the basis of their configurations or, if we use psychological terms, act from the point of view of their interests, goals and motivations. The compatibility concept is based on the fact that connections between agents in any natural system are a form of cooperation, or competition, oriented on the outcome of their activity, whether this activity is intentional or spontaneous. These synergy conditions occur in numerous examples in natural systems and are the focus of study of synergetics [8]. They were described at the cellular level within the theory of functional systems by Anochin [1], and appear more obviously at the level of individuals, groups, organizations and states.

In order to describe diversity and compatibility formally in modeling, we can take all possible traits, configurations, factors and characteristics of the agents and order up a vector space of these traits. We could imagine then the complete vector of interests for each agent, which characterizes the individuality of this agent within the space of these traits. If every agent has such "summarized" individuality then we could formally compare agents using their vectors. This permits us to quantitatively define a difference in configuration as a distance between "individuality vectors". We do not need to know the exact nature of each "interest", configuration or trait corresponding to some vector. We need only know the number of traits or dimensions, through which the differences between members of a group could be analyzed (dimension of the vector space of individual differences). This presentation of compatibility of configurations is easy to operate with mathematically in EVS-modeling. We have to note that compatibility of interests does not mean complete similarity of the agents; it is the result of the "synchronicity" of their configuration to have an interaction and connection.

The relationships between these three formal environmental parameters of a population (sociability, diversity and compatibility) appear to have some subtleties which have an impact on the behavior of individual elements and contribute to the development of a system. Let's start with sociability.

2. Global games on the cooking table

2.1. SOCIABILITY TAKES OVER

Paul Erdos and Alfred Renyi started from a population of isolated nodes, randomly adding links between nodes, creating random pairs. During this addition of more and more links to randomly chosen nodes some pairs started to be connected with other pairs, and at some critical number of added links the whole population of nodes became a big interconnected cluster. Thus, they found a first order phase transition effect in clustering behavior. A similar effect was observed in various physical phenomena and was called percolation.

Our Compatibility model[1] [14] demonstrated a similar effect: with an increase of

---

[1] The Compatibility model contained features reminiscent of spin glass models, considering an ensemble of N cells, each of which possesses a "resource of life" R, and a k-dimensional "vector of traits" v, where each component can equal +/- 1. Each cell forms connections with other cells, and both the maximal number of connections per cell S, termed the 'sociability' and the rate of connection attempts a are fixed and identical for all cells. At every time step each cell i attempts a random connections, and its life resource R_i is adjusted according to the "quality" of its current connections. If $R_i > 0$, the cell dies, to be replaced by a new cell having the maximal life resource R and a random vector of traits. The quality of a contact between i-th and j-th cells is evaluated according to the traits of both cells:

$$q(i, j) = \sigma(T_{ij})(v_i, v_j) = \tanh(\gamma T_{ij}) \sum_{m=1}^{k} v_{i_m} \cdot v_{j_m} \qquad (1)$$

where (,) denotes the inner product of two vectors, T_{ij} is the duration of the contact between the cells i and j, and $\sigma(T)$ is the "efficiency of the contact" - for small T it linearly increases, and after several time steps saturates at $\sigma = 1$. The quality varies between σk for aligned trait vectors, to $-\sigma k$ for anti-aligned trait vectors. For cell i having n_i connections and connection set $\{i_m\}$, the value of the life resource at the next step is

$$S_i(t+1) = S_i(t) - \delta(n_i, \{i_m\})$$

$$\delta(n_i, \{i_m\}) = \frac{\delta_0}{1 + \alpha n_i} - \sum_{m=1}^{n} q(i, i_m) \qquad (2)$$

At each time step, a cell i is randomly chosen and a possible contact cell j, which is not a member of the connection set of i is randomly selected. The possible profit of this connection for both i and j is determined by calculating $\delta(n_i, \{i_m\})$ and $\delta(n_i, \{j_k\})$. If the effect is beneficial to both i and j then the connection is formed.

agent's sociability (number of contacts that a given agent can check/hold per step), there was a critical value (critical sociability), above which a population created huge clusters of interconnected agents, and below which it represented a number of small groups clustered by interests. A first order phase transition was observed as a function of sociability with the critical point $S_c=P^{0.6}$, where P is the population size. As we said, this S_c was a threshold between having a population organized into a large number of small clusters as opposed to a small number of large clusters (Fig.2). When sociability exceeded a critical value, it not only forced self-organization of a population into a few large clusters, it seemed to prohibit "small groups of common interests", as there were almost no small clusters left.

There are several comments that should be made about the role of sociability in such clustering. Our first point is something missed in random graphs theory: *the flexibility of the structure of connections in natural systems*. In graph theory links, once established, stayed without change. The rigidity of the links in random graph theory started partially to ease after the introduction of the so-called "small-world" architecture of networks. Thirty years ago Granovetter proposed a concept of weak and strong links, and a clustering coefficient as a ratio between actually established and possible links [Grannovetter, 1973]. This meant that the links, once established, also stayed, but with various weights of importance, and it was implemented in network modeling. In contrast, in our EVS models links were far from being constant. In our EVS, the agents dropped the links for various reasons, simulating "traveling interactions" of particles or bodies in a real world. Even after the phase transition, i.e. after a population demonstrated existence of a structure, relationships between elements in these large, "totalitarian", clusters of interconnected agents were generally dynamic, altering in time while still preserving the global functionality of the structure.

Second, it seems that *the value of maximal sociability existing within the population is more important than the distribution of sociability values within a population* (i.e. how many agents are with low, average or high sociability within it). The most interesting result in Erdos and Renyi's work was that the critical number of links to add for such a phase transition was equal to the number of nodes, i.e. on average one link per node. It was difficult to judge what exactly was the range of sociability (maximal number of links, which a single node could have), as only the overall number of added links was counted. However Erdos proved that the more a graph is growing, the more even is the distribution of links among the nodes, i.e. nodes are having approximately the same number of links. Does this mean that the critical sociability value (i.e. the number of holding links, that causes a large population to be unified into a large cluster) in the random graph theory would be $S_c=1$? No, it does not, as S_c is not the average number of links, but rather the maximum number of links that an agent could have in a population.

Investigations of actual clustering behavior in naturally occurring systems showed that a small number of elements that have a sociability dramatically larger than the rest of the population can hold everybody together. "Networkers" called them "*hubs*". This ability to hold a large number of links makes hubs play the <u>role of structuring elements within a population</u>. The sociability of the Internet's hubs obviously exceeds a critical sociability many times over, resulting in a clustering of the population into large clusters organized around these hubs.

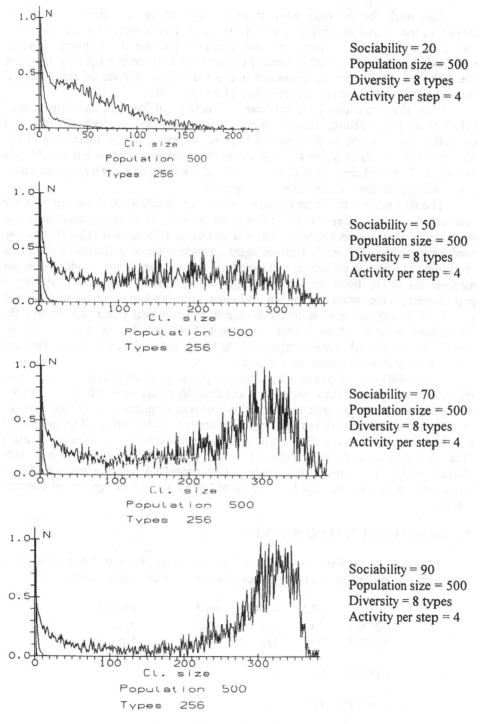

Sociability = 20
Population size = 500
Diversity = 8 types
Activity per step = 4

Sociability = 50
Population size = 500
Diversity = 8 types
Activity per step = 4

Sociability = 70
Population size = 500
Diversity = 8 types
Activity per step = 4

Sociability = 90
Population size = 500
Diversity = 8 types
Activity per step = 4

Figure 2. Phase transition in clustering behavior.

This might be the case when there is only one agent within a population exceeding the critical sociability, while others might have a sociability only equal to one, and it would be sufficient to have a phase transition in clustering, holding everybody together. On the other hand, if the elements have on average five links, not one link, they might not be organized into a big cluster if none of them have a sociability at or above the critical value for a phase transition.

What if a population does not have the "luxury" of having these "organizing" hubs? How does self-organization emerge in a population in which the maximal sociability value is not as dramatically high as, for example, the sociability of the Internet's hubs? Natural systems emerge before the development of such hubs. How do they do it? That is where a study of the critical values and the terms of a phase transition in clustering behavior become especially important.

The third point is that there is a *dominant role of sociability in clustering behavior over other formal parameters*. We investigated the roles of diversity, sociability and resource arrangements in clustering behavior in various EVS models [13-18]. Thus we used a wide range of diversity between agents, a wide range of sociability and various conditions of resource exchange, including "no resource parameter" conditions in our models. So far in these models sociability values mainly determined whether a population of agents would be unified into some virtual system or would stay in small pieces. Sociability appeared to be the dominant environmental factor, determining the affiliational behavior of an individual element. Compatibility and diversity of the population had a much weaker impact on clustering, but at the same time influenced sociability and clustering behavior indirectly.

If sociability dominates in determining the phase transition in clustering behavior, why do we worry about the compatibility and diversity of agents in this clustering? We do so because in natural systems connections between elements of a structure are not only flexible and temporal, but in many cases are not established at all because of an incompatibility of elements. That is why a real system eventually develops a hub, many times exceeding some formally sufficient number of connections: to hold together sub-clusters of diverse elements that would not be connected with the majority of the population unless they belonged to a chain of "small worlds" or "groups of common interests".

2.2. COMPATIBILITY MAKES A MOVE

Suppose that we have a population P of various agents, each having individual configurations, presented as a set of values given by D "traits"-components.

	trait 1,	trait 2,	trait k,	trait D
Agent 1:	v_{11},	v_{12},	$,...v_{1k}...$	v_{1D},
Agent 2:	v_{21},	v_{22},	$,...v_{2k}...$	v_{2D},
			....	
Agent j:	v_{j1},	v_{j2},	$,...v_{jk}...$	v_{jD},
			....	
Agent P:	v_{P1},	v_{P2},	$,...v_{Pk}...$	v_{PD}
Range of values:	$0-t_1$	$0-t_2$	$0-t_k$	$0-t_D$

Thus, the first agent has an individuality expressed by components: $v_{11}, ..., v_{1D}$, and the last agent has an individuality expressed by components: $v_{P1}, ..., v_{PD}$. For simplicity let's imagine than we are able to use the same number D of traits in order to describe a configuration of agents.

These agents wander around in a neighborhood of S_j agents, which is their individual limit of accessibility to others, randomly choosing S_j number of agents and "checking them out". We do not go into details of the specifics of what exactly should be synchronized between elements for the establishment of a connection, or what was not good enough when a connection was terminated or not established at all - we trust their judgments.

When their configuration is more compatible with some agent than with the one they currently keep a connection with, they switch to from a connection with the agent with maximal compatibility. In order to compare their configurations and find the optimal connection, EVS are calculating the difference $\sigma_{ijk} = |v_{ik} - v_{jk}|$ between each two agents on each trait, summing these sigma over the traits and finding for each agent j the minimal value of these overall sums of the differences:

$$\hat{\sigma}_j = \min_i \sum_{k=1}^{D} \sigma_{ijk}, \qquad k = 1,...,D; \quad i,j = 1,...,P. \qquad (1)$$

We agreed that our agents, or elements have not only one, but S_j number of links to check/hold per step, which reflects their individual sociability. In this case the agents deal with compatibility through a search minimizing their differences not just with one, but with S_j agents:

$$C_j(S_j) = \sum_{1}^{S_j} \sum_{k=1}^{D} \sigma_{ijk}, \qquad (2)$$

where $C_j(S_j)$ is a compatibility value of j-th agent with its S_j links.

First, let us consider a "sterile" case which, although more attractive to computer scientists or mathematicians than to natural scientists, illustrates some of the subtleties of dealing with compatibility and diversity notions. If:

1) all agents in the population are different, $\min \sum_{1}^{D} \sigma_{ijk} > 0$, and

2) the range of values t_k of all traits has the same "step" u between values, i.e. $\min \sigma_{ijk} = u$, and

3) this range of values is the same for all traits, i.e. $t_1 = ... = t_k = ... = t_D = un$, and

4) the population represents all possible types and values of the traits, so
$P = (t_{max}/u)^D = n^D$, where t_{max} is the maximal value by a trait,

then the optimization of the structure of connections has a chance (after certain time steps) to find a stable solution. The compatibility value between an agent i and an agent j would be $\sum_{k=1}^{D} \sigma_{ijk}$, with a minimum value equal u. The number of combinations for a most compatible link with the minimum difference between two agents would be at the range of D to $2D$. In general there are $n^d(n^d-1)$ compatibilities pairs to consider, as we exclude self links. The mean compatibility is given as:

$$\bar{\sigma} = \frac{\sum_{i=1}^{n^d} \sum_{j=1}^{n^d} \sum_{k=1}^{n^d} |i_k - j_k|}{n^D(n^D - 1)} = \frac{D}{3}\left(\frac{(n-1)n^{D-1}}{n^D - 1}\right)(n+1) \qquad (3)$$

For $D=1$, this mean equal $(n+1)/3$. In the case in which all traits behaved a fully independent, one would expect that mean would be $D(n+1)/3$. Clearly, in general, these traits do not act as if they are independent. For very large D, the mean tends to:

$$\bar{\sigma} = \frac{D}{3}(1-1/n)(n+1), \qquad (4)$$

which is slightly less than the independent value. However in the case of large n, the mean tends to $D(n+1)/3$. Thus the traits do tend to behave as if independent in the case in which they have a large range of values.

The dynamics of the links depends upon the actual compatibility between the agents forming the link, and the preferred links are those with the smallest compatibility values. The probability of holding a link will therefore depend upon its compatibility value. The range of compatibility values for a given agent i will depend, at least in this model, upon the specific choice of i. For any given agent i with sociability S_i, and any compatibility value m, one can calculate the probability that the minimum compatibility value over S_i, links is m according to the rather cumbersome formula:

$$prob(\hat{\sigma}_i = m) = \left[\frac{1}{n^D} \sum_{\substack{k_1+...+k_D \geq m \\ k_j \leq \max(n-i)(i-1)}} 2^{\gamma_n(i_1,...,i_D;k_1,...,k_D)}\right]^{S_i} \qquad (5)$$

We agreed first to consider a luxury case when in our population one can find an agent with any value of any trait. At every step the agents would chose a smaller difference value and update the structure of their connections. Eventually all agents in this situation would find the most compatible agents to associate with, making a nice difference-lattice of connected and optimally compatible agents, with the difference for each trait equal to just one unit u in D-dimensional space. It would be a "correctly defined problem with a final solution" for specialists on difference lattices. The system would find its optimum and would stay around it without much dynamics.

Unfortunately for mathematicians and fortunately for the rest of the world the first three conditions do not hold in real natural systems: the ranges of configuration parameters and the "steps" of these ranges are different, and often in a different order (i.e. the ranges t_i of all traits are different, and the units of the measurement of the traits

u_i are trait-specific), and the population has a limited number of combinations of types and values.

Let's look at what would happen in this situation to the structure of connections. In game theory we would find many optimal solutions with even a small increase of diversity of options. A similar effect could be observed here. In eq. (3), the numerator would include a much wider range of possible values and so the average difference between agents would increase.

In general, each trait has its own unique range, so that the total number of possible configurations will be $\prod_1^D \dfrac{t_i}{u_i}$. The populations of elements in real systems do not represent all possible values of configurational traits, i.e. $P << \prod_1^D \dfrac{t_i}{u_i}$. With a limited number of types within the population the probability to get a refusal for a connection (not finding a compatible type) increases, no matter how many steps the optimization of the connection structure runs. It means that agents often do not have a chance to find the "best case scenario" and should oscillate between several "as good as it gets", trying to find a compatibility optimum.

What is good about this "mess"? This "everything is different" supports a dynamics of compatibility search, forcing elements to change/update their structure of connections more often and to not become frozen at a single optimum point or narrow region of values. It increases the degrees of freedom (number of combinations) in the structure of connections, complicating the compatibility search and making it more dynamic.

In this sense, a *uniqueness principle* (unique ranges of values for unique traits of unique configurations) makes natural systems more adaptive. It helps them to develop a region of optimal states to choose from under various conditions versus a rigidity of the lattices with a narrow optimum around a fixed point of values.

2.3. SOCIABILITY CONFRONTS DIVERSITY

With even a slight increase in the range of traits the structure of connections starts to become very volatile, but, as we know, natural systems demonstrate a great range on most of their configurational signs. How do they not decompose and manage to actually "play a system"? Here sociability is having a hidden "fight back" with the diversity. As you can see from equation (3), with the increase of sociability of an agent the average difference between agents decreases, and so does their compatibility (eq. (2)).

It can be easily seen from both empirical observations and mathematical presentation that the higher the individual sociability of some agent, the higher its compatibility, as it is the sum of the differences with all the agents it is associating with.

With this optimization algorithm and from the law of large numbers a resulting compatibility value would oscillate around the expectancy values: $E_j = min\ Cj$ and the dispersion of differences between the j-th agent and his S_j number of contacting agents $B_j^2 = \Sigma(\Delta\sigma_{ijk})$. The central limit theorem would tell us that the probability of:

$$\alpha B_j < Cj - E_j < \beta B_j \quad \text{has the limit with } S_j \rightarrow \infty :$$

$$\frac{1}{\sqrt{2\pi}} \int\limits_{\alpha}^{\beta} e^{-\sigma^2/2} d\sigma \qquad\qquad (4)$$

With an increase of sociability the probability to optimize the structure of connections increases. The agents have a chance to choose "best from the best" in terms of their compatibility, so that the value of C_j is approaching the expected minimum E_j and the range of σ becomes more localized. In such a way sociability eases a compatibility search, but also <u>decreases the diversity</u> of agents within the local clusters of interconnected agents. The simple selection of the most compatible agents would lead to a range of configurational values within a cluster that is much narrower than the range of such values within the population. But this selection rule is not only the way to decrease the diversity in a compatibility search.

In addition, at some point evolution provided systems with the ability to adapt their own configuration to their environment, an ability evolving from simple mechanical changes to complex adaptive intentional behavior. In natural systems once connected to beneficial elements, agents tend to demonstrate such adaptation behavior, developing traits that make them more compatible with their associates and decreasing traits that distinguish them from their groups. This can be observed in a wide range of systems, but is especially visible in social, economic and psychological systems. Whether the cluster is a social group with its rules, an industry with its technical regulations or an interaction between family members, all of these clusters tend to develop a common language, common infrastructure and common rules, making the contacts between the members of the cluster easier, but also making the members of the cluster more similar to each other. "*The small worlds*" architecture appears and develops with the development of such clustering, supporting "group norms" and "group language" within these small worlds of often connected agents (so-called "*strong links*"[6]) and increasing the isolation of a cluster from the rest of population with a resultant decrease in the cluster's internal diversity.

This tendency of "regression to the mean" does not succeed in the complete isolation and localization of the cluster's configurational values around some narrow range. The increase of sociability increases the cluster size of associated agents and in a diverse population it means an increase in dispersion and thus an increase of the average differences, and again volatility in connections. At some point agents end up being connected with so-called "*weak links*", which are less compatible with their major configuration than the members of their small world, but it seems that in both real life [6] and in our modeling it is something natural and beneficial. That is how diversity controls sociability, and how sociability controls diversity.

2. 4. HOLDING TIME DOES NOT HOLD IT TOGETHER.

Although the inclusion of compatibility and diversity brings these models closer to reality, they are difficult to analyze formally. Even this simple complexity makes it difficult to fully appreciate the influence that sociability has over the clustering phenomenon. One day, Bill Sulis suggested a way to simplify the models even further and to "forget this diversity-compatibility mess". The compatibility structure was eliminated leaving sociability behind. In its place was substituted a "holding time", a maximum duration over which an agent would hold a connection irrespective of the nature of the agent to which it was connected. Choice was eliminated from the model, all actions being random. At each time step, agents would eliminate connections held past their holding time, then randomly search for free agents to connect with. New connections would be formed up to the agent's individual sociability limit, but the search extended only up to this limit, so sometimes connections would remain unsatisfied. If anything, the model was expected to remain unsaturated. Disconnecting on the minimum of the two holding times ensured volatility in the connections. Short of making purely random disconnections, it seems difficult to think of a simpler system that still supports sociability in a non-trivial manner.

Even in such a simple model one could observe an interesting and subtle phase transition in the clustering behavior. At very low sociabilities, the system formed only isolated two element clusters. At slightly larger sociabilities, clusters of all sizes could be observed to form. At slightly larger sociabilities, yet again mostly small clusters formed with occasional very large clusters. The spreading of the cluster distribution for intermediate sociabilities was reminiscent of a second order phase transition, yet it occurred on a narrow range of sociabilities which is more typical of a first order phase transition. Of course this could be just a finite size effect. But in addition to this first transition a second transition appeared. As the sociability was further increased large clusters involving most of the system began to appear even more frequently. The cluster distribution was bimodal, one peak centered on the small clusters, one peak centered over very large clusters, with nothing in between. The maximum of the peaks shifted, as the sociability was increased, from being centered over the very small clusters to over the very large clusters.

Thus two phase transitions were observed. At low sociabilities we went from a cluster distribution which was unimodal (peaked over small clusters) to uniform across the range, to bimodal (maximally peaked again over small clusters). As the sociability increased a second transition occurred, the distribution remaining bimodal, but shifting from a maximum over small clusters to one over very large clusters. Two order parameters were selected to reflect these two transitions. The first was defined as the size of the modal cluster divided by the size of the largest cluster. The second order parameter was defined as the longest region of consecutive zeros in the cluster distribution.

The onset of these phase transitions varied as a function of the holding time. One result of this Holding Time model was a very important effect: the more "stickiness" a population has, i.e. the longer agents are stuck with certain connections because of their holding time obligations, the smaller clusters are within the population. A self-organizing system does not like when somebody forces its elements to be connected longer than it needs.

3. Stages of development of a system

The game between diversity, compatibility, sociability and size of population seems to contribute to the development of natural systems. Let us briefly summarize a couple of possible scenarios of such development (Table 2).

As a prerequisite for a system to be developed on our cooking table, at a minimum two conditions should be met that are surprisingly similar to thermodynamic concepts. Actually this is not surprising at all as phase transitions and critical exponents were first described in detail in the thermodynamics literature [11]. In thermodynamics temperature, density and pressure are the parameters determining the changes of systems. With an increase of pressure (and so density) the access of elements to each other increases. In the Van der Waals' non-ideal gas model, an interaction term sums up all binary interactions between molecules and is proportional to the square of the density, i.e. the square on the number of particles in a given volume. It seems obvious that the requirement of the density condition, i.e. the distance between elements sufficient for an interaction during the functional period, is universal in all natural systems.

Another control parameter for a phase transition in clustering behavior in thermodynamics is temperature, which is an increase in the number of contacts of an element with others per time unit. This parameter can find an analogy with our sociability parameter. We saw that with an increase in sociability a phase transition in clustering occurs, and a system appears to be interconnected even while having very volatile connections and contacts. Examples of the emergence of systems arising from an increased rate of interactions, or an increase of potential access to other agents can be found in the dynamics of emergence and growth of a social group, of a city, or of business corporations.

3.1. STAGE ONE: ADAPTATION AND DIVERSITY INCREASE

Conditions of sociability and density provide elements with opportunities for more or less intense interactions. One problem is that up to this point there is no speciation of the population accumulated in these interactions, and so the range of configurational traits of interacting agents can be very broad, sort of "whoever comes we greet".

As mentioned above, agents in natural systems tend to adapt their configuration to the configurations of other agents that interact with them. This initial adaptation increases the "fractality" of configurational traits, diversifying their units u_i. In this sense the diversity of interacting elements increases, thus reducing their success in holding a compatible connection. Wandering around each other, agents do establish temporary connections, which are unstable because of low compatibility and moderate sociability. It leads to the emergence of a conditional population of weakly connected (weakly synchronized) agents.

3.2. STAGE TWO: SOCIABILITY INCREASE AND POPULATION INCREASE

The diversity of agents and the instability of their clusters lead to a situation in which agents actively participate in "other worlds" or "other populations" interacting time to time with those agents. The usual strategy of a system against diversity is an increase in sociability.

If we use an analogy from thermodynamics, an increase of diversity among interacting agents leads to more frequent changes in contacts, and so to more intense interactions, which in thermodynamics is equal to an increase in temperature. This might affect the intensity of interactions between agents, which are not usually interacting frequently due to low density/far distances. In physical terms an increase of temperature might give an agent sufficient energy to fly for a longer distance and to reach a more distant agent, and in social terms an increase in the sociability of an agent might be associated with the bringing into a community its "relatives and friends" to participate in its functioning.

In this sense an increase in sociability might lead to the increase of a population of agents that regularly interact. As we described above, an increase of sociability and population size allowed interacting agents to establish connections with more compatible agents. An improvement in compatibility leads to clustering behavior, and there is a chance for a population to switch from a number of small temporary groups to large clusters of interconnected "small worlds".

3.3. OPTIMIZATION OF THE STRUCTURE OF CONNECTIONS

In section 2 of this article we saw that an increase in compatibility, induced by an increase of sociability, helps to optimize the compatibility search and to find the local minimum of the difference function. This could be done under the condition of some "stickiness" in the population, allowing it to hold preferences and the most compatible associations.

This stage might be very short if there is little perturbation from the environment. We might call it "the spring" of the system, as it has the best adaptivity and the best potential. Its diversity and sociability are high, the population is growing and the compatibility search makes the structure of connections more and more perfect. Unfortunately this short period ends as the system starts to adapt to the fine structure of connections resulting from the compatibility search and starts to "cut" contacts that are not used for a while.

3.4. TOTALITARIAN CLUSTERING

Under stable environmental condition the adaptivity of some systems appears as a re-weighing of the connections, giving higher preferences to the most compatible agents. High sociability develops a tendency of "adaptation to the majority" in agents. It allows the system to make a "renormalization" of the configurational traits, decreasing the differences between connected agents.

The decrease of degrees of freedom (diversity of states), which are not used in the system within the most stable clusters, and high sociability create conditions for the development of uniformity within the whole population. If a change in an agent's configuration is not stopped by environmental change, the population has a chance to develop stable clusters, structuring an increasing a rigidity of the system.

3.5. LIFE IN A MESS OR A PURE DEATH?

Table 2 shows two different scenarios for the next stage of development. The condition for this stage is simple, universal and unavoidable: the change in environment. Under one scenario (A) we satisfy the system's wish to have the highest compatibility between agents - very small differences between them within the clusters and small differences between clusters. The population has high homogeneity, everything "flies with the same speed", and everybody "speaks the same language". Is this not a dream of any young manager, teacher or politician?

Suppose we have a perturbation of this system arising from its from environment. This happens all the time in natural systems. Such a perturbation, i.e. an impact of some factors or elements causing a deviation from the established phase space, requires the system to respond with certain configurational traits, which in previous conditions were not required or were not popular.

Condition of development	Stage of development	Popul. Size	Socia-bility	Diver-sity	Compa-tibility
Density and sociability increase.	0. Wondering, gathering as separate elements or unstable small groups	Low	Higher	Low	Low
Diversity increase as adaptation	1. Establishing unstable connections with low compatibility.	Higher	Same	Higher	Low
Access to distant elements	2. Sociability (and so population) increase, compatibility increase and clustering	High	High	Higher	Higher within the population
Optimization of structure of connections	3. Finding and holding the most compatible elements.	High	High	Lower	High within the population
"Renormalization" and "structuring"	4. Decrease of degrees of diversity, which are not used in the system and stabilizing the most compatible connections.	Association of "small worlds" clusters	Lower	Same	High within the cluster, low within the population
Environmental request for a change, low diversity	5A. Decomposition and death due to involvement of elements in other systems.	Decompo-sition	Low	Lower	High within the population
Environmental request for a change, high diversity	5B. Remodeling of structure of connections	Ensemble architecture		Higher	High within the cluster, low within the population

Table 2. "Death of rigid" scenario (includes stage 5A) and "Survival of an ensemble" scenario (includes stage 5B).

Elements continue doing what they were doing. There are no miracles here: they continue calculating their differences and searching for the most compatible arrangements. A perturbation by definition causes a deviation in the configuration of some agents in some segment of the population. These deviant agents try to find compatible partners, but nobody prefers to establish a connection with them because of the resulting differences between their configuration and the rest of population. Thus, in totalitarian systems any ill agents as well as any novelty will die if the perturbation were is not very significant.

If we have a significant perturbation, it requires an adjustment of the structure of connections to the configurational change resulting from this perturbation. What does a system with limited diversity and strong ties between agents do in such a situation? It starts to decompose very quickly and dies. The elements can not longer find compatible members within this system due to the change in their configuration and start to participate in the clusters of the other systems.

Scenario 5B also has a population (association) of "small worlds" with the diversity within these world-groups probably lower than that within the population. The diversity of the population in this scenario still remains high. This means that the links between these "small worlds" are weak, and these world-groups are not super-stable .Also, from time to time their members could establish connections with some average-compatible agent, and a member of another small world could temporarily occupy the vacant space. This is an "*ensemble*" architecture of the population, when diversity is not only allowed, but is required both between the clusters and within the clusters.

These two scenarios show again the conditional benefits of unification and diversity. Unification, which happens under high sociability conditions and regression to means is beneficial for the functioning of a system under unperturbed conditions. The diversity of the connections, and imperfections in compatibility, create the set of just "good enough" connections, producing an ensemble of diverse elements, ready to establish connections with even more diverse elements. The diversity of an ensemble is its life saver, as it makes it more adaptive to a change of environment. The most stable, rigid and non-adaptive natural systems are those which have little diversity of elements, and the most unstable, but the most adaptive systems have a high diversity of agents.

4. Conclusions

We discussed that:

- Sociability is the major factor affecting clustering behaviour in a diverse population;
- Diversity and compatibility have ways to control sociability, and sociability has ways to control the diversity;
- Diversity, compatibility and sociability could be considered as global factors affecting the development of a system, as its interaction within the developmental stages defines the specific of these stages;
- Diversity of agents and an ensemble architecture of connections are beneficial for the survival of a natural system functioning in a changing environment, while unification is beneficial in stable conditions;
- Establishment of interactions between agents of a population on the basis of

compatibility of their configurations is associated with a first order phase transition (in clustering behaviour), common in physical systems;

- Stickiness of agents decreases the possibility of a 1st order phase transition, but leads to a second order phase transition, common for biological systems.
- Compatibility of interests in making a connection makes a phase transition from a population of small clusters to an all-unified population smooth. Absence of compatibility makes this transition sharp.
- Artificial holding of a connection instead of compatibility condition delays the phase transition in size of population and sociability conditions, but then makes the phase transition very sharp.

6. Acknowledgments -- The author is very grateful to Dr. William Sulis for his constructive editing and useful recommendations, which led to the final revision of this paper.

5. References

1.   Anochin P.K. (1975) Biology and Neurophysiology of the Conditional Reflex and its Role in Adaptive Behavior. Oxford Press.
2.   Arbib M.A. (Ed) The handbook of brain theory and neural networks. 1995. MIT Press.
3.   Bak, P. Tang C. and Wiesenfeld K.: *Self Organized Criticality*, Phys Rev A 1988. **38 (1)**, pp. 364-374.
4.   Burks, A.W. (Ed.) (1970). Essays on Cellular Automata. Urbana, IL: University of Illinois Press.
5.   Barabasi, A-L. (2002). Linked: The New Science of Networks. Perseus Publishing: Cambridge, Massachusetts.
6.   Grannovetter, M. (1973) The Strength of Weak Ties". American Journal of Sociology 78, pp. 1360-1380.
7.   Grimmett G. (1989) Percolation. Berlin: Springer-Verlag.
8.   Haken H. (1983) Advanced synergestics. Instability hierarchies of self-organizing systems and devices, Springer, Berlin.
9.   Palmer E.: *Graphical Evolution*. 1985. New York, Wiley Interscience.
10. Trofimova I.N., Mitin N.A., Potapov A.B. Malinetzky G.G. (1997) Description of Ensembles with Variable Structure. New Models of Mathematical Psychology. Preprint N 34 of KIAM RAS.
11. Stanley H.E. (1971) Introduction to Phase Transitions and Critical Phenomena. Oxford, England: Oxford University Press.
12. Trofimova I. (1997) Individual differences: in search of universal characteristics. In: M.A.Basin, S.V. Charitonov (Eds.). Synergetics and Psychology. Sanct-Petersburg.
13. Trofimova I., Potapov A.B. (1998). The definition of parameters for measurement in psychology. In: F.M. Guindani & G. Salvadori (Eds.) *Chaos, Models, Fractals*. Italian University Press. Pavia, Italy. Pp.472-478.
14. Trofimova, I., Potapov, A., Sulis, W. (1998) Collective Effects on Individual Behavior: In search of Universality. *International Journal of Chaos Theory and Applications*. V.3, N.1-2. Pp.53-63.
15. Trofimova I. (1999). Functional Differentiation in Developmental Systems. In: Bar-Yam Y. (Ed.) Unifying Themes in Complex Systems. Perseus Press. Pp. 557-567.
16. Trofimova I. (2000) Modeling of social behavior. In: Trofimova I.N. (Ed.) (2000). Synergetics and Psychology. Texts. Volume 2. Social Processes. Moscow. Yanus Press. (in Russian). Pp. 133-142.
17. Trofimova I. (2000) Principles, concepts and phenomena of Ensembles with Variable Structure. In: Sulis W., Trofimova I. (Eds.) Nonlinear Dynamics in Life and Social Sciences. IOS Press, Amsterdam.
18. Trofimova I., Mitin N. (2001) Self-organization and resource exchange in EVS modeling. *Nonlinear Dynamics, Psychology, and Life Sciences*. Vol. 6 N.4 Pp. 351-562.

TETRAHYMENA AND ANTS-
SIMPLE MODELS OF COMPLEX SYSTEMS

WENDY A.M. BRANDTS
Department of Physics
University of Ottawa
150 Louis Pasteur, Ottawa, Ont.
Canada K1N 6N5
wbrandts@ca.inter.net *

TETRAHYMENA -
You single-celled creature
with universal developmental features
Can you tell me before
the end of my run
and swear by your genius genes
Why three is between
two and one
and what in biology
does it mean?

ANT ANT

You can't be late
you must confer
with your colony
and choose quickly
tasks you prefer
before they allocate

(W.Brandts, 2002)

1. Introduction - Simple Models of Complex Systems

The advent of high power computing and sophisticated software packages such as e-cell
have encouraged the proliferation of a brand of modelling known as *in silico* in theoretical

* Partial funding provided by NSERC

J. Nation et al. (eds.), Formal Descriptions of Developing Systems, 249–267.
© 2003 *Kluwer Academic Publishers. Printed in the Netherlands.*

biology and other sciences. In this kind of modelling, everything known about a system is put into the model in detail, including experimental and field data and a plethora of parameters, established dynamics are patched together with hypothetical models to fill in the gaps, and then simulations are run. The assumed justification for this kind of modeling is the complexity of biological systems: diversity is rampant, many details are suspected to be crucial, thus everything at the lower level must be included to generate the higher level behavior. The problem with this approach is that the internal workings of the model do not become understood, i.e. a simulation remains a black box and little in the way of strategic relationships beween variables, or the effect of the numerous parameters on the dynamics, can be uncovered. We may simulate nature, but we will gain little insight.

In this paper, we hope to demonstrate the value of simple models at the appropriate 'high' level. This approach to modelling is based on the conviction that the complexity of biological organization forbids a completely reductionistic attack. Complex systems demand that different levels of explanation be employed (Brandts [3]). It acknowledges the emergence of new phenomena at each level of description, while still permitting bridging between levels.

We will sketch two simple models of complex biological systems:

1. A phenomenological formalism, non-linear fields, for modelling spatial patterns in a continuous developmental system. We apply the model to pattern formation in the single cell organism *Tetrahymena*.

2. Temporal patterns in a discrete system. We describe a network model with state-dependent coupling coefficients of task allocation in an ant colony.

Both these systems (models) are self-organizing, i.e. there is no central, hierarchical control. Order at the system level emerges from simple interactions. We look for patterns, similarities and differences, ordering principles and laws hidden beneath the observed biological complexity. We seek 'the essence' and explore to what extent simple mechanisms or laws can explain complex data and behaviors. This approach is not exhaustive or exclusive, but is susceptible to 'add ons' to the basic principle, e.g. higher order corrections, refinements, the effect of environment, etc..

We examine what each model can and can not do. This is the role of a model (as opposed to a 'black box'). What a model can *not* do may be as enlightening for understanding complex systems as what it *can* do.

2. Pattern Formation in Tetrahymena

2.1. REDUCTIONISM AND GLOBAL MECHANISMS

While physics has been regarded as a reductionistic science in which matter can be understood in terms of the molecules and successively of the atoms, nucleons, quarks, etc. that compose them, it has been argued that living systems cannot be arbitrarily reduced or decomposed in this way without losing their vital characteristic properties, and thus that biology is wholistic. Nevertheless, physics does have intrinsically non-reductionistic laws, such as the Pauli exclusion principle which crucially affects the behavior of two electrons (determining for example, atomic orbitals and hence chemistry) and is in no way derivable from the properties of a single electron. On the other hand, modern biology has obtained successes using reductionistic tactics, for example, in molecular genetics some large scale behaviors seem to be accounted for by the function of a few constituent molecules (genes).

In the area of pattern formation, scientists from many different disciplines have been concerned with such questions as the dynamics of growth, formation of pattern boundaries, occurrence of phase transitions, changes in symmetry, etc.. The search is ongoing, not only for the solutions to dynamical equations, but even for appropriate basic models that capture the essence of pattern formation mechanisms. The challenge of biological systems is particularly great, because the elementary 'constituents' of a pattern, as well as the fundamental 'interactions' of the processes are often unknown; moreover, the great degree of complexity often found in biological systems makes it difficult to choose the most fruitful and tractable simplifications at any of the many levels of modeling (e.g. molecular, supramolecular, tissue, macro-structure, etc.).

As it applies to pattern formation, the "degree to which the properties of a 'higher' (more inclusive) level are generated by mechanisms evident at the 'lower' level," is discussed by Frankel [13] in terms of whether patterns originate from local interactions or from global mechanisms. (In the case of *Tetrahymena*, he argues for a dualistic view in which large-scale patterning can be achieved both by additive extensions of local mechanisms and by a more global 'pattern factor').

While the biological distinction between local and global interactions can often be made clear, and tested in a laboratory situation (e.g. by severing communication between parts of the system and observing the effects), in the mathematics of modeling the distinction between global or local control can be ambiguous. This is because dynamical systems often have alternative mathematical representations which may be entirely equivalent but lead to seemingly different interpretations. An example from classical physics is mechanics, in which a particle can be described as moving according to Newtons laws, which are local, or so as to minimize action over any segment of its future path (non-local in both space and time).

In the particular case of our field model, the mathematical expression of pattern control is formulated in terms of minimization of total energy, a global property. However, the model could in principle have been formulated in terms of a differential equation, which expresses field values and interactions locally. In fact, there are many different dynamical processes (equations) which could give rise to the same steady state minimum energy solutions of the field. In this essay we will outline the field model of large-scale patterning in *Tetrahymena*. It is a non-reductionistic model that describes the organism at a high or macroscopic level of the structural hierarchy, organizing large and various collections of experimental data, and predicting developmental configurations and pathways. Thus it should stand on its own as a useful and testable model. Nevertheless, we also discuss how this macroscopic approach to modeling development can potentially be bridged to more familiar reductionistic microscopic mechanisms of physics and chemistry.

2.2. MORPHOGENETIC FIELD CONCEPTS

The use of fields that we make here has its formal origins in the role that fields play in physical theory. The mathematical properties of physical fields inspire their adoption to biological problems, not only because there already are well developed formal mathematical structures pertaining to physical fields which make possible quantitative predictions of dynamical behaviors, but because biological fields actually exhibit qualitative similarities to physical (mathematical) fields (e.g. smoothness, continuity and differentiability).

The organization in biological organisms implies that cells 'know where they are'. The field captures the idea that within a region of a developing system, activity is coordinated and may be controlled by a distribution of some property or quantity. Continuity in the biological field implies, for example, that two cells close together are more similar than two distant cells.

The global aspect of the field corresponds closely to the wholistic nature of physical fields. In electrostatics for example, a unique electric field everywhere in a region of space can be determined given sufficient information about the field in a restricted region (e.g. given the appropriate Neumann or Dirichlet conditions on the boundary of the region). This ability to interpolate the field is analogous to regeneration in organisms, where a whole can be re-formed from the information contained in some of the parts.

The field model applied to *Tetrahymena* is intended to show global pattern formation. Patterns are coherently organized over the entire cell surface. The information and control necessary to form the pattern is present in the surface as a whole.

In models of morphogenesis, several interpretations of the expression of morphogenetic information have arisen: positional information, prepattern, mechanical forces. The essence of positional information (Wolpert [31], Lewis *et al* [21]) is that there is a cell parameter related to the cell's position in the developing system. In the most literal interpretation, the cells have their position specified with respect to some coordinate system and then interpret their positional value by differentiating in some particular way. True positional values are independent of the structures that they form and thus position is decoupled from differentiation. This idea is most clearly exemplified in transplant experiments where the transplanted material differentiates into a structure characteristic of its origin but appropriate for the position specified by its host environment (e.g. Bryant [11]). In the prepattern scheme, the biological system differentiates in response to differing threshold values of a nonhomogeneously distributed parameter, without the system actually having any internal geometric representation of its position (e.g. Maini *et al* [23], Murray [24], Nagorcka [25], Nijhout [28]). In a more directly physical interpretation, mechanical forces such as stresses and strains can be shown to generate spatial patterns (Goodwin & Trainor [19], Odell *et al* [29], Oster *et al* [30]).

The field approach we take in this paper does not limit itself to one interpretation. We do not need at this point to specify whether the field values represent *positions* which are characteristically associated with the formation of certain structures (positional information interpretation) or on the other hand whether the field values themselves directly represent or result in the formation of *structures* (prepattern interpretation). The physical basis of the morphogenetic field, like positional information, is a largely open question because it is generally not known how pattern is maintained and controlled in biological systems such as the cell cortex. The simplest interpretations of positional information rely on scalar fields, which could represent densities or concentrations of, for example, morphogens or cell surface structures. Molecular interpretations may seem the most obvious or simple, given the present trend for much of biology to be interpreted in molecular terms. For example, in a reformulation (Lewis [22]) of a well established model of positional information, the Polar Coordinate Model (French *et al* [15]), it was inferred that a system of diffusible chemical signals determining the growth and pattern of positional values could anticipate the phenomena codified by the Polar Coordinate Model. Other possible physical bases for the morphogenetic field, such as cell surface structures, mechanical effects such as elasticity, or more abstractly, order parameter fields (Brandts [4]) are not excluded. In the case of *Paramecium*, a relative of Tetrahymena, it is thought that distribution of Ca2+ dependent mechanoreceptors could provide the positional information on the cell cortex (Frankel [14]).

In (Brandts [4]), we demonstrate how the energy functional of the field in our model, treated in thermodynamic terms, may be used to derive dynamical equations of the field. One resulting interpretation is that of a nonlinear diffusion–like process exhibiting both diffusive and aggregative behavior. A more abstract interpretation of the field, such as an order parameter field, is also possible. Order parameters are useful in biochemical models, for example, to describe calcium regulated fluid and crystalline phases in lipids (Lara–Ochoa [20]). A visco–elastic field version of the morphogenetic field is also an intuitively natural candidate for our present field model, because of the 'stretching' and 'crowding' behavior of the morphogenetic field in *Tetrahymena*. Before persuasive arguments about the role of specific cell components and their physicochemical behavior can be advanced, the three cellular regions in which pattern organization might be located, the cell membrane, the underlying cytoskeleton (containing microfilaments and microtubules, which in other cellular systems contribute to the maintenance of cell polarity), and the outermost layer of the subcortical cytoplasm, will require further study (Frankel [14]).

2.3. CYTOGEOMETRY OF TETRAHYMENA

Tetrahymena is a ciliated protozoan. It has a number of cell surface features that are discernible after fixing and staining, and which maintain a stable characteristic organization, both in intrinsic structure and in placement at the cell surface, generation after generation. A typical cell has 18 to 21 longitudinal rows of cilia, which can be used like a grid for measuring its size and the circumferential location of cell surface features. The normal (wildtype) configuration has one *oral apparatus* (which we will denote by OA), making it by definition a *singlet*. The asymmetry in the organization of membranelles that compose the OA makes it possible to assign a 'handedness' to the OA, which for the normal configuration is defined to be a right–handed OA. In situations where pattern reversals occur, left–handed OA's can be identified. To the cell's right of the (right–handed) OA, located near the posterior end of one, two, or three adjacent ciliary rows are the *contractile vacuole pores* (CVP's). A group of CVP's on adjacent ciliary rows is referred to as a CVP *set*.

Homopolar *doublets* are fusions of two cells, joined side by side with the longitudinal anterior–posterior (A–P) axes parallel and are obtained experimentally by various manipulations of cells or cell pairs (Nelsen & Frankel [26]). Doublet configurations are not stable and they tend to down regulate in size. A plethora of forms can be observed as these doublets 'shrink' and regulate back to single cell forms along several different possible pathways.

Doublet experiments begin with a population of cells all having a double complement of all cell features; in particular, approximately double the usual number of ciliary rows, and with two OAs and two CVP sets. These *balanced doublets* have the OAs on longitudinally opposite sides of the cell. Reversion to singlets involves a loss of ciliary rows, the

254 WENDY A.M. BRANDTS

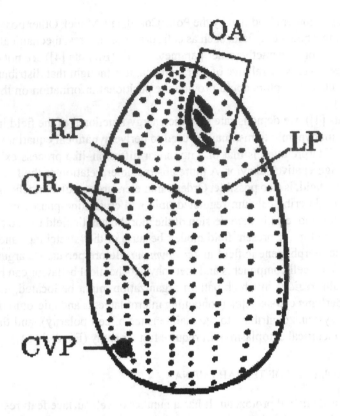

Figure 1. Singlet configuration of *Tetrahymena* showing distinctive features which can be used as positional markers or pattern elements in describing the regulation of cell configurations. Cell markers shown are: OA, oral apparatus; CR, longitudinal ciliary rows; RP, LP, right and left postoral ciliary rows, respectively, extending from the oral area; CVP, contractile vacuole pores, which appear to the cell's right of the OA in the right-handed cell.

disappearance of one of the OAs, and of one of the CVP sets. In the conversion of a doublet cell bearing *two* oral structures back to a singlet cell with *one*, a surprising triplication state with *three* oral structures, the middle one of inverted symmetry, sometimes appears. This intermediate state with three oral structures is difficult to understand from traditional biological ideas, such as lateral inhibition, because the third oral structure appears close to the original structures and hence would correspond to the presumptive region of highest inhibition. Usually the loss of ciliary rows entails a shift in the relative positions of the OAs so that the cell becomes 'unbalanced' before the transient state with three OAs appears. These additional oral structures have a highly variable location but are found only in the narrower of the two circumferential arcs spanned by the two original right–handed OAs. They are always of abnormal shape, resembling a left–handed version of the normal right–handed OA. In addition to the triplication forms described here, throughout development various other configurations of doublets and singlets appear. These have been modelled in detail and are discussed in (Brandts & Trainor [9]).

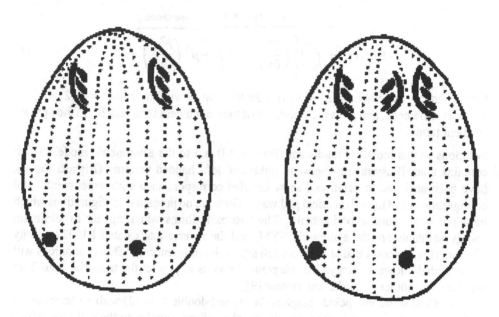

Figure 2. Configurations of *Tetrahymena*: Doublet(a) and Triplet(b). Right-handed doublets are created experimentally by the fusion of two right-handed singlets and thus have a double complement of all cell features. During regulation, a triplet configuration sometimes appears, with a third OA having opposite handedness.

2.3.1. *Field Model*

The idea of describing patterns on the cell surface with this model is to take the simplest field which is single-valued, smooth and continuous about a closed circumference, and interpret different values of the field as different cell structures. We chose a two dimensional vector field, where different values of the vector's orientation in morphogenetic space, $\theta(x, t)$, represent different cell structures. Two scalar fields could in principle also have been used.

The dynamical development of the field is governed by an energy functional, $E(\theta)$, so that the equilibrium configurations of the field, which correspond to stable biological patterns, are minimum energy configurations (Brandts and Trainor [8]). The term 'energy' as used here is not necessarily the same as the physical energy of the system. We assume however that it is a dynamical quantity which the system minimizes. The energy incorporates two competing tendencies in the field behavior:

(1) adjacent field vectors interact so as to favour an *optimal gradient* of winding. This term reflects the idea that there is an optimal spacing of positional values or pattern features for the cell, and was inspired by the qualitative biological scheme, the Cylindrical Coordinate Model (Nelsen and Frankel [26]), a variant of the Polar Coordinate Model

(French *et al* [15]; Bryant *et al* [10]).

(2) *smoothness*– changes in direction of winding cost energy.

The simplest form for an energy functional which meets these criteria is a quartic in the field gradient, combined with a second derivative smoothing influence.

$$E(\theta(x)) = \int_0^L \{ \overbrace{\left[\left(\frac{\partial \theta}{\partial x} \right)^2 - 1 \right]^2}^{\text{optimal gradient}} + \overbrace{\beta^2 \left(\frac{\partial^2 \theta}{\partial x^2} \right)^2}^{\text{smoothness}} \}dx \qquad (1)$$

With boundary conditions $\theta(L) = \theta(0) + 2\pi W$; W = winding number = 1,2.
L = circumference, and β^2 is the weighting of the smoothness term relative to the optimal gradient term.

Solutions to this model (Brandts and Trainor [9]) and to the asymmetric model, which contains a small asymmetry between right- and left- handed winding (Brandts [5, 6]), have been analysed in detail and show detailed correspondance with experimental cell configurations and their dynamical pathways. There are no predicted configurations which have not been found experimentally. The two basic forms which appear as minimum energy solutions are the symmetric SYM, and the reverse-intercalated RI. The energy diagram (Fig.1) shows that at different cell sizes (circumferences), different solutions will attain globally lowest energy. This diagram allows us to predict the transition size from one form to another (Brandts and Trainor [9]).

The SYM solutions include singlets (W=1) and doublets (W=2) with a normal set of cell features, arranged in sequence at the usual positions, relative to the cell size (which varies in a doublet experiment from 2x down to 1x the singlet circumference). The RI solution (Fig.4) includes both singlet and triplet configurations.

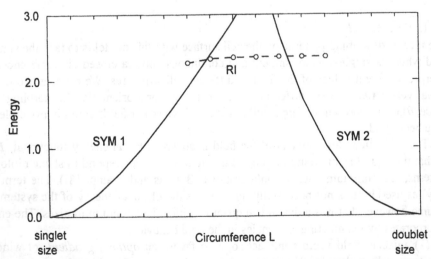

Figure 3. Energy Diagram. The total energy versus the cell circumference L, from singlet to doublet size, for SYM and RI solutions. As the cell reduces from doublet to singlet size, the sequence of (minimum energy) configurations is: SYM(W=2) → RI → SYM(W=1), which corresponds to pathways with triplet states.

The model contains only one free parameter, β^2, which can be fit to experimental data. Since β^2 determines both transition points on the energy diagram simultaneously, we can choose its value by fitting to the location (circumference value L) of one of the transitions; this gives at the same time a prediciton of where the other one falls. In the analysis of cell configurations below, the value $\beta^2 = 0.2$ is used.

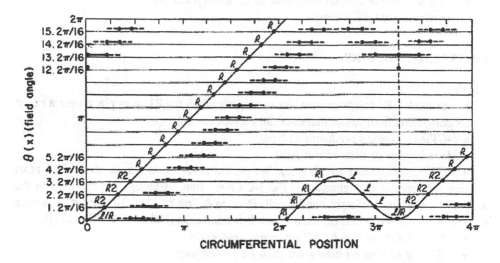

Figure 4. RI SOLUTION. The field value (angle) $\theta(x)$ versus circumferential position about the cell, for a cell circumference ($L = 3.250\pi$) that has a minimum energy solution RI. The right-hand portion of this figure ($3.25\pi - 4\pi$) repeats the left portion($0 - 0.75\pi$). Example cell configurations can be read along the horizontal lines (see Brandts and Trainor [9] for details). Right-handed OA's are denoted R, left-handed L. R semi-cells of unequal size are denoted with $R1$, $R2$. The domain of CVP formation is indicated by solid and dashed lines. Reading from the top down, cell configurations shown are: (2x) singlet with three CVP; (1x) singlet with two CVP, one broad; (1x) singlet with two CVP, one narrow; (8x) singlet with one CVP; (3x) triplet with one CVP; (1x) fusing triplet with two CVP. Solution graphs like this give dynamical paths of individual configurations as L decreases, and can be used to generate samples for comparison to population data.

2.3.2. *Solutions and Predictions*

In the proceeding sections, we compare, in summary, the configurations given by our model to the many experimental configurations which have been documented and measured. Predictions appear in italics followed by parenthetical indication of whether they have been tested experimentally.

The experimentally recognized division of cells into two main classes, the 'doublets' and the 'singlets' is *not* the natural grouping for our model. Our model has two classes of solutions, SYM and RI, which reflect differences in their underlying symmetry. Both SYM and RI solutions can be manifested as SYM(W=2) doublets, or RI triplets not completely expressed, or as singlets. Conversely, doublets may be represented by SYM(W=2) or by RI solutions, and singlets may be represented by SYM(W=1) or by RI solutions.

SYMMETRIC (SYM) SOLUTIONS

Singlets and Doublets
The SYM solutions give the simplest ('most symmetric') forms for doublets and singlets in our model.
- *Doublets and singlets should be plentiful (yes).*
- *CVP location in singlet proportional to circumference (yes).*
- *CVP2 location in doublet proportional to semicell sc2 (yes).*
- *Distribution of CVP's (yes).*

REVERSED (RI) SOLUTIONS

Singlets
In our model, RI solutions give rise to singlets when the RI takes place over a range of morphogenetic field values that does not include the OA.
- *1-CVP-set singlets predominate (yes).*
- *More singlets as L decreases (yes).*
- *3-CVP-set singlets possible (yes).* We predict several new singlet states. The most striking new singlet configuration is the 3-CVP-set state. Although not reported in the original doublet study of right-handed cells (Nelsen & Frankel [26]) 3-CVP-set cells were subsequently observed in experiments on left-handed cells (Nelsen & Frankel [27]).
- *A broad CVP-set in a 2-CVP set singlet (maybe).*
- *≈ 25% of RI singlets have more than 1 CVP-set (no).*

CVP Distributions
- *Most 2-CVP-set singlets have CVP at typical singlet location (yes).*
- *Additional CVP's to the right of OA (yes).*

Our theoretical distribution of CVP's in doublets and 2-CVP-set singlets is qualitatively similar to the experimental distribution:
- *CM2 distribution is wider than CM1 (yes).*
- *Doublets and 2-CVP-set singlets have same CM1 (yes).*
- *CM2 of 2-CVP-set singlets halfway between doublet CM's (yes).*
- *CM2 of 2-CVP-set singlets wide (yes).*
- *Doublet distributions sharp (yes).*

Triplets
- *L only in narrower semicell sc1 (yes).*
- *L position anywhere within sc1 (yes).*
- *L position uncorrelated with sc1 width (yes).*
- *L position uncorrelated with width sc2 (yes).*
- *Absolute and relative widths of sc1 decrease with L (yes).*
- *Cell shrinkage uniform (yes).*
- *Average relative width sc1/c = 0.33 (yes).*
- *Maximum 20% triplets (yes).*
- *Most triplets have 1 CVP set (yes).*
- *CVP2 location is proportional to sc2, independent of sc1 (yes).*

• *Triplet with 2-CVP sets has large relative width, scl/c = 0.37 (yes).*

Left handed cells, dynamical pathways, and another pattern, mirror reversed solutions, have also been studied (Brandts [5, 6]).

The biological system that we have chosen to apply our morphogenetic field model to is, as a single celled system, atypical. The traditional application of positional information models is to multicellular organisms. In multicellular organisms, individual cells might be thought of as responding genetically to their local information, and effecting nuclear changes which lead to a particular differentiated state in that cell. Hence, which genes get turned on determines what kind of cell develops. We extend the idea of positional information to systems wherein genetic changes cannot be the sole repository of positional cues, namely to the single cell.

2.3.3. Conclusions on the Morphogenetic Field Model

In this part of the essay, we have seen the translation and extension of an important biological concept, positional information or pattern specification, into a rigorous and explicit formalism: a vector field governed by a non-linear energy density functional.

One of the more profound insights offered by our model refers to the relationship between observed biological states and underlying field configurations (reflecting the presumptive physical processes). Similar observed states may arise from fundamentally different field solutions, and conversely, one type of field solution can give rise to very different biological patterns. For example, the observed 1-CVP-set singlets can result from either a symmetric solution or a RI solution; conversely, the RI field solution can give rise to triplets, 1-CVP-set singlets, and singlets with 3 CVP sets. The lesson to be learned here is that 'taxonomy' based only on the superficial expressed patterns may not provide sufficiently complete insight into the fundamental underlying processes or the dynamical relationships between patterns. In particular, the fundamental classes of patterns in this problem may turn out not to be 'singlets' and 'doublets', nor 'balanced' and 'unbalanced' doublets, but something more closely related to the symmetry of the field solutions. Our model suggests new relationships between patterns, with a simple underlying logic but a vast diversity of observables.

Our very broad interpretation of positional information, viz. as a general means of distinguishing locations in an ordered system, suggests its widespread applicability, not only to pattern formation problems of direct biological relevance but also to patterns in purely physical systems. The difference between pattern specification or positional information in biological vs. physical systems is primarily in its expression[8], e.g. the physical or chemical complexity of the fields and the substances in which the patterns are manifested. Biological systems have more complex parts in which to show the effects of positional ordering (e.g. the genetic switching of each cell in Wolpert's [31] positional information scheme) but the underlying general principles of local activity leading to global ordering may be identical, whether they be, for example, magnetic particles interacting electromagnetically in a ferromagnet, or cells in tissue interacting by series of complex physiological means.

The proof of this argument lies in developing explicit models of generalized morphogenetic information for pattern formation problems, and testing them in individual systems, both physical and biological, such as we have done for *Tetrahymena*. The success

of our program and its intuitive simplicity, argues strongly for the value of developing
general high-level models of biological processes, in addition to the detailed molecular
approaches typically undertaken. While the molecular approaches might be imagined to
supply a mechanism to support positional information and morphogenetic fields, they
generally do not by themselves have the power (the tractability) at such a microscopic
level to make the necessary high-level predictions for relatively large-scale patterns.

We turn now to a brief description of our other example of a 'high-level' model of a
complex biological system, a network model of task allocation in ants.

3. Modeling Self-Organization in Ants

3.1. COLLECTIVE BEHAVIOR AND SELF-ORGANIZATION

Collective activities in social insects, such as ants, bees, and termites lead to complex
spatio-temporal patterns. We are prompted to ask, does complexity at colony level imply
that individuals are complex? i.e. do ants measure many parameters and perform complex
computations to adjust their behavior? In Physics and Chemistry (e.g. phase transitions,
chemical mixtures)we also see the emergence of complex spatio-temporal patterns. These
pattern formation processes have been described by the theory of self-organization. In this
view, complex macroscopic patterns emerge from *interactions* among individuals with
simple behaviors.

Self-organization explains *how* collective behavior arises. It does not explain *why*
collective behavior arose in evolution. However, the *mechanisms* of collective behavior
influence the *path* of evolution and thus we need to understand the mechanisms.

3.2. TASK ALLOCATION IN ANT SOCIETIES

Ants communicate by pheromones and antennation and they interact with their environment. They perform many different tasks, such as nest cleaning, foraging, patrolling,
and waste management. How are these tasks allocated? In this essay, we consider such
complex behavior in ant societies which exhibits no central control, i.e. there is no 'boss.'
In some species (polymorphic), tasks are determined by caste. In monomorphic species,
where all ants are identical, other mechanisms must be at work. In our model of task
allocation we aim to determine to what extent interactions between identical individuals,
rather than intrinsic differences between individuals or interactions with the environment,
account for group behavior. Other models of task allocation in biology include (Gordon
et al. [18]; Deneubourg and Franks [12]; Bonabeau *et al.* [1, 2]).

The study of the dynamics of a collection of interacting units such as cells or agents
poses great challenges. Models which are complex enough to show a range of dynamical behaviors are typically too complex to characterize fully analytically or numerically.
There are many examples of other networks - neural, genetic, immunological, societal-
where the dynamics are not well understood, either because the couplings are asymmetric,
time- or state-dependant, or because units can have more than two states. One result of
our work is that stochastic processes known as birth-death jump processes with state-
dependent transition probabilites are useful for studying non-hierarchical task allocation
(Brandts *et al.* [7]).

Interactions between units in network models depend on coupling strengths, which are usually assumed constant. For example, in classical neural networks, the rule that governs switching between neuron states uses a weighted sum over the states of other neurons, and the coupling strength between two neurons does not depend on the state of these neurons. In this essay, we focus on the dynamical behavior of a population of units whose coupling coefficients *does* depend on the state of those units. Task allocation was recently modeled using this state-dependant coupling coefficient approach (Gordon *et al.* [18]). The model ants in this network change categories through deterministic switching rules based on fields, which can be interpreted as summed chemical or contact cues from individuals. They considered eights tasks and hence eight categories of ants. Each ant was sensitive to three different fields. They employed three matrices of category-dependent coupling coefficients (3X8X8 parameters, with some reductions due to symmetries). The network showed very complex dynamics, including multistability. It did reproduce some observed features of task switching, such as fluctuating yet stable populations, and perturbations to one category that propagate to other categories. But the model is too complex to analyze fully and categorize behaviors or even gain good intuition about its properties.

3.3. TWO-CATEGORY NETWORK MODEL

The Two-Category Model (Brandts *et al.* [7]), inspired by the above mentioned eight-category model, has a simple enough architecture to allow us to determine analytically its rich repertoire of dynamical behaviors and classify all possible interaction matrices into a limited number of category interaction types, which in turn determine the resulting attractors and their basins. Our analysis provides much needed insight into the nature of the solutions and complexity of many-category models. It reveals the types of category interactions that lead to stable populations that recover quickly from perturbations, and that allow the equilibrium populations to be chosen by suitable parameter choises.

In our Two-Category Network Model, we consider identical units (ants) that belong to either of two categories. These categories or tasks can be interpreted as, e.g., active vs inactive, or as outside vs inside the nest. We model how the population gets divided into categories dynamically and focus on ant-ant interactions (while later adding ant-environment interactions).

The model contains N identical units, each with a state S_i. $S_i = 1$ denotes category 1. $S_i = -1$ denotes category 2.

n_i is the population of category i; $n_1 + n_2 = N$.

The interaction between the i-th and j-th unit is *category-dependent* and given by the interaction matrix:

$$\alpha_{IJ} = \begin{pmatrix} \alpha_{11} & \alpha_{12} \\ \alpha_{21} & \alpha_{22} \end{pmatrix}, \tag{2}$$

3.4. NETWORK DYNAMICS

The dynamics of the network were generated with asynchronous random updating of the state of individual units (which is the only source of noise). A unit in category 1 sees the field F_1:

$$F_1 = n_1 \alpha_{11} S_1 + n_2 \alpha_{12} S_2 - \alpha_{11} S_1 \qquad (3)$$
$$= n_1(\alpha_{11} + \alpha_{12}) - \alpha_{11} - N\alpha_{12} \qquad (4)$$

A unit in category 2 sees the field F_2:

$$F_2 = n_1 \alpha_{21} S_1 + n_2 \alpha_{22} S_2 - \alpha_{22} S_2 \qquad (5)$$
$$= n_1(\alpha_{21} + \alpha_{22}) - (N-1)\alpha_{22} \qquad (6)$$

We can define the critical populations:

$$n_a = \frac{\alpha_{11} + N\alpha_{12}}{\alpha_{11} + \alpha_{12}}, \qquad n_b = \frac{(N-1)\alpha_{22}}{\alpha_{22} + \alpha_{21}}. \qquad (7)$$

The dynamics proceeds as follows: choose an ant at random, then switch its state according to the rules below.
A unit in category 1 will switch to category 2 if $F_1 < 0$, i.e. if

1. $n_1 < n_a$, for $(\alpha_{11} + \alpha_{12}) > 0$, or if
2. $n_1 > n_a$, for $(\alpha_{11} + \alpha_{12}) < 0$.

Similarly, a unit in category 2 will switch if $F_2 > 0$, i.e. if

1. $n_1 > n_b$, for $(\alpha_{22} + \alpha_{21}) > 0$, or if
2. $n_1 < n_b$, for $(\alpha_{22} + \alpha_{21}) < 0$.

3.5. DYNAMICAL BEHAVIORS

The dynamics show attractors (e.g. see Fig.5 and Fig.6), repellors, and lines of fixed points (e.g. see Fig.5). These behaviors are either deterministic or stochastic. We classify interaction types (see Table 1) according to whether the effect one category has on another is inhibiting or activating; this leads to 16 possible matrices, 8 symmetric, 8 asymmetric.

TABLE I.   Interaction Types

type	self-	cross-
activator	a	a
inhibitor	i	i
bigot	a	i
loner	i	a

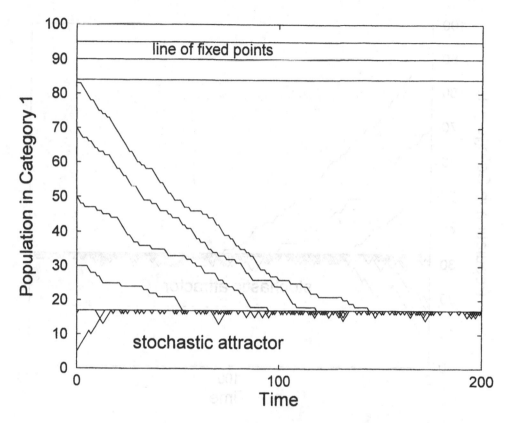

Figure 5. Category 1 population versus time for various initial conditions from model simulations of an activator-inhibitor interaction. The system exhibits bounded fluctuations on a stochastic attractor for $n_1 < n_b$, monotonically decaying solutions for $n_b < n_1 < n_a$ that lead to the trapping region $n_1 < n_b$, and constant solutions for the line of fixed points $n_a < n_1 < N$. $N = 100$, $a_{11} = 1$, $a_{12} = 5$, $a_{21} = -5$, $a_{22} = -1$. Here $n_a = 83.5$ and $n_b = 16.5$.

The interaction type determines what dynamical structures can arise (see Brandts *et al.* [7] for details).

3.6. CANDIDATE COUPLINGS FOR TASK ALLOCATION

Four sign patterns produce stable but fluctuating (stochastic) population dynamics.

1. Loner-loner
2. Inhibitor-inhibitor
3. Inhibitor-loner
4. Loner-inhibitor

The model thus makes predictions about the coupling patterns (i.e. loner or inhibitor) required for ants to show stable but fluctuating behavior among the various categories.

The model can also be treated analytically. The evolution of the model can be formulated in two possible ways, as an iterated function system and as a birth-death jump

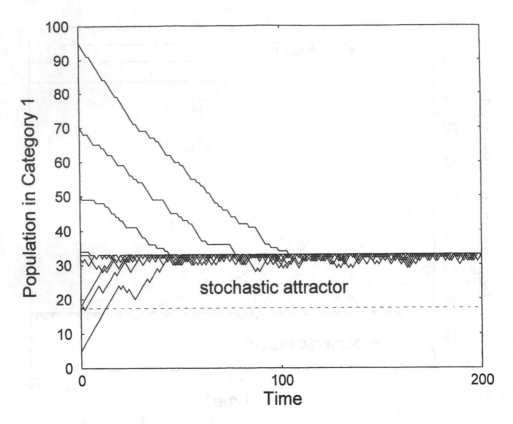

Figure 6. An asymmetric stochastic attractor arises for loner-loner interaction with matrix elements $a_{ii} = -5, a_{12} = -1, a_{21} = -10$. The critical population values $n_a = 17.5$ and $n_b = 33$, are not centered on $N/2$. All initial conditions converge to solutions that fluctuate persistently near n_a. $N = 100$.

process. The trajectories of the model can thus be viewed as random walks with state-dependent biases in population space. A master equation with state-dependent transition probabilities can be derived and the probability density of the population calculated. This formalism also allows computation of decay rate of the transients following perturbations. Task allocation with more than two tasks could also be analyzed in this way.

Interactions with the environment can be examined by adding an environmental stimulus in two ways: altering the threshold associated with a given field, or altering the field with a bias term. If the stimulus strength is independent of the category population, the category population follows the time variation of the stimulus (e.g. circadian activity). The critical parameters n_a and n_b, which determine attractor boundaries and type, also follow the stimulus. Accordingly, the attractors move around in time and qualitative changes may occur if n_a, n_b cross.

If the stimulus strength depends on category population (e.g., foragers deplete new food source) there are more opportunities for n_a, n_b to cross and qualitatively different dynamics can arise, depending on the initial stimulus level.

Evolution of task allocation can also be studied. In this view, evolution selects the set of dynamical rules or parameters of the model, i.e. instead of natural selection fixing ratios of individuals in each task (e.g. caste or threshold), it may select an optimal way to switch tasks. For example, there may be a propensity for an outside worker to switch to foraging when a new food source appears rather than selection for more foragers (Gordon [16, 17]).

It is possible to study how cross-interactions and switching categories evolves in this model by observing the effect of changing interaction coefficients.

For example, if units initially exert control only over their own population (self-interaction is inhibitory and there is no cross-interaction) wide fluctuations about the median population occur. As cross-interaction is increased, initially, the effect is small (safe, gradual evolution); but as the magnitude of the cross-interaction approaches the self-interaction, there is a dramatic narrowing of the attractor, i.e. tighter control.

3.7. CONCLUSIONS ON ANTS

We have analyzed the rich dynamical behavior of a two-category network model of task allocation without hierarchical control (for details see (Brandts et al. [7]). We were able to determine the full repertoire of dynamical behaviors as a function of the interaction matrices. It turned out that the 8 symmetric and 8 asymmetric interaction matrices could usefully be classified into a limited number (10) of category interaction types using the notion of 'activator' and 'inhibitor'. Moreover, particular category interaction types determine the characteristices of the population behavior, i.e. the attractor behavior. Intrinsic fluctuations are one prediction of our model with respect to the sources of variablity observed in measurements of population number of ants in different tasks.

We described and analyzed the attractors of the model: features such as their stability, boundaries, deterministic vs stocastic nature, location, width, the montonicity of transient solutions, the rate of decay of perturbations. All these features can be obtained from time series data of category population numbers and thus direct comparisons between data and model are possible. We also studied a number of general features of our model.

We further determined the conditions on the interaction matrix which lead to stable behavior of the populations and isolated a number of category interaction types as candidates for task allocation models.

4. General Conclusions

We have argued in this essay for the value of 'simple' models for complex biological systems, drawing on the examples of a field model of pattern formation in *Tetrahymena*, and a network model of task allocation in ants. In contrast to complex *in silico* models, where everything known about a system is put into a simulation (and which leads to a big 'black box' but little insight), we recognize that the emergence of new phenomena at each level of description -a feature of complexity already present in purely physical systems (e.g. hydrodynamics vs. molecular dynamics)- necessitates a multi-level approach to modeling in complex biological systems. Strategic relationships between variables (and parameters) must be discovered, and subsequently, the model and its indigenous concepts bridged to other levels.

References

1. Bonabeau, E., G. Theraulaz, J.L. Deneubourg, S. Aron and S. Camazine (1997). "Self-organization in social insects," *Trends Ecol. Evol.* 12, 188–193.
2. Bonabeau, E., G. Theraulaz, J.L. Deneubourg (1998). "Fixed Response Thresholds and the regulation of division of labor in insect societies," *Bull. Math. Biol.* 60, 753–807.
3. Brandts, W.A.M. (1997). "Complexity: A Pluralistic Perspective," in *Physical Theory in Biology: Foundations and Explorations,* edited by C.J. Lumsden, W.A. Brandts, and L.E.H. Trainor. World Scientific Publishing, London.
4. Brandts, W.A.M. (1995) "Relevance of field models to global patterning in ciliates," in *Interplay of Genetic and Physical Processes in the Development of Biological Form.* edited by E. Beysens, G. Forgacs and F. Gaill. Singapore: World Scientific.
5. Brandts, W.A.M. (1993a) "A Field Model of Left-Right Asymmetries in the Pattern Regulation of a Cell," *IMA J. Math. Appl. in Medic. & Biol.* 10, 31–50.
6. Brandts, W.A.M (1993b) "A Field Model of Symmetry Reversals in the Pattern Regulation of a Cell," in *Experimental and Theoretical Advances in Biological Pattern Formation,* edited by H.G.Othmer et al., Plenum Press, New York, 1993.
7. Brandts, W.A.M., A.Longtin, and L.E.H.Trainor (2001). "Two-category Model of Task Allocation with Application to Ant Societies," *Bull. Math. Biol.* 63, 1125–1161.
8. Brandts, W.A.M, and Trainor, L.E.H (1990a) "A Non-Linear Field Model of Pattern Formation: Intercalation in Morphallactic Regulation," *J. theor. Biol.* 146, 37–56.
9. Brandts, W.A.M, and Trainor, L.E.H (1990b) "A Non-Linear Field Model of Pattern Formation: Application to Intracellular Pattern Reversal in *Tetrahymena*," *J. theor. Biol.* 146, 57–87.
10. Bryant, S.V., French, V., and Bryant, P.J. (1981) "Distal Regeneration and Symmetry," *Science* 212, 993–1002.
11. Bryant, P.J. (1974) "Pattern Formation, growth control, and cell interactions in *Drosophila* imaginal discs," in *The Conal Basis of Development* edited by S. Subtelny and I.M. Sussex. Academic Press, New York.
12. Deneubourg, J.L. and N.R. Franks (1995) "Collective control without explicit coding: The case of communal nest excavation." *J. Insect Behav.* 8, 417-432.
13. Frankel, J. (1995) in *Interplay of Genetic and Physical Processes in the Development of Biological Form.* edited by E. Beysens, G. Forgacs and F. Gaill. Singapore: World Scientific.
14. Frankel, J. (1989) *Pattern Formation: Ciliate Studies and Models,* Oxford University Press.
15. French, V., Bryant, P.J., and Bryant, S.V.(1976) "Pattern regulation in epimorphic fields," *Science* 193, 969-981.
16. Gordon, D.M. (1996) "The organization of work in social insect colonies," *Nature* 380, 121-124.
17. Gordon, D.M. (1999) *Ants at Work: How an Insect Society is Organized,* New York: The Free Press.
18. D.M. Gordon, B.C. Goodwin, and L.E.H. Trainor (1992). "A parallel distributed model of the behavior of ant colonies," *J. theor. Biol.* 156, 293.
19. Goodwin, B.C. and Trainor, L.E.H. (1985) "Tip and Whorl Morphogenesis in *Acetabularia* by Calcium–Regulated Strain Fields," *J. theor. Biol.* 117, 79-106.
20. Lara–Ochoa (1991) "A Dynamical Model for Phase Transitions of Lipids Induced by Calcium Ions," *J. Theor. Biol.* 148, 295-304.
21. Lewis, J.H., Slack, J.M.W. and Wolpert, L. (1977) "Thresholds in Development," *J. Theor. Biol.* 65, 579-590.
22. Lewis, J.H. (1981) "Simpler Rules for Epimorphic Regeneration: The Polar–Coordinate Model Without Polar Coordinates," *J. Theor. Biol.* 88, 371-392.
23. Maini, P.K., Benson, D.L., Sherratt, J.A. (1992) "Diffusion Models with Spatially Inhomogeneous Diffusion Coefficients," *IMA J. Math. Appl. in Medic. & Biol.* 10.
24. Murray, J.D. (1981) "A pre–pattern formation mechanism for animal coat markings," *J. Theor. Biol.* 88, 161-199.
25. Nagorcka, B.N . (1989) "Wavelike isomorphic prepatterns in development," *J. Theor. Biol.* 137, 127-162.
26. Nelsen, E.M., and Frankel, J. (1986) "Intracellular pattern reversal in *Tetrahymena thermophila*. I. Evidence for reverse intercalation in unbalanced doublets," *Dev. Biol.* 114, 53-71.
27. Nelsen, E.M., and Frankel, J. (1989) "Maintenance and regulation of cellular handedness in *Tetrahymena*," *Development* 105, 457-471.

28. Nijhout, H.F. (1980) "Pattern Formation on lepidopteran wings: determination of an eyespot," *Devl. Biol.* 80, 267-274.

29. Odell, G.M., Oster, G., Alberch, P., & Burnside, B. (1981) "The Mechanical Basis of Morphogenesis," *Dev. Biol.* 85, 446-462.

30. Oster, G.F., Murray, J.D. and Maini, P.K. (1985) "A Model for chondrogenic condensations in the developing limb: the role of extracellular matrix and cell tractions," *J. Embryol. exp. Morph.* 89, 93-112.

31. Wolpert, L. (1989) "Positional information revisited," *Development Supplement* 3-12.

EMBRYOGENESIS AS A MODEL OF A DEVELOPING SYSTEM

O. P. MELEKHOVA

M. V. Lomonosov Moscow State University

1. Introduction

Embryogenesis provides a unique chance to concurrently perform an experimental investigation and theoretical treatment of a purely natural model of a complex developing system. Embryogenesis comprises the earliest life period of a multicellular organism *(Metazoa)*. In this period, a hierarchical structure of organs and a diversity of cell types arise from a single cell - a fertilized egg. All these events are controlled by a single genetic program and supported by the intrinsic resources of the egg. From the very beginning, the embryo is a complex system: the egg itself possesses an intrinsically nonuniform cytoplasm, which, later on, ensures the self-organization of the complicated structure of the organism in total. In other words, formation of an embryo organism needs no special influence from outside and, hence, may be considered a self-organizing system.[1] Hereafter, we essentially focus on the low-vertebrates case.

For generations, the embryogenesis of any living species of animals in general reproduces itself with high accuracy. Nevertheless, each individual group of cells may behave in a way that is not strongly determined. Right after fertilization, not only does the embryo-space organization begin, but also the individual time of life starts. In embryogenesis, this individual time counts according to the cell division (pendulum-type count of the individual time) and to the irreversible processes (sand-glass count of time) of cell differentiation (specialization), growth, and morphogenesis. The occurrence of special types of cells obviously results from the differential activity of genes.

In recent years, developmental biologists have generally employed molecular-genetic analysis in their study of the problem of spatial organization. In the space-time aspect, embryogenesis appears to result from the interaction of the genetic program of cells with the so-called *morphogenes*, the special regulators of gene expression, distributed nonuniformly over the embryo. In some cases, investigators detected genes and their products that controlled certain morphogenetic effects. Thereby, the detailed mosaic pattern of cause-consequence connections was partially discovered for some animals as regards these processes at the molecular and cellular levels.

However, one should also take into account the general macroscopic dynamics of development, which is similar for different kinds of animals. We believe (Melekhova [7],

---

[1] According to H. Haken [4], a system experiences self-organization as long as, free of any special influence from the outside, by itself it acquires its own structure

J. Nation et al. (eds.), Formal Descriptions of Developing Systems, 269–276.

[9], [10]) that the site and time that start the events of morphogenesis and differentiation, i.e., that give rise to the coordinated hierarchy of organs, follow a single epigenomic space-time program. Therefore, it seems fruitful to emphasize some characteristic features of embryogenesis as a model of a complex developing system.

2. Characteristic features of embryogenesis as a model of developing system

Reading out of genetic information in embryogenesis occurs gradually and, at the beginning, is governed by the egg architecture. An as-fertilized egg already has a complicated structure, which encodes the pattern of the future organism. In this pattern, the coordinate axes and body domains are outlined, thus predicting the location of the general acts of differentiation. Beginning from the single-cell stage, the inner space of an embryo already appears to be a nonuniform dynamic system. Metabolic gradients are known, such as the gradients of sensibility to damage (Child [1]), structural heterogeneity of the cytoplasm, and nonuniform distribution of the regulatory molecules and energy substrates in the egg.[2] These egg properties provide a ground for the nonlinear behavior of a developing embryo, as well as for its equivocal response to external and internal influences.

Fast division into a cell multitude, which terminates at a certain critical number of cells, is a necessary phase for all multicellular organisms *(Metazoa)*. These resultant cells are expected to possess a wide choice for differentiation (large potencies of development).

Qualitative diversity within this cell population grows gradually, and the process of early differentiation is of a bufurcational nature. This diversity first shows itself once the cells become quantitatively different in nonspecial characteristics like the rate of dividing, response to the intrinsic and extrinsic influences, and the general level of oxidation processes.

The process of differentiation (specialization) is only possible with a group of cells whose amount has attained a certain critical number. The differentiation begins from a latent period of determination, in which the "cell fate" becomes predetermined at the gene level; here, the development potential reduces whereas neither biochemical nor morphological changes are visible. Typical of early embryogenesis, synthesis of matrices, needed for a coming phase of development, occurs in advance. The above matrices, capable of fast activation, are for a while preserved in an inactivated form.

The state of competence is a typical precursor of the determination period. In that state, the cells are extremely susceptible to a weak stimulus and ready to respond by passing to differentiation. In a competence state, the cells possess wide-range potencies to development. Moreover, the cells become extremely liable to damage effects so that the call fate is uncertain and can change depending on the close environment (e.g., with transplantation).

The state of competence, of a particular sensibility to a stimulus inducing differentiation, is an endogenic property. This arises and dies out as strictly connected with the age of the cells, i.e., at definite stages of the embryo development, and exists not synchronously

---

[2] It is at the expense of the mother matrices and intrinsic energy substrates that embryogenesis is maintained during the period of time when the embryo-environment interaction consists only of diffusion-assisted gas exchange. Therefore, the embryo makes a somewhat less open system as compared with an adult organism.

in all of the embryo rudiments.[3] The same signal initiates different reactions and specific kinds of synthesis in the cells of different "ripeness", i.e., of different biological age. Currently, the molecular-genetic aspects of those phenomena are being thoroughly investigated. Most likely, those phenomena are conditioned by membrane receptors as well as by the metabolic factors ensuring the signal transmission within a cell, and by the readiness of the cytoskeleton for morphological reconstructions. With all that knowledge, we are still not quite realize the nature of the endogenic processes that make cells competent with respect to either kind of differentiation.

Differentiation events, resulting in the occurrence of different kinds of cells and leading to the development of organs, are merely coordinated through induction: in the mutual contacts, the "older" rudiments induce (initiate) differentiation in the "younger" rudiments. In early embryogenesis, an important role also belongs to the phenomenon of *positional information* (Wolpert [14]). Thanks to the latter, the cells experience differentiation according to their location in the embryo. The distribution gradients of the "morphogenes" (regulatory elements of the gene expressions) between the cells, as well as the difference between the latter in the threshold sensitivity to those "morphogenes", are believed to be responsible for the different directions of differentiation.

Typical of the early phases of development, an embryo exhibits a capability of *embryonal regulation*, which shows itself in the recovery of a normal way of developing next to the partial removal, stirring, or adding to the cell material. Indeed, a half of a multicell embryo of sea urchin can develop to a quite fair larva (Driesch [2]). It is important, therefore, that the positional information is contributed before the steady determination of the cell material has occurred.

In embryogenesis, the *development phases*, as well as the governing homeostatic systems, sequentially replace one another. Those phases provide the "reversibility barriers" and sequentially determine the biological age, i.e., the following respective morphophysiological states of an embryo.

- Within the first phase of cleavage, cell dividing brings about the occurrence of a multicell embryo, called a *blastula*.
- Cell populations of different quality appear within the next phase, *gastrulation*.
- An axial rudiment appears, in the vertebrate case, at the *neurulation* phase.
- Within the next phase - *organogenesis* - differentiation occurs, which terminates with the formation of organs.

Next to it, low vertebrates hatch out from the capsule to start their larva period of life with its active motion and feeding.

Each transition from one phase to another is associated with passing through the state of instability against external influences and with increased susceptibility to damage. Therefore, those transitions may be referred to as the critical periods of development (Svetlov [12]). C.H. Waddington [13] imaged the development issue as a *morphogenetic landscape*, in which the periods of numerous potencies, determination, and sensibility to the internal and external effects were alternated with the periods of the steady-state programmed development.

---

[3] A group of embryo cells, from which an organ begins to develop.

3. Embryogenesis from the viewpoint of the self-organization theory

According to the self-organization theory, one can consider the embryo interior as an active nonlinear environment, whose sensitivity to external excitation is nonuniform and which demonstrates threshold effects and typical metabolistic parameters. Those parameters experience fluctuations, induced by such excitations and capable of spontaneously strengthening or dying out (dissipating). In such environments, irreversible processes are possible, which can be localized into space-time structures.

Synergetics proposes certain rules of and indications to the bifurcational ways of development. One can comprehend by intuition the applicability of those rules to embryogenesis, in particular, to the dichotomic occurrence of the early differentiations. *Order parameters* (Haken [3]) is another synergetic concept used when describing a hierarchy of arising structures. That term defines the limited number of degrees of freedom, peculiar to a continuous complex system; all the other parameters align according to those parameters.

The stability states, which a system tends to approach and in which it for a while exist, are defined as "attractors". The latter seem close to an embryological term *pattern*, describing an expected ("planned") space organization of an embryo.

Cells behave according both to the specific chemical signals, which influence quite certain molecular reactions, and to the energistic regulation, which determines whether the reception of a governing signal and the cell response are at all possible. Therefore, in the early individual development, i.e., before the function of the own genome of the embryo is turned on, the likelihood of the governing role of the energy parameters is high (Melekhova [7]). We have chosen one of the energistic indicators of cell metabolism, the level of free-radical reactions. Free radicals (FR) are formed as intermediate products of various oxidation-reduction reactions. This indicator is of an integrated nature, i.e., representative of the total concentration of FR in a test object. Thus, free radicals seem to provide a measure of the rate of energy exchange (storage and consumption) in a cell. We regularly did our measurements during the sequential phases of embryo development. An autoradiography technique was used to reveal the distribution of FR reactions. Finally, we could compare the individual regions of an egg or the groups of cells in terms of the FR distribution over the hystological section of a test embryo (Melekhova [6], [7]).

We found that fertilization resulted in extremely nonuniform distribution of FR reaction (track density). Then, a region (multicell) emerged, in which the FR reactions looked self-accelerated (Fig. 1). In that region only, the FR relative concentration exhibited sharp random (not related to a certain cell) fluctuations. Quite soon, the FR density over this region increased tenfold. Next to it, the cell cycles slowed down and the gene expression began, indicating to the initiated differentiation (Fig. 2). In the region of still "low-energy" of the embryo, the above effects came later but also began with the growing rate of energy exchange (FR level).

At the peak of energy exchange within the determination time (when one In any of the next phases of the further development, the determination (bifurcation) of the cell types always begins with the separation into the lower-energy and higher-energy regions (Fig. 3). In the organogenesis phase, embrional induction becomes a primary mechanism, coordinating the localization of differentiating cell types relative one another. Besides, a similar *energy pattern* was revealed in various inducing systems: the inductor appeared

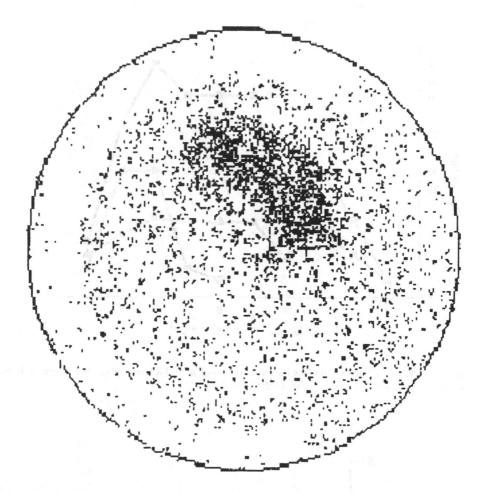

Figure 1.   Regional arrangement of the FR reactions in the embryos of *Anura*, the autoradiography data (Melekhova [7]). On the left: hystological sections (schematic) of the embryos, with the zones (labeled) dotted according to the observed track density. On the right: the frequency diagrams (schematic), showing the track density per 25 $\mu\, m^2$ of the section area.

the older (regarding the biological age) rudiment, characterized by a higher energy level. Upon the contact with the inductor, the "younger" rudiment experienced increasing of its energy level to thus prepare itself for differentiation.

At the peak of energy exchange within the determination time (when one of the alternative ways of differentiation is being selected), the cells become unstable and highly sensitive. They respond to even weak damaging influences (including those of the contaminated environments) by the disturbance of differentiation. Consequently, specific anomalies (terates) may occur. Based on such effects, we invented a method of indicating contaminants in a water environment (Melekhova [8]). It looks like, in normal development, such an enhanced liability and instability of cell populations is a critical condition of a competence phenomenon. The latter arises spontaneously at a certain individual time as a capability of receiving a signal to begin differentiation.

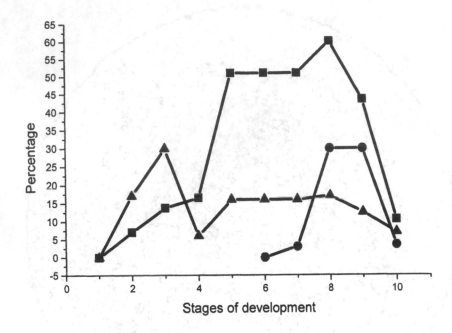

Figure 2. (■) density fluctuations of the FR-tracks, (●) growth of the FR level, and (▲) RNA synthesis in the nuclei of the embryo cells in zone 5 of Fig. 1 (Melekhova [7]).

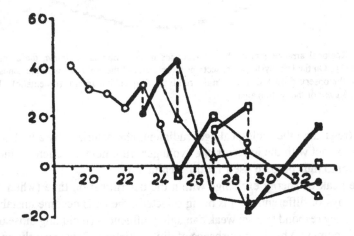

Figure 3. Bifurcation events in the lens of an amphibian embryo: sequential occurrence of the irregularities in the FR concentration, next to which differentiation of the respective groups of cells begins (Melekhova [5]). $\Delta C / C$ is the local FR concentration as referred to the average FR level over a $25\ \mu m^2$ area of the hystologic section.

4. Conclusion

Judging from the above experimental data, the instability states, lawfully arising in embryogenesis, are a necessary condition for the differentiation and general development of an embryo. Energetic metabolism provides a factor that controls the onset of the instability state. It looks like the instability regions are localized in the individual time and inner embryo space according to an epigenomic space-time program. The latter must be inherent in the egg cytoplasm by the nonuniform distribution of substrates, promoters, and inhibitors of the FR reactions. This program is realized through the relationship between the rates of the energy exchange, peculiar to the different regions of a developing embryo.

One can expect that the critical amounts of and proportions between the substrates, promoters and inhibitors of the energy exchange must put into all the individual time, intrinsically reserved for the embryogenesis of a certain species. The individual ages of particular cell populations is determined by the consumed fractions of the energosubstrates (Melekhova [9]).

Note again the condition for entering the differentiation period: the energy metabolism must achieve its peak level and thus bring the system to a highly unstable and liable state. This condition predicts certain unpredictability of each cell fate, which definitely correlate with the concept of chaotic state (Ruelle [11]). Moreover, an appreciable chaotic component appears a necessary element in the phenomenon of transition between two different homeostatic (determined) states of a developing system.

Whichever uncertainty, a normal development still appears in total a highly determined and exceptionally reproducible issue. Therefore, a temporary state of uncertainty as regards the choice from the alternative ways of further development, which all of the cells are born to make, suggests an idea of *determined chaos*. Indeed, in the lawful course of development, both the occurrence of that *necessary* chaotic state and the ways of passing out of it by a fair choice are believed to be strictly controlled by the epigenomic program as an issue of a higher hierarchic level. Furthermore, the rate of FR oxidation-reduction reactions appears an *order parameter*, which is to control the starting point of differentiation events in embryogenesis (Melekhova [10]).

Acknowledgement. The author is grateful to Dr. Yu.P. Liberov for the helpful discussing and translating of the manuscript.

References

1. Child, C.M.(1941) *Patterns and Problems of Development*, Univ. of Chicago, Chicago.
2. Driesch, H. (1894) *Analytische Theorie der Organischen Entwicklung*, W. Engelmann, Leipzig.
3. Haken, H. (1980) *Synergetics*, translated into Russian under the title *Sinergetica*, Mir, Moscow.
4. Haken, H. (1991) *Information and Self-organization*, translated into Russian under the title *Informatsiya i samorganizatsiya*, Mir, Moscow.
5. Melekhova, O.P.(1976), Nekotorye fiziko-khimicheskie aspekty differentsirovki glaznogo zachatka u Anura (Some Physicochemical Aspects of the Differentiation of the Eye Rudiment of Anura), Cand. Sci. (Biol.) Dissertation, Moscow State University.
6. Melekhova, O.P. (1976) Free-Radical Processes in the Embryogenesis of Anura, *Ontogenez* 7, 131–140.

7.   Melekhova, O.P. (1990) Physicochemical Characteristics of the Space-Time Arrangement of Embryoge-
     nesis in Anura, in *Kletochnaya reproduktsiya i protsessy differentsiatsii* (The Cell Reproduction and the
     Processes of Differentiation), Nauka, Leningrad, pp. 30–51.
8.   Melekhova, O.P. (1994) Evaluation of the Embryotoxicity of Water Environment, *Izv. Russian Akad Nauk,
     Ser. Biological*, no. 4, pp. 661–666.
9.   Melekhova, O.P. (1998) Time as a Development Factor, in *Prostranstvenno-vremennaya organizatsiya
     ontogeneza* (Space-Time Organization of Ontogenesis), Moscow State University, Moscow, pp. 24–39.
10. Melekhova, O.P. (2000) Self-organization Phenomena in Embryogenesis, in *Sinergetika-3* (Synergetics-3),
     Moscow State University, Moscow, pp. 319–325.
11. Ruelle, D. (2001) Applications of Chaos, in W. Sulis and I. Trofimova (eds.), *Nonlinear Dynamics in the
     Life and Social Sciences*, NATO Science Series A: Life Sciences, JOS Press, Amsterdam, pp. 3–12.
12. Svetlov, P.G. (1978) *Physiology (Mechanics) of Development*, Leningrad (in Russian).
13. Waddington, C.H. (1962) *New Patterns in Genetics in Development*, Columbia University, New York.
14. Wolpert, L. (1983) *From Egg to Embryo: Determinative Events in Early Development*, Cambridge
     University, Cambridge.

Chapter 5.

Presentations
for Discussion

CLINICAL RESEARCH HISTOMARKERS FOR OBJECTIVELY ESTIMATING PREMORBID VAGAL TONE CHRONOLOGY IN GULF WAR VETERANS' ILLNESSES AND IN ACUTE STRESS REACTION

H.S. BRACHA, J.M. YAMASHITA, T. RALSTON, J. LLOYD-JONES, G.A. NELSON, D.M. BERNSTEIN, N. FLAXMAN, & F. GUSMAN
National Center for PTSD, Department of Veterans Affairs
Spark M. Matsunaga VA Medical and Regional Office Center, Honolulu, Hawaii

Abstract: While laboratory techniques for estimating genetic susceptibility for adult psychopathology have received much recent attention, laboratory techniques for objectively estimating a person's early stress related autonomic nervous system perturbations have been under-researched. Biological psychiatric research on heart rate variability suggests that early life episodes of low vagal tone may predict poor stress resilience in adults. This chapter will detail a research technique for retrospectively estimating in adults the chronology of low vagal tone episodes experienced prior to age ten. This technique makes use of the developing enamel matrix, one of very few tissues that cannot recover after being stressed. This is likely to be clinically useful in understanding the etiology of several disorders, especially Posttraumatic Stress Disorder (PTSD) as well as post-deployment disorders of unclear etiology such as Gulf War Veterans' Illness (GWVI). Finally, our proposed technique may be useful in the research on Acute Stress Reaction, Acute Stress Disorder, and on stress resilience (hardiness) in populations expected to be exposed to high levels of cumulative stress (e.g., special forces and other active duty personnel during deployment).

1. Relevance to the department of defense and veterans affairs mission

Post-deployment disorders such as Gulf War Veterans' Illnesses (GWVI) has become a topic of considerable concern. Extensive research over the last twelve years strongly suggests that the most likely mechanism in the etiology of these disorders in most veterans is neurochemical brain changes resulting from stress sensitization. Additionally, there is extensive indirect evidence that pre-enlistment extreme stress is a predisposing factor for GWVI [8, 23, 29, 43, 55]. However, it has been very difficult for the scientific and federal community to convince GWVI patients and other stakeholders that the above neurobiological mechanism underlying GWVI makes it akin to Combat Related-Posttraumatic Stress Disorder (CR-PTSD) and thus probably amenable to evidence-based treatments currently established for PTSD.

We have reasoned that the current generation of patients and veterans is increasingly accustomed to expecting confirmation or refutation of medical decisions by laboratory tests. Laboratory tests are perceived as tangible physical evidence. Psychiatry has only recently embarked on including biomarkers to enhance clinical decision making. Therefore, an evidence-based biological laboratory estimate of pre-enlistment stress may be of interest for research in GWVI. Additionally, a biological laboratory estimate of pre-enlistment stress may be of interest for research on stress resilience and

J. Nation et al. (eds.), Formal Descriptions of Developing Systems, 279–288.

hardiness. Understanding resilience and hardiness is a new and growing area of emphasis at the DoD [44].

2. The parasympathetic nervous system in acute stress reaction

Extreme autonomic nervous system perturbations during early life can produce long-term deleterious effects. For reviews of recent important work see Porges et al. [34, 45], McEwen [18, 28, 52] and Sapolsky [47, 48].

Stress research has focused on the adverse effects of adreno-cortical activation. Following Porges' important theoretical contributions, there has been growing attention to low vagal tone as an important final common pathway leading to the adverse effects of the acute stress reaction [34, 45].

The vagus is a complex bi-directional system with a left and right myelinated branch. Each branch has two source nuclei with fibers originating either in the dorsal motor nucleus or the nucleus ambiguus. As Porges (1995) points out, while previous research has focused on the dorsal motor nucleus of the vagus, less attention has been given to the motor pathways originating from the vagal nucleus ambiguus.

The nucleus ambiguus is the more anterior and rostral (limbic) of the two vagal nuclei. Furthermore, a potentially important hemispheric lateralization exists in the ambiguus motor neurons. As articulated by Porges (1995, p. 228):

"In the study of acute stress and emotional expression, vagal pathways originating in the right nucleus ambiguus are critical. The right nucleus ambiguus provides the primary vagal input to the sino-atrial node to regulate heart rate ...and to the larynx to regulate the vocal intonation. Acute stress, especially during painful procedures, is associated with high heart rate and high pitch vocalizations and cries (e.g., the high pitched cries of infants in severe pain). Both characteristics are determined by a withdrawal of vagal efferent outflow originating in the nucleus ambiguus. These vagal afferents can act instantaneously to change heart rate and the pitch of vocalizations. Unlike the involuntary and often prolonged, characteristic pattern of vagal outflow from the dorsal motor nucleus, the outflow from the nucleus ambiguus may exhibit rapid and transitory patterns associated with perceptive pain or unpleasantness. The central nucleus of the amygdala, implicated in emotional lability [and fear], directly communicates with the nucleus ambiguus. Thus, the branch of the vagus originating in the nucleus ambiguus is closely linked to the rapid expression and regulation of emotional state..."

3. Nucleus ambiguus control of enamel secretion by ameloblasts

The nucleus ambiguus has been implicated in several physiological markers of Combat Related-Posttraumatic Stress Disorder (CR-PTSD) such as heart-rate variability (respiratory sinus arrhythmia) [33, 46]. There has been little attention given to the fact that vagal cholinergic neurons originating in the nucleus ambiguus control the trophic parasympathetic regulation of blood flow to the enamel secreting ameloblasts. The nucleus ambiguus neurons ending in the ameloblast layer travel by a circuitous route through other cranial nerves [21, 32]. The dramatic slowing of enamel secretion by the ameloblast is produced by an abrupt drop in ambiguus cholinergic activity (vagal tone) which also produces the change in vocal intonation, facial expression, and especially saliva secretion and thus the symptom of "dry mouth." All of these are well-documented physical signs of acute stress. These neurons affecting the rate of enamel secretion are of the same extended amygdala origin as the neurons that coordinate increased heart rate, breathing, and peri-laryngial constriction (high-pitch vocalizations and the feeling of a "lump in the throat") which are part of acute stress reactions [33].

As a rule, during acute stress reaction and acute stress disorder, the trophic parasympathetic functions slow or cease transiently [15]. Prior to age ten, the trophic "luxury" parasympathetic functions that typically slow or cease during stress include secretion of the still developing dental enamel matrix [19, 22, 51, 54].

4. "DDE-SH Rings": a proposed technique for estimating early vagal tone chronology

Our technique uses some of the histological markers known in the fields of anthropology and paleopathology as Developmental Defects of Enamel or DDE. In this review, we use the term DDE-Stress Histomarker Rings (DDE-SH Rings; pronounced "desh rings") primarily to differentiate them from other DDE which are mineralization defects but also to emphasize the recent understanding of the etiology of these histomarkers. DDE-SH Rings have been developed as a Histomarker based upon carefully conducted research by paleopathologists, forensic dental anthropologists, and bio-archeologists. For recent work see Dean, Leakey, and Reid, (*Nature* 2001).

DDE-SH Rings are dome-shaped malformations demarcating the layer of dental enamel secreted by ameloblast cells during periods associated with low vagal tone [2, 3, 9, 30, 41]. Both acute and more chronic stress can be estimated via DDE-SH Rings of various widths.

There are various grades of histological features which together we call DDE-SH Rings. These include microscopic histomarkers (Circadian Cross Striations, Circaseptan Striae of Retzius, Accentuated Striae of Retzius, and Wilson Bands) as well as clinically detectable histomarkers (Linear Enamel Hypoplasias and other Gross Enamel Hypoplasias) [16, 17, 20]. There is growing consensus in the paleopathology and dental anthropology literature that "Selyean stress" lasting one-week or longer will produce DDE-SH Rings [3, 22], and that the transient disruption of ameloblast activity during severely depressed vagal tone can be detected as long as the tooth remains largely intact [17, 26]. Studies by Wright [60] suggest that the DDE-SH Rings called Accentuated Striae of Retzius, may be especially useful and can be produced in as little as one or two days of extreme stress.

At present it may be too early to determine which histomarker will be the most useful in clinical stress research in humans. Nevertheless, based on the above literature and additional literature cited below, we have proposed that human enamel is an untapped resource for detecting periods of low vagal tone that occur during early brain development [5, 6, 7].

DDE-SH Rings are conceptually akin to tree rings that mark periods of environmental adversity during a tree's development. The developing enamel matrix is one of very few tissues that cannot recover after being stressed. [1, 11, 12, 13, 22, 36,

37, 38, 38, 39, 40, 51, 53, 59]. Thus, the developing enamel has promise as an accessible repository of information on vagal tone chronology prior to age ten when human 3rd molars (the last developing teeth) complete crown formation [19, 54].

As one of the very few premorbid physiological stress markers, DDE-SH Rings may be important for research on the etiology of anxiety and mood disorders. DDE-SH Rings may be useful in researching disorders such as Combat Related-Posttraumatic Stress Disorder (CR-PTSD), Chronic Fatigue Syndrome (CFS), and Fibromyalgia Syndrome (FS). Additionally, DDE-SH Rings may be useful in researching medically unexplained syndromes, such as GWVI, often seen in military veterans in which it is hard to determine whether pre-enlistment stress sensitization has an etiological role in the patient's current illness.

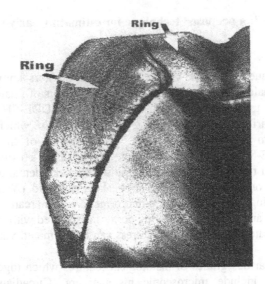

Figure 1: A single microscopic DDE-SH Ring in a combat veteran with a post-deployment disorder (Combat Related-PTSD). The stress Histomarker provides indelible evidence of stress sensitization occurring prior to enlistment, in this case at age 8. Incidentally, this particular tooth also demonstrates evidence of bruxism providing objective evidence of recent symptoms of anxiety.

5. Current clinical and forensic uses

Earlier DDE-SH Rings research in contemporary populations has focused almost entirely on deciduous dentition. DDE-SH Rings in deciduous teeth were found to be useful tools for ruling in, or ruling out, late prenatal, intranatal, and infancy distress in research participants [24, 25].

Several early studies of DDE-SH Rings conducted with deciduous teeth suggest that clinical DDE-SH Rings may be several times more common in patients with neurodevelopmental disorders than in controls. This finding is consistent with the role that low vagal tone during infancy, intranatal and prenatal period plays in the etiology of these disorders [10, 58]. Additionally, poorer clinical outcomes were shown to be associated with DDE-SH Rings [30]. Early stressors, as estimated by the location of

DDE-SH Rings, were associated with a more severe impairment [30]. These studies lend support to the potential utility of DDE-SH Rings.

In a forensic investigation, Skinner and Anderson [54] reported in detail the pattern and timing of a series of DDE-SH rings. They have demonstrated that the rings could be matched to a premortem record of physiological stress experienced. This chronological determination is possible since the rate of enamel elongation is well known. Prior studies observed that, although there is a certain amount of known non linearity in this rate, permanent enamel elongates in the cervical (root) direction at approximately 2.60 μm/day (about one millimeter/year). This information was obtained by measuring the mean inter-striae distance for a first permanent molar from a terminally ill child given timed injections of tetracycline which marked the enamel with orange striae [27].

6. Prevalence of DDE-SH rings in molars in non-clinical populations

Our current research specifically focuses on DDE-SH Rings in permanent molars. Permanent molars have been the least studied human teeth as far as vagal tone is concerned. The most severe form of DDE-SH Rings is Dental Enamel Hypoplasias which are relatively uncommon in permanent molars [31, 57]. Goodman and Rose [19], in their review of five large studies of contemporary populations (over 1,000 subjects), concluded that approximately 11% of first molars and 5% of second molars show enamel hypoplasias. In 3rd molars the prevalence is unknown. The one published study of 24 healthy subjects over age 16 [49], reported enamel hypoplasias in none of the 3rd molars.

The prevalence of less severe (subclinical) DDE-SH Rings is also unclear, although generally believed to be higher. Wright found that out of 43 permanent teeth with subclinical DDE-SH Rings, only 51% manifested clinical hypoplasias [60]. The first study of subclinical DDE-SH Rings specifically focusing on molars is currently being completed by our research team.

7. Evolutionary reasoning for research on DDE-SH rings

A research program investigating enamel markers of vagal tone perturbations is a step towards developing a clinical tool (biomarker) for researching the autonomic nervous system's response during Acute Stress Reaction. Evolutionary biological reasoning has led us to take this approach. Extensive research suggests that survival during extreme stress, throughout human evolution, depended primarily on blood supply to the brain and heart. Several other organs, such as skin, intestines, other mucousae, nails, hair, and bone were of lower priority and grow predominantly during spans of low stress such as sleep. We have reasoned that the anatomical structures of lowest survival priority may be a neglected indicator to the negative effects of stress. While little research has been done on the topic, amelogenesis of the still erupting teeth is one luxury trophic function likely to be among the lowest survival priorities during extreme stress.

8. Few bio-markers are indelible and unaffected by adulthood stress

One unique strength of our DDE-SH Rings technique for estimating premorbid vagal tone chronology is that enamel (the most durable human tissue and is the only human tissue that can withstand cremation) cannot naturally remodel or undergo repair after its initial formation in the way that other human tissues, including dentin, bone, and brain tissue can [19, 22, 42, 51, 54]. DDE-SH Rings therefore, cannot be affected by a stress reaction that occurs after the age at which enamel secretion ceases (shortly after age 10). Since enamel has the permanence of stone, DDE-SH Rings are indelible indicators of pre-adulthood experiences. In other words, DDE-SH Rings are a specific marker of infancy and childhood stress and are completely unaffected by post-pubertal and adult stress. Specifically, DDE-SH Rings cannot be produced by research confounding factors after age 10 such as current or recent exposure to alcohol, drug abuse, smoking, medical illness, medications, poor nutrition, or head injuries. Similarly, DDE-SH Rings are not affected by the research participants' quality of self-report as influenced by current mood, cooperativeness, current psychopathologies, or cultural factors.

9. Conclusions

The possible role of stress sensitization and kindling-like mechanisms in the etiology of anxiety and mood disorders suggests a wide range of disorders to which this technique may be applicable [4, 14, 35, 50, 56]. A histological marker that can retrospectively estimate vagal tone chronology during the first ten years of life would be valuable for clinical research on both PTSD and GWVI. Additionally, such a technique may be valuable for research on Acute Stress Reaction, Acute Stress Disorder, and for understanding stress resilience and hardiness. Finally, DDE-SH Rings may be useful in researching any post-deployment disorder in which it is difficult to convincingly determine whether or not pre-enlistment stress sensitization has an etiological role in the person's current illness.

10. Acknowledgements

This material is based upon work supported in part by the Office of Research and Development, National Center for PTSD, Department of Veterans Affairs, Spark M. Matsunaga Medical and Regional Office Center. Support was also provided by a National Alliance for Research on Schizophrenia and Depression (NARSAD) independent investigator award, a NAMI Stanley Foundation award, and a Japan Ministry of Education and Technology - Basic Research Projects Award. The authors thank Col. Donald A. Person MD and Andrew Williams MA for comments on sections of this manuscript.

11. References:

1. Antoine D, Dean C, Hillson S. 1999. Dental Morphology 1998. In *Dental Morphology 1998*, ed. JT Mayhall, T Heikkinen, pp. 48-55. Finland: Oulu University Press
2. Ash MM. 1993. Dental Anatomy, Physiology and Occlusion Philadelphia: W.B. Saunders Company.
3. Aufderheide AC, Rodriguez-Martin C. 1998. The Cambridge Encyclopedia of Human Paleopathology Cambridge: Cambridge University Press.
4. Bell IR, Schwartz GE, Baldwin CM, Hardin EE. 1996. Neural sensitization and physiological markers in multiple chemical sensitivity. *Regulatory Toxicology and Pharmacology* 24:S39-S47
5. Bracha HS, Lopez HH, Flaxman NA, Lloyd-Jones JL, Bracha AS, Ralston T. 2002. Can Enamel Serve as a Useful Clinical Marker of Childhood Stress? *Hawaii Dental Journal* 9-10
6. Bracha HS, Ralston T, Yamashita J, Nelson G, Lopez HH, Cummings T. 2003. Stressful experiences in children and adolescents: Initial report from the PSEI-NCPV Honolulu study. *Hawaii Med. J.* 62:54-9
7. Bracha HS, Lloyd-Jones J, Flaxman N, Sprenger M, Reed H. 1999. A new laboratory estimate of cumulative autonomic perturbations during early brain development. *Clinical Autonomic Research* 9:231
8. Chalder T, Hotopf M, Unwin C, Hull L, Ismail K, David A, Wessely S. 2001. Prevalence of Gulf war veterans who believe they have Gulf war syndrome: questionnaire study. *BMJ* 323:473-6
9. Clark Spencer Larsen. 1999. Bioarchaeology: Interpreting behavior from the human skeleton Cambridge: Cambridge University Press.
10. Cohen HJ, Diner H. 1970. The significance of developmental dental enamel defects in neurological diagnosis. *Pediatrics* 46:737-47
11. Dean C. 1999. Hominoid tooth growth: using incremental lines in dentine as markers of growth in modern human and fossil primate teeth. In *Human growth in the past: studies from bones and teeth*, ed. RD Hoppa, CM Fitzgerald, pp. 111-127. Cambridge: Cambridge University Press
12. Dean MC. 2000. Progress in understanding hominoid dental development. *J. Anat.* 197:77-101
13. Dean MC, Leakey MG, Reid DJ, Friedeman S , Schwartz GT, Stringer C, Walker A. 2001. Growth processes in teeth distinguish modern humans from *Homo erectus* and earlier hominins. *Nature* 414:628-31
14. Essex M, Klein M, Cho E, Kalin N. 2002. Maternal stress beginning in infancy may sensitize children to later stress exposure: effects on cortisol and behavior. *Biol. Psychiatry* 52:776
15. Fagius J. 1997. Syndromes of autonomic overactivity. In *Clinical autonomic disorders evaluation and management*, ed. PA Low, pp. 777-790. Philadelphia: Lippincott-Raven
16. FitzGerald C.M. 1998. Do enamel microstructures have regular time dependency? Conclusions from the literature and a large-scale study. *J Hum Evol* 35:371-86

17. FitzGerald C.M., Rose JC. 2000. Reading Between the Lines: Dental Development and Subadult Age Assessment Using the Microstructural Growth Markers of Teeth. In *Biological Anthropology of the Human Skeleton*, ed. MA Katzenberg, SR Saunders, pp. 163-186. New York: Wiley-Liss, Inc.
18. Goldstein DS, McEwen B. 2002. Allostasis, homeostats, and the nature of stress. *Stress* 5:55-8
19. Goodman AH, Rose JC. 1990. Assessment of systemic physiological perturbations from dental enamel hypoplasias and associated histological structures. *Yearbook of Physical Anthropology* 33:59-110
20. Guatelli-Steinberg D. 2001. What can developmental defects of enamel reveal about physiological stress in nonhuman primates? *Evolutionary Anthropology* 10:138-51
21. Harati Y, Machkhas H. 1997. Spinal cord and peripheral nervous system. In *Clinical autonomic disorders evaluation and management*, ed. PA Low, pp. 25-46. Philadelphia: Lippincott-Raven
22. Hillson S. 1996. Dental anthropology Cambridge: Cambridge University Press.
23. Lee HA, Gabriel R, Bolton JP, Bale AJ, Jackson M. 2002. Health status and clinical diagnoses of 3000 UK Gulf War veterans. *J R Soc Med* 95:491-7
24. Levine RS, Turner EP, Dobbing J. 1979. Deciduous teeth contain histories of developmental disturbances. *Early Hum. Dev.* 3/2:211-20
25. Leviton A, Needleman HL, Bellinger D, Allred EN. 1994. Children with hypoplastic enamel defects of primary incisors are not at increased risk of learning-problem syndromes. *ASDC J Dent Child* 61:35-8
26. Liebgott B. 2001. The Anatomical Basis of Dentistry St. Louis: Mosby.
27. Massler M, Schour I, Poncher HG. 1941. Developmental pattern of the child as reflected in the calcification pattern of the teeth. *Am. J. Dis. Child.* 629:33-67
28. McEwen BS, Lasley EN. 2002. The End of Stress as We Know It Joseph Henry Press.
29. Morgan CA, III, Grillon C, Southwick SM, Davis M, Charney DS. 1997. Exaggerated acoustic startle reflex in Gulf War veterans with posttraumatic stress disorder. *Am. J. Psychiatry* 153:64-8
30. Murray GS, Johnsen DC, Weissman BW. 1987. Hearing and neurologic impairment: insult timing indicated by primary tooth enamel defects. *Ear Hear.* 8:68-73
31. Murray JJ, Shaw L. 1979. Classification and prevalence of enamel opacities in the human deciduous and permanent dentitions. *Arch. Oral Biol.* 24:7-13
32. Nieuwenhuys R, Voogd J, van Huijzen C. 1981. The Human Central Nervous System. A Synopsis and Atlas Berlin: Springer-Verlag. 1-253 pp.
33. Porges SW. 1995. Cardiac Vagal Tone: A Physiological Index of Stress. *Neurosci. Biobehav. Rev.* 19:2:225-33
34. Porges SW. 2001. The polyvagal theory: phylogenetic substrates of a social nervous system 3. *Int. J. Psychophysiol.* 42:123-46
35. Post RM, Weiss SRB, Smith MA. 1995. Sensitization and kindling: implications for the evolving neural substrates of post-traumatic stress disorder. In *Neurobiological and clinical consequences of stress*, pp. 203-224. Philadelphia: Lippincott - Raven

36. Reid DJ, Beynon AD, Ramirez Rozzi FV. 1998. Histological reconstruction of dental development in four individuals from a medieval site in Picardie, France. *Journal of Human Evolution* 35:463-77
37. Reid DJ, Dean MC. Brief communication: the timing of linear hypoplasias on human anterior teeth. American Journal of Physical Anthropology 113(1). 2000. Wiley-Liss, Inc. Ref Type: Generic
38. Reid DJ, Dean MC. 2000. The timing of linear hypoplasias on human anterior teeth. *Am. J. Phys. Anthropol.* 113:135-9
39. Risnes S. 1998. Growth tracks in dental enamel. *J Hum Evol* 35:331-50
40. Risnes S, Moinichen CB, Septier D, Goldberg M. 1996. Effects of accelerated eruption on the enamel of the rat lower incisor. *Adv. Dent. Res.* 10:261-9
41. Roberts C, Manchester K. 1997. The Archaeology of Disease Cornell University Press.
42. Rose JC, Armelagos GJ, Lallo JW. 1978. Histological enamel indicator of childhood stress in prehistoric skeletal samples . *Am. J. Phys. Anthropol.* 49:511-6
43. Roy M. 1994. *Fed. Prac.* In press
44. Rumsfeld DH. 2002. Transforming the Military. *Foreign Affairs* 81:20-32
45. Sahar T, Shalev AY, Porges SW. 2001. Vagal modulation of responses to mental challenge in posttraumatic stress disorder 4. *Biol. Psychiatry* 49 :637-43
46. Sahar T, Shalev AY, Porges SW. 2001. Vagal modulation of responses to mental challenge in posttraumatic stress disorder. *Biol. Psychiatry* 49:637-43
47. Sapolsky RM. 1997. McEwen-induced modulation of endocrine history: a partial review. *Stress* 2:1-12
48. Sapolsky RM. 2002. Chickens, eggs and hippocampal atrophy. *Nature Neuroscience* 5:1111-3
49. Sarnat BG, Schour I. 1941. Enamel hypoplasia (chronologic enamel aplasia) in relation to systemic disease: a chronologic, morphologic and etiologic classification. *J. Am. Dent. Assoc.* 28:1989-2000
50. Sax, Strakowski. 2001. Behavioral sensitization in humans. *J Addict Dis* 20:55-65
51. Scott GR, Turner CGI. 1997. The anthropology of modern human teeth: dental morphology and its variation in recent human populations Cambridge. Cambridge University Press.
52. Seeman TE, McEwen BS, Rowe JW, Singer BH . 2001. Allostatic load as a marker of cumulative biological risk: MacArthur studies of successful aging. *Proc. Natl. Acad. Sci.* 98:4770-5
53. Simpson SW. 1999. Reconstructing patterns of growth disruption from enamel microstructure. In *Human growth in the past: studies from bones and teeth*, ed. RD Hoppa, CM Fitzgerald, and pp. 241-263. Cambridge: Cambridge University Press
54. Skinner M, Anderson GS. 1991. Individualization and enamel histology: a case report in forensic anthropology. *J. Forens. Sci.* 36:939-48
55. Smith TC, Smith B, Ryan MA, Gray GC, Hooper TI, Heller JM, Dalager NA, Kang HK, Gackstetter GD. 2002. Ten years and 100,000 participants later: occupational and other factors influencing participation in US Gulf War health registries. *J Occup Environ Med* 44:758-68
56. Stam R, Bruijnzeel AW, Wiegant VM. 2000. Long-lasting stress sensitisation. *Eur. J. Pharmacol.* 405:217-24

57. Suckling GW, Pearce EIF, Cutress TW. 1976. Developmental enamel defects in New Zealand children. *New Zealand Dent. J.* 72:201-10
58. Via WF, Churchill JA. 1957. Relationships of cerebral disorder to faults in dental enamel. *Am. J. Dis. Child.* 94:137-42
59. White TD. 1991. Dentition. In *Human Osteology*, ed. AC White, pp. 101-110. San Diego: ACADEMIC PRESS
60. Wright LE. 1990. Stresses of conquest: a study of Wilson bands and enamel hypoplasias in the Maya of Lamanai, Belize. *Am. J. Hum. Biol.* 2:25-35

LIMITS OF DEVELOPING A NATIONAL SYSTEM OF AGRICULTURAL EXTENSION

FELIX H. ARION
*University Assistant of Management and Extension at
University of Agricultural Sciences and Veterinary
Medicine Cluj-Napoca, Department of Management and
Extension.*

Abstract: The concept of extension is extremely wide and complex, being developed in national contexts that differ in many characteristics: beneficiaries, objectives, financial resources, and so on. At the moment it is interesting for government to know the extent to which it can give up its power and responsibility without incurring negative effects, knowing both that most of the beneficiaries are not able to sustain the cost of being helped and that agricultural extension follows a number of national government objectives. These objectives include domestic food security, reducing poverty, decreasing the disparity between urban and rural living standards, and many others.

Modern agricultural extension is considered to have begun during the Irish potato blight between 1847 and 1851, and since then has grown into one of the largest efforts for development and knowledge dissemination in the world. World Bank studies estimated that more than 750,000 extension agents were employed all over the world by Ministries of Agriculture, universities, or research institutions. Hundreds of thousands have been trained, and hundreds of millions of farmers have had contact with agricultural extension services. Approximately US$ 6 billion was been spent in 1988 for extension activities (Alex [2]). The demand for agricultural producers to be more competitive at the local, national or international level is increasing all the time, with a corresponding need for more knowledge on which to base their decisions.

This increased demand for information results in producers seeking information from a variety of sources, including institutions and organizations from diverse areas of activity: social, banking, legal, education, and many more. These resources also include other farmers, farmers' organizations, private companies for agricultural inputs and/or outputs, some non-governmental organizations, and mass-media. For many of these organizations, agricultural information is only a secondary or tertiary activity, and without uniform coverage during the year. The strategic source of information remains the agricultural extension service.

All those actors combine at the national level to form an interconnected and very complex system that cannot be understood only in terms of its separate elements (Drygas [5]). Figure 1 offers a simplified image of the national extension service, including only the general components of the system. Acting as part of a society that is developing and changing all the time, extension is under the influence of the social, economic and political evolutions of society. The change is more obvious not regarding the elements of

289

J. Nation et al. (eds.), Formal Descriptions of Developing Systems, 289–298.
© 2003 *Kluwer Academic Publishers. Printed in the Netherlands.*

the system, which more or less remain the same, but with respect to the relations among them, and especially the relative place and the importance of each one. The institutions that were originally created as a response to acute agricultural and food problems as producers associations, starting with the first decades of the last century gradually came under the authority of government, especially the Ministry of Agriculture. Thus government became the main element of the system for a long period of time. One particular reason for this is that government will invest in agricultural extension when it considers that extension represents an efficient instrument for achieving national objectives through its agricultural policy.

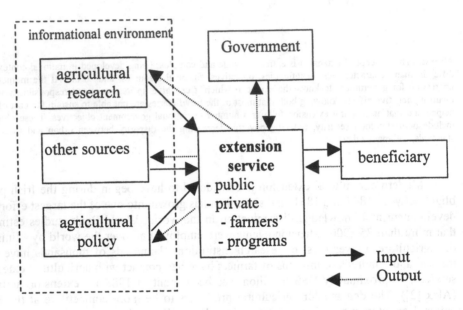

Figure 1. System of agricultural extension at national level.

There are a large variety of objectives that must or could be included here (van der Ban [3] and Szalczyc [7]):

• Increasing agricultural production (in diversity as well as volume), as a result of increasing demand. Imports of food products could be a serious problem for the balance of trade, because food is a primary necessity, and the stability of the government depends directly on the level of covering the food request of the population.

• Stimulation of economic growth, which is closely related to the previous objective, but with an increased importance attributed to production cost and to capacity to compete on the international markets.

• Increasing the welfare of families from rural areas and reducing the disparities between rural and urban population. The rural population is an important source for votes, and the fact that they are generally poorer means that they represent a good opportunity for election campaigns. Besides those political aspects, keeping the tradition and culture of the rural areas, and stopping migration to richer areas, must represent a priority for the state.

• Offering technical information to rural populations is an important objective, especially when this information is designated for small or poor farmers, who are not able to pay for it. This is a means of promoting special agricultural extension channels, desired by the state, too.

• Encouraging sustainable agriculture, thereby reducing environmental degradation (erosion, level of salinity, water or soil pollution, etc.), even if the government has to face, at least temporarily, the dissatisfaction of the farmers who could be affected in a negative way.

Some of these objectives are not so clear as they seem to be at first view, and could interfere, in a negative way, with agricultural extension beneficiaries. For instance, increasing agricultural production in a short period of time will usually have as a consequence ignoring the sustainable aspects of production, and even more, could generate a reduction of prices and of revenues for producers. It is obvious that the efficiency of extension services is greater in cases in which governmental objectives coincide with farmers' particular ones (e.g., introduction of new, more productive seeds; plant rotation for erosion reduction; promoting new technologies). In that case such objectives constitute very important instruments for government. With little effort, extension agents could gain the confidence of the beneficiaries, the confidence that can be used, in the future, for achieving other objectives that are not so popular with farmers. In that sense, it is necessary to appreciate the role of the state extension agencies for supporting those objectives that do not directly coincide with the farmers' needs, even at the risk of losing, partially, the confidence of the farmers, with a potential effect in future promoting of farmers' needs. This is a result of previous reticence linked to the actions that were only for promoting the interest of the state, and ignoring the private interest of the farmers, as the main beneficiaries of the extension activity.

Practically, in some states, entire extension services are part of a central or local administration, and are financed by the state budget. Government investments in agricultural research, as a whole, and for extension as an instrument for dissemination, are made not only for promoting some national objectives, but also for achieving positive financial results. It is appreciated that the average internal rate of return for these investments is about 40%, quite high when compared to other areas of agricultural investment. Studies have shown a significant impact of extension in developing countries, where the rate of return on extension investments has generally ranged between 5% and 50%. A recent media study of 289 studies of economic returns on agricultural research and extension found an average rate of return of 80% for extension investments, equal to the return on research investment (Alex [2]). Of course, the level of results depends on many factors, including the structure of the extension service, the level of bureaucracy, quality of the agents, and so on. At the same time, there are external factors: market for agricultural extension, changing of technology, raw material available, marketing chain, transport facilities, etc. Anyway, raising the level of knowledge and information of the agricultural producers is one of the most important factors that lead to such results, and the extension service could be the instrument that intermediates the information transfer to them. As a result, the activity of the state agents must be combined with other policy instruments that are designed to improve the external factors, which means that extension agents have to share their work and time among different activities. This is one of the reasons that in recent years there has been a powerful tendency in the direction of privatizing the governmental extension, even though, in this situation, farmers would have to assume new responsibilities for decision making and, nevertheless, have to pay a fee for having access to extension.

All over the world, the role of government in the system of agricultural extension is undergoing a serious debate, initiated in 1980s with critical moments in 1990s, because of the incapability of governments to find sufficient funds for agricultural development and of the unsatisfactory results. The public sector in this field is seen as inefficient and unsustainable and, maybe more important, under continuous budgetary pressure. In this direction, emphasis on limiting government expenditures and interventions is reinforced by changing attitudes towards agriculture. The general public, being increasingly urbanized and better-fed, is loosing its direct link with agriculture. As long as food security was a priority, taxpayers agreed to fund public agricultural extension and programs, but having achieved food security, the public interest changes to better use of resources, better products, and greater consumer choice. Of course, reducing the government role involves giving the private sector opportunity to prove its resilience and strength in promoting new production technologies and providing goods and services to farmers. In many areas, public extension is not considered anymore to be essential where there exists a private alternative. In fact, trade liberalization that provides export opportunities to farmers, imports and consumers, reinforces the government role in establishing norms and regulations rather than direct intervention in production and markets.

As a conclusion, public extension services are under pressure for their own poor performance and are criticized as being inefficient and ineffective, lacking clear objectives, motivations and incentives; being poorly managed and not being accountable to clients; and lacking relevant technologies (Alex [2]). Factors affecting their performance include the increasing inability of government to fund adequately and unwillingness of donors to support them; increasing transformation of agriculture in several parts of the world from subsistence to commercialized agribusiness; gradual change of technology from being largely a public good to private good, and so the incentives of private sector to invest in its dissemination and adoptions by clients; technological developments in mass-media and increased specialization among farmers (Sulaiman [8]); and budget deficits that make it almost impossible for some governments to pay for extension services. It is desirable that agents work to be remunerated according to beneficiaries' satisfaction, for improving the quality of their activity.

In some countries farmers are not convinced that state agents serve their interests, but the interests of friends, politicians, and consumers (van der Ban [3]). Even more, many public extension services have low coverage, often working with no more than 10 percent of potential clients, of which a minority are women. Accountably to clients is lacking because top-down bureaucracy prevents farmers from influencing extension agendas, with the result that they are short of relevance for farmers.

Maybe the most used argument is that is not fair for taxpayers to pay for a service that is designated for the own benefit of a small number of agricultural producers. The last claim must be analyzed before being accepted. A competent agricultural extension service will have as a result the improvement of agricultural production efficiency. This in turn will lead to decreasing food prices (if these prices are not controlled by government, or by external markets), and the impact will be considerably higher for reducing the cost of living for beneficiaries than for increasing the revenues of producers. In other words, a good extension seems to be more effective for consumers than for producers. That could mean that consumers should be the ones to pay for extension.

It becomes clear that a privatized extension service will have not only advantages, but disadvantages too. A detailed analysis of the present and future local situation is the only way to determine if the advantages are higher than the disadvantages, and whether privatization is desirable or not. The general rule for this analysis is that there is no general rule that could or must be followed in every case. Because it excludes beneficiaries from the process of establishing the objectives of extension work, a commercial extension system cannot substitute entirely for the role of the state in agricultural extension activity. Factors limiting their efficiency include confused legislation in the field of agriculture, an unstable target population, an insufficient infrastructure, production that is used mainly for farmers' needs, etc.

An alternative to governmental or privatized extension is one financed and directed by a farmers' organization. Even though in poor countries the information available is limited, the models that exist in various countries proved to be efficient and viable. The crucial initial effort involves finding and organizing local groups that have the potential to develop an organization capable of offering extension services. Extension services that were built by different local groups (churches, volunteers, etc.) play a special role because they have a good and close relationship with the local population. The disadvantages are that they operate only at a local level, depend on the level of involvement of members and/or volunteers, and their work is limited to discovering problems and passing the responsibility of solving them to other services, which can be ineffective or inefficient.

However, when such organizations succeed in becoming real farmers' organizations, with a viable operation, recognized at a local or national level, their results in extension activity are quite positive. Advantages of such organizations include, among others: (Albrecht et. al., [1] and Ciurea, Lăcătuşu, Puiu, [4]) continuity in time for agents in the same post, resulting in a more powerful identification and attachment for solving the problems; evaluation of extension is made continuous, because farmers could express at anytime their opinion about the quality of extension work; a better motivation of agents, because the accent is on the local situation, their part in the process of establishing objectives, and increased schedule flexibility; a concentrated view of extension, with more attention to local resources and solving problems with limited funds; more interest from agents for using their initiative because their professional prestige depends directly on their performance; extension activity is designated for non-favored target groups (small farmers, women, etc.) who are not included in governmental programs for extension; elimination of rigid economical thinking in terms of costs and benefits, and a better emphasis on ecological and sustainable agriculture.

Such organizations normally have success because their aim is more orientated toward beneficiaries, and their effort is concentrated towards motivating people to act. In that direction they could successfully collaborate with other local organizations, and their efficiency could be improved by this means. The main problem is that it requires a long period of time and considerable efforts to create an organization capable of offering real extension services, or to hire an agent, especially at the beginning. These organizations are confronted, anyway, with a number of disadvantages that must not be ignored, including: agents are solicited by bigger extension companies that offer them higher salaries; even if extension is paid for uniformly by the members of the group, it is offered non-uniformly, as a result of different distances from office to farms, and due to internal fluctuations of farms personnel (on what are, generally, family farms); membership could be restricted to farmers that can afford to pay the fee, in which case farmers who are not members could benefit by paying for consultation through simple visits; in the event of financial problems it may become impossible to pay for extension. In spite of the fact that they have proved to be superior, from many points of view, to public extension services, the development of associations at a regional or national level is extremely difficult, especially in less developed countries.

There are many efforts to avoid the inefficiency of state extension services, making and developing some specialized organizations that are led and financed on the frame of diverse programs. Unfortunately, this kind of extension services works properly only as long as there are enough external personnel and financial resources. That is why autonomous projects in extension are, generally, under powerful pressure to succeed, and the external personal is applied over a short period of time. As a result, these organizations tend to pursue short-term success, ignoring the long-term effects. The consequence of this approach is the crash of the entire activity when the funds are finished or when the experts leave the organization.

Worldwide there were different experiences in the transformation from a public extension system to a privatized one, in diverse commercial forms, in order to increase the role of privatized agricultural extension services. Though in New Zealand public extension is completely privatized, there are other modalities for transferring a part of government responsibilities to other actors. In countries from South America, Asia and Africa, these include sub-contracting and a voucher system (Chile), a voucher system targeted at small farmers to contract private extension (Costa Rica), share-cropping between farmers and extension staff for a profit (Ecuador), contracting of subject matter specialists by farmers' groups (China), extension associated with contract out-grower schemes (Kenya), and privatized service centers (Ethiopia).

In European countries with a more developed agriculture, they include cost-sharing of advisors (Turkey), cost-recovery from users (The Netherlands), Extension services rendered by farmers' associations (Denmark), and the more complicated system from Germany, which combines many models in different areas: completely privatized, semi-privatized, subsidized farmer associations, voucher system. (Kidd, [6]) They suggest a rapid evolution of extension systems in different parts of the world, and they offer a lot of possibilities and mechanisms for government to give up some of its prerogatives in this field. At present, the active models of extension services start from an integral privatized one, to ones in which the state has kept a substantial involvement.

In cases where there exist different types of organizations that offer extension services, the state loses part of its control and administration of the activity involved. If we agree with the argument above, the state should still subsidize those costs associated with achieving certain national objectives, even though a commercial organization is in charge. As a result, the budgetary effort is reduced, but not entirely eliminated, in order to increase the efficiency of their utilization.

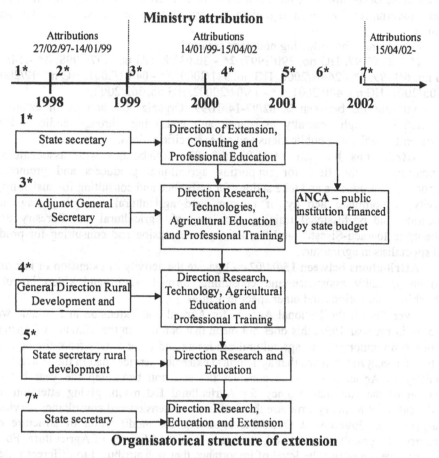

Ministry attribution

Attributions
27/02/97-14/01/99

Attributions
14/01/99-15/04/02

Attributions
15/04/02-

1* 2*   3*   4*   5* 6*   7*

1998   1999   2000   2001   2002

1*
State secretary → Direction of Extension, Consulting and Professional Education

3*
Adjunct General Secretary → Direction Research, Technologies, Agricultural Education and Professional Training

ANCA – public institution financed by state budget

4*
General Direction Rural Development and → Direction Research, Technology, Agricultural Education and Professional Training

5*
State secretary rural development → Direction Research and Education

7*
State secretary → Direction Research, Education and Extension

Organisatorical structure of extension

Figure 2. Evolution of extension in the structure of Ministry of Agriculture in Romania
(27 July 1997 – September 2002)

Agricultural extension, like many other activities, is under strict regulations, which are created mainly by the Government and reflect its view about the issue. In Romania, a country for which agriculture is very important in the context of national economy (18.5% of GDP in 1997), a "clear" national system of extension was created on 30 September 1998 by Government Decision no. 676/1998 regarding creation, organization and function of National Agency for Agricultural Extension (ANCA), published in Official Monitor no. 381 from 06.10.1998. But setting up the National Agency for Agricultural Extension is a result of international support, almost an obligation, in the frame of 3 PHARE Programs: RO 9505-01-01 "Development of National Extension Service" (NES) with the German company AHT International; RO 9505-01-02 "Creating Demonstrative Farms and Producers Groups" (DEMO) and RO 9505-01-03 "Strategical Analysis of Agricultural Knowledge Informational System" (AKIS) in collaboration with the Scottish company SAC International. The legal support on which the extension activity is based in Romania includes 11 normative acts: 1 law, 1 Presidential Decree, 7

Government Decisions, 1 Government Ordinance, 1 Order of Ministry of Finance and 4 other Government Decisions regarding the organization and function of Ministry of Finance.

For Figure 2 the following notations are used:

1* - 27/07/97, HG no. 390/1997; **2*** - 30/03/98, HG no. 197/1998; **3*** - 14/01/99, HG no. 6/1999; **4*** - 26/04/2000, HG no. 331/2000; **5*** - 04/01/2001, HG no. 12/2001; **6*** - 3/05/2001, HG no. 440/2001; **7*** - 15/04/20002, HG no. 362/2002;

Attributions between **27/02/97-14/01/99**: Organize the activity of extension of information through specialty assistance and consulting, through audio and video instruments, publishing publications and other specific actions.

Attributions between **14/01/99-15/04/02**: Collaborates with associations and foundations on the field for supporting agricultural producers and promotes the supporting of producers and take actions for assistance and consulting for sustaining their activity; Co-ordinate activity of research and agricultural technology, agricultural education and professional training, Academy of Agricultural and Forestry Science "Gheorghe Ionescu-Şişeşti", and the activity of extension and consulting for producers and specialists in agriculture.

Attributions between **15/04/02-**: Organize the activity of extension of information through specialty assistance and consulting, through audio and video instruments, publishing publications and other specific actions

Even though the National Agency for Agricultural Extension in Romania was set up only in the year 1998, this does not mean that before then the Ministry of Agriculture accorded no importance to agricultural producers and other agents from the agricultural sector. To analyze the place that agricultural extension services had in the structure of the Ministry of Agriculture, let us consider the situation beginning the year before the creation of the National Agency for Agricultural Extension, giving attention to any modification of Ministry structure that influenced extension and consultancy services. By analyzing the Government Decisions regarding the modification of structure of the Ministry of Agriculture and Food, respectively of the Ministry of Agriculture, Food and Forestry, we can notice the level of importance that was attributed to different aspects of activity of the Ministry (programs, SAPARD, different agencies, etc.) It could be observed that agricultural extension granted to agricultural producers and other economic agents from agriculture remains largely unchanged, with the major exception of the creation of the National Agency for Agricultural Extension, as a result of re-organization of the former Direction of Extension, Consulting and Professional Education that was in direct sub-ordination to a secretary of state (Figure 2).

Once ANCA was set up, the Ministry's interest in professional training did not entirely disappear, even though this primarily became the obligation of ANCA. It created the Direction Research, Technologies, Agricultural Education and Professional Training, avoiding the "extension" from its name until this year. There is another notable change in the vision of Ministry of Agriculture, Food and Forestry regarding the agricultural extension. Though until January 2001 the Counties Offices of Agricultural Extension (OJCA) were under direct (hierarchical and financial) authority of ANCA, in the period January 2001 – April 2002, they were placed under the authority of the General Direction for Agriculture of each county (an institution that represents the Ministry in the territory). After that they were put at the disposal of local authorities, being financed from their budget – a first step of the Government towards placing the problem of financing extension activities in the hands of extension agents, because it is probable that a local community, especially if is a small one, will not want to pay for services that are not addressing their needs.

The problem of privatizing agricultural extension in Romania is, at least at first glance, difficult for two reasons. Firstly, ANCA and its offices in the territories were created by Government with workers who spent almost their entire careers in the structures of state organizations, and their mentality is difficult to change. *Volens nolens,* until they are changed it is difficult to set up a service not financed from the state budget. Moreover, it is difficult for private firms to break the monopoly created by ANCA, with the notable exception of big companies that offer a "pseudo-extension" in the moment of selling their products (mainly inputs for agricultural producers), and that lack of competition will lead to a decrease of efficiency for activities of ANCA. Secondly, the structure of Romanian agriculture at the moment (with 22.5 million inhabitants there are 5-5.2 million land properties; 3.1 million agricultural households; 1.0 million family farms – totaling 8.1 million hectares, with only 0.24% with a surface more than 10 hectare and a average surface for exploitation of 2-2.4 hectares) make the possibility of setting up a commercial system of agricultural extension extremely vulnerable, as an obvious result of the very limited financial resources of agricultural producers.

In conclusion, the increasing role of the private, non-governmental sector in agricultural extension is an actual trend all over the world. This movement requires state involvement, and is realizable over the long term with costs that are not only financial, but also social and cultural. It must be kept in mind that this movement in the direction of privatizing the national services of agricultural extension appeared in countries with viable, commercial farms (Germany, Holland), that can afford to pay for information and services. Privatized agricultural extension was then begun on an experimental basis in countries with a less developed commercial agriculture – but where the idea was easily adopted by Government as result of lack of money for paying for extension services from the state budget, and of financial involvement in these experiments of important international financial organizations (World Bank in Costa Rica and Mexico, for example).

Experience showed, anyway, that an entirely public extension system is not a viable solution, especially on long term, and particularly if a powerful and efficient agricultural sector is a priority. At the same time, it proved that there is not a common model, that can be applied everywhere. From both the economic and social point of view, the involvement of the state in extension activity, at least financially, is justified. There are extension functions that can not be taken by the private sector, and many farmers cannot be efficiently served by a commercial extension they cannot afford. This does not means necessarily that only the public extension part of responsibilities can be transferred to farmers' organizations, to non-governmental organizations, or to other institutions, but it must be sure that they are offered and delivered to beneficiaries. Government could maintain only the role of developing extension strategies that to take into consideration the modalities of covering the costs, the implementation and evaluation.

References

1. Albrecht H., H. Bergmann, G. Diederich , E. Großer, V. Hoffman, P. Keller, G. Payr, R. Schulzer (1989) *Rural Development Series: Agricultural Extension. Volume 1: Basic Concepts and Methods*, Deutsche Gesellschaft für Technische Zusammenarbeit (GTZ) GmbH, Federal Republic of Germany, Eschborn, United Kingdom

2. Alex, G., 1/23/2001, *Agricultural Extension Investments: Future Options for Rural Development*, World Bank Papers, US

3. Ban A. W. van den; H. S. Hawkins (1999) Agricultural Extension, Second Edition, Blackwell Science, Oxford, United Kingdom

4. Ciurea I.V.; G. Lăcătuşu, I. Puiu (2000) *Consultanţă Agricolă*, Iasi, Romania

5. Drygas, M. (1996) *Objectives, goal and structure of the Agricultural Knowledge System*, in International Conference: Agricultural Extension as a Link of the Agricultural Acknowledge System in the Process of Modernizing Rural Areas and Agriculture and in the Integration Process with the European Union, Poświętne, June 12-14, 1996. Drygras, M., Duczkowska-Malysz, K., Siekierski, C. and Wiatrak, A.P. (eds.), The Ministry of Agriculture and Food Economy, The University of Agriculture Warsaw, Foundation for Assistance Programmes for Agriculture; Centrum Doradztwa i Edukacji w Rolnictwie, Poznań, Poland, 1997, 21-28

6. Kidd, A.; J. Lamers, V. Hoffman (1998) *Towards pluralism in agricultural extension*, in Beratung im ländlichen Raum. Begleitordner für Kurs-Trainer und Kurs-Teilnehmer, Institut für Sozialwissenschaften des Agrarbereichs Fachgebiet Landwirtschaftliche Kommunications und Beratungslehre, Universität Hohenheim, Stuttgart, Germany

7. Szalczyk, Z. (1996) *Agricultural extension services – the domain of the state or farmers?*, in International Conference: Agricultural Extension as a Link of the Agricultural Acknowledge System in the Process of Modernizing Rural Areas and Agriculture and in the Integration Process with the European Union, Poświętne, June 12-14, 1996, Drygas M. (ed.), The Ministry of Agriculture and Food Economy, The University of Agriculture Warsaw, Foundation for Assistance Programmes for Agriculture; Centrum Doradztwa i Edukacji w Rolnictwie, Poznań, Poland, 1997, 162-163

8. Sulaiman, R.V.; Sadamate, V.V. (2000) *Privatising Agricultural Extension in India*, National Centre for Agricultural Economics and Policy Research (NCAP), New Delhi, India

AN ACCUMULATION MODEL FOR THE FORMATION OF MINI BLACK HOLES

GÖKSEL DAYLAN ESMER
University of Istanbul
Faculty of Science, Physics Department
Istanbul, TURKEY

Hawking [2] and Zel'dovich and Novikov [6] have suggested that mini black holes could have been produced during the very early stages of the cosmological expansion of the universe. Hawking's model is based on the assumptions of chaotic cosmology [4], which imply that the universe is not homogeneous and isotropic at the beginning.

In our study (inspired by a suggestion of W.H. McCrea) called "A new model for the formation of mini black holes" [1], the author and Sehsuvar Zebitay have proposed an alternative model to describe the formation of mini (primordial) black holes. The model is based on an idea that the fluids which fill the universe at the beginning, just after the Big Bang, are separated into mini clusters, moving with supersonic velocities and colliding with each other under the forces of gravitational attraction. We considered the time era from 10^{-43} to 10^{-8} sec., during which about 36 various generations of mini black holes with masses ranging in the interval from 10^{-5} to 10^{30} gm are possible.

First, let us give a non-technical description of the model; see [1] for more details. Our model has the following assumptions.

I)   The universe is homogeneous and isotropic at the beginning era and evolves according to the standard cosmogenic model described by the Friedmann equations [4].

II) The fluid that fills the universe at the beginning era conforms with the Basic Particle model [5].

III) The fluid that fills the universe is separated into clusters by any means in the time interval of 10^{-43} – 10^{-8} sec. On the other hand, we do not consider times after 10^{-8} sec.

IV) All structural properties of the clusters and fluid are the same.

V)   The clusters collide with each other under the influence of their gravitational field. The velocity of accumulation of the clusters in the cross-section region is greater than the velocity of sound.

VI) Mini black holes are formed as a result of Jeans condensation in a common accumulation volume of the clusters after the collision.

In the time period under consideration, the conditions for the collapse of clusters under their own gravitational field are not satisfied. The velocity of sound in the beginning era is given by $v_s = c / \sqrt{3}$. Thus the velocities of supersonic clusters will be

J. Nation et al. (eds.), Formal Descriptions of Developing Systems, 299–302.
© 2003 *Kluwer Academic Publishers. Printed in the Netherlands.*

in the interval $v_s < v < \sqrt{3}\, v_s$, and collisions of the clusters will be possible before their dispersion through the expansion of the universe. In order for these clusters to reach supersonic speed, their radii must be of the same order as the Jeans radius, the critical radius for collapse. After the head-on collision of two identical, supersonic, spherical clusters, an accumulation region is formed; see Figure 1. The collision produces one, and only one, mini black hole in the accumulation region, with a radius and mass of the same order as the original colliding clusters; see Figure 2. Moreover, during the expansion of the universe, some parts of the fluid which do not share in the condensation will be accrued by the mini black hole. Of course, not all clusters will take part in collisions, nor will all clusters satisfy the necessary supersonic velocity condition, nor will all collisions be head-on. Most of this remaining material will be dispersed in the expanding universe, though near-head-on collisions may result in mini black holes with a spin.

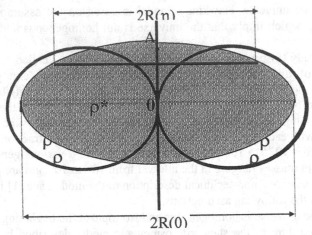

Figure 1. Head-on collision of two spherical clusters under their gravitational field. The accumulation region that occurred in a common region after the collision is shown as shaded.

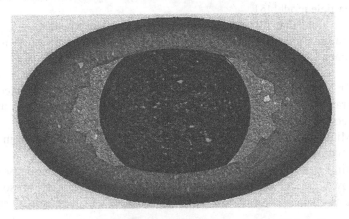

Figure 2. The accumulation region that occurred after the head – on collision of two spherical clusters. Only one condensation occurs in the accumulation region.

Thus we obtain that the head-on collision of the two identically spherical clouds with supersonic velocities ends with the creation of one mini black hole in the accumulation region, with a radius of the same order as that of the original clusters. One can also show that the mass of mini black holes which occur in the accumulation region will be about Jeans mass. This shows that the mass of condensation, which appears after the collision, is the same order as each of the colliding masses.

Furthermore in the time interval that we consider (10^{-43} to 10^{-8} sec.) it is shown that there would have occurred 36 consecutive generations of primordial black holes. Because the Jeans mass (the critical mass for collapse) increases with the evolution of the universe until the end of the radiation era, the masses of the mini black holes that are created increase during the expansion of universe.

When we look at Table I, we see that the life span of the first 35 generations of black holes is not enough to allow detection at this time. However, when we consider the contribution of the 36th generation, we see that these black holes have a sufficiently long lifespan to be observed. So the later generations of mini black holes that are predicted by this model might survive up to the present time. Thus this model naturally explains the concept of missing mass.

t (sn)	ρ (gr/cm^3)	r (cm)	m (gr)	τ (sn)
10^{-43}	$1{,}5.10^{93}$	3.10^{-33}	4.10^{-5}	$6{,}4.10^{-42}$
10^{-41}	$1{,}5.10^{89}$	3.10^{-31}	4.10^{-3}	$6{,}4.10^{-36}$
10^{-40}	$1{,}5.10^{87}$	3.10^{-30}	4.10^{-2}	$6{,}4.10^{-33}$
10^{-38}	$1{,}5.10^{83}$	3.10^{-28}	4	$6{,}4.10^{-27}$
10^{-36}	$1{,}5.10^{79}$	3.10^{-26}	4.10^{2}	$6{,}4.10^{-21}$
10^{-34}	$1{,}5.10^{75}$	3.10^{-24}	4.10^{4}	$6{,}4.10^{-15}$
10^{-32}	$1{,}5.10^{71}$	3.10^{-22}	4.10^{6}	$6{,}4.10^{-9}$
10^{-30}	$1{,}5.10^{67}$	3.10^{-20}	4.10^{8}	$6{,}4.10^{-3}$
10^{-28}	$1{,}5.10^{63}$	3.10^{-18}	4.10^{10}	$6{,}4.10^{3}$
10^{-26}	$1{,}5.10^{59}$	3.10^{-16}	4.10^{12}	$6{,}4.10^{9}$
4.10^{-24}	10^{54}	10^{-13}	$1{,}5.10^{15}$	$3{,}4.10^{17}$
10^{-23}	$1{,}5.10^{53}$	3.10^{-13}	4.10^{15}	$6{,}4.10^{18}$
10^{-22}	$1{,}5.10^{51}$	3.10^{-12}	4.10^{16}	$6{,}4.10^{21}$
10^{-20}	$1{,}5.10^{47}$	3.10^{-10}	4.10^{18}	$6{,}4.10^{27}$
10^{-18}	$1{,}5.10^{43}$	3.10^{-8}	4.10^{20}	$6{,}4.10^{33}$
10^{-16}	$1{,}5.10^{39}$	3.10^{-6}	4.10^{22}	$6{,}4.10^{39}$
10^{-14}	$1{,}5.10^{35}$	3.10^{-4}	4.10^{24}	$6{,}4.10^{45}$
10^{-12}	$1{,}5.10^{31}$	3.10^{-2}	4.10^{26}	$6{,}4.10^{51}$
10^{-10}	$1{,}5.10^{27}$	3	4.10^{28}	$6{,}4.10^{57}$
10^{-8}	$1{,}5.10^{23}$	3.10^{2}	4.10^{30}	$6{,}4.10^{63}$

Table 1. Relation between the time (t) and the density (ρ), radius (r), mass (m), and lifetime (τ) of mini black holes.

I am planning to do further research.

1) I will try to determine the total mass of the universe which transformed into mini black holes within the estimated time span.

2) I will try to explain the black holes remaining to the present time, taking Hawking's process related with the evaporation of black holes into consideration together with this model, and in this way calculate the contribution of these black holes to "the missing mass" that isn't observed now and which is postulated by General Relativity.

References:

1. Daylan Esmer, G. and Zebıtay Ş. (2000) A new model for the formation of mini black holes, *General Relativity and Gravitation* 32, 1241-1254.
2. Hawking, S.W. (1971) Gravitationally collapsed object of very low mass, *Mon. Not. R. Astron. Soc.* 152, 75–78.
3. Hawking , S.W. (1975) Particle creation by black holes, *Commun. Math. Phys.* 43, 199 – 220.
4. Misner, C.W., Thorne, K.S., and Wheeler, J.A. (1973) *Gravitation*, ISBN 0-7167-0334 -3, Freeman, San Francisco.
5. Weinberg, S., (1972) *Gravitation and Cosmology : Principles and Applications of the General Theory of Relativity*, ISBN 0-471-92567-5, John Wiley, New York.
6. Zeldovich, Ya. B. Novikov, I.D. (1967) The hypothesis of cores retarded during expansion and the hot cosmological model, *Soviet Astronomy – A.J.*, 10, 602 – 603.

This work was supported by the Research Fund of the Istanbul University, project number Ö908/12122000.

SUBJECT INDEX